LAND RESOURCES

NAGI Series proposed Volumes

- Resources: Concepts, Classification, and Role/ **R.P. Misra**
- Human Resources/ **S. Nangia** (JNU), **N.C. Jana** (Burdwan) and **R.B. Bhagat** (Mumbai)
- Land Resources/**H.S. Sharma** and **S. Padmaja** (Jaipur)
- Water Resources/ **R.B. Singh** (Delhi)
- Biotic Resources: Forests and Animals/ **A.K. Pandey** (Allahabad)
- Soils of India/ **Arun Chaturvedi** (Nagpur)
- Mineral Resources/ **Rana Pratap Singh** (Bodh Gaya)
- Energy Resources/ **Rasbihari P. Singh** (Patna)
- Marine Resources/ **S. Chattopadhyaya** (Trivandrum)
- Managing Our Resources: Problems and Prospects/ **R.P. Misra**

NAGI SERIES-3

LAND RESOURCES

Processes, Degradation and Evaluation

Series Editor: R.P. Misra

Edited by
H.S. Sharma
S. Padmaja

Published in Collaboration with
NATIONAL ASSOCIATION OF GEOGRAPHERS, INDIA
by
CONCEPT PUBLISHING COMPANY PVT. LTD.
NEW DELHI-110059

ISBN-13: 978-93-88937-53-5

First Published 2020

Published and Printed by

Concept Since 1974

Concept Publishing Company Pvt. Ltd.
A/15-16, Commercial Block, Mohan Garden,
New Delhi-110059 (INDIA)
T : 91 11 25351460, 91 11 25351794
E : publishing@conceptpub.com; *W* : www.conceptpub.com

Cataloging in Publication Data--Courtesy: D.K. Agencies (P) Ltd. <docinfo@dkagencies.com>

Land resources : processes, degradation and evaluation / edited by H.S. Sharma and S. Padmaja.
pages cm. — (National Association of Geographers India series ; 3)
Contributed articles.
Includes bibliographical references and index.
ISBN 9789388937535

1. Land use—India. 2. Natural resources—India. 3. Geomorphology—India. 4. Land degradation—India. I. Sharma, H. S. (Hari Shanker), 1945- editor. II. Padmaja, S., editor. III. National Association of Geographers, India, issuing body. IV. Series: NAGI series ; 3.

LCC HD876.5.L36 2020 | DDC 333.73130954 23

Foreword

The National Association of Geographers, India (NAGI) in its annual meeting held on November 6, 2004 at MD University, Rohtak (Haryana), decided to publish a series of ten volumes on The Natural and Human Resources of India'. This volume is the third in the series and focuses on land resources, which, in my view is the basic resource of all resources. Mother Nature does present itself in the form of land, water, energy, air and open space (Kshiti, Jal, Pavak, Gagan, Samir) which we use as resources for development, it is, however, the human mind that perceives them as resources.

The idea of the series emerged when the General Body (GB) of the Association took up the assessment of progress the discipline of Geography has made during the year. Members of the Association were highly concerned about the unscientific and callous use of many of the natural resources of India such as land, water, air, energy, flora and fauna, and consequent danger of Global Warming and Climate Change. They were alarmed at the neglect, under-utilization, and even mal-utilization of country's human resources. While the destruction and depletion of natural resources threaten food security and welfare of the future generations, lack of appropriate policy to develop the human resources has led to mass poverty, ignorance, and debasement of human values. Together, they have brought the world in the grips of Global Warming and Climate Change on the one hand, and escalating violence and hopeless socio-economic development scenario on the other.

The members felt that India, like many other countries of the world is likely to face a number of intractable problems due to the overuse and misuse of its resources, both natural and human. Its land and soil resources are being depleted so fast that the country may face food shortage once again in not too distant a future. Fertile agricultural land is being swallowed by SEZ, which now number over 200 in the country. Overuse of chemical fertilizers and pesticides has reduced the natural fertility of the soils to a great extent. It is petroleum, an energy resource that India imports from West Asia especially that is keeping its agricultural production growing at around 3 per cent per annum. In fact, we are living from petroleum to mouth. The country faces serious food security problem in the future.

India is depleting and polluting its ground and surface water resources so fast, that in the next few decades it will not be able to supply potable water to its growing population. To increase agricultural productivity in a monsoon climate regime, irrigation is a must. While there are vast land areas yet to be irrigated, the present irrigation system (flood irrigation) is so wasteful that the ground water level is going down too fast to be depended upon for future supplies. The situation is much worse with regard to the surface water. Most of the lakes and tanks are encroached upon, and the rivers are not only robbed of their waters for capal irrigation but also for meeting the drinking water needs of the ever growing cities. They do not have enough water to remain perennial during summer months. Further, they are highly polluted by the untreated effluents released by the municipal governments.

The situation with regard to other natural resources such as forests, animals, minerals, is equally disturbing. The country has never been rich in conventional energy resources, except in coal. Burning of the coal, however, is counterproductive as it leads to air pollution which threatens all life on planet earth, humans included. The prospects of augmenting the energy resources are meager. The situation need not, however, be hopeless. Hydrogen, found so abundantly in the atmosphere may open a new vista in the energy sector. What is needed is the right and timely application of science and technology in which India is not behind the advanced countries of the world. But that is not enough. India must evolve a style of life that is truly Indian. It has to be based on the Gandhian principle of 'living simply so that other can simply live'. It must embark on a major programme of resource conservation and scientific management.

The Association wanted to highlight the emerging problems associated with the availability, use, and development of the natural and human resources of the country so that the future generations neither suffer from resource constraints, nor from the environmental problems, the exploitation of natural and human resources entails. In order to make the people aware of the impending problems, the Association decided to bring out this series of books, which look at India's natural and human resources and the problems associated with them.

The task of editing the series was entrusted to me. The titles of the volumes and the names of respective volume editors are as under:

- Resources: Concepts, Classification, and Role: R.P. Misra
- Human Resources: S. Nangia (JNU), N.C. Jana (Burdwan) and R.B. Bhagat (Mumbai)
- Land Resources: H.S. Sharma (Jaipur), S. Padmaja (Hyderabad)
- Water Resources: R.B. Singh (Delhi)
- Biotic Resources: Forests and Animals: A.K. Pandey (Allahabad)
- Soils of India: Arun Chaturvedi (Nagpur)
- Mineral Resources: Rana Pratap Singh (Bodh Gaya)
- Energy Resources: Rasbihari P. Singh (Patna)
- Marine Resources: S. Chattopadhyaya (Trivandrum)
- Managing Our Resources: Problems and Prospects: R.P. Misra

This volume is the Third in the series. It centers on Land Resources and has been edited by one of the most distinguished Geographers of India—Professor H.S. Sharma of Department of Geography, University of Rajasthan, Jaipur with Professer S. Padmaja Department of Geography Osmania University, Hyderabad.

The volume, as apparent from the contents, covers almost all aspects of Land resources of India; it fully meets the standards set by the NAGI.

Allahabad
30th November, 2018

R.P. Misra
Series Editor

Preface

This is the third volume in the series, The State of Natural and Human Resources of India' sponsored by the National Association of Geographers, India (NAGI). As a first step in this direction, the Association resolved to publish compendiums putting together articles, reports and other materials relevant to the theme in a manner that would present a total and comprehensive scenario of the state of natural and human resources of India. However, in the present volume original unpublished papers have been included on general land use, geomorphology, land degradation and models of mapping. The general editor of the series, Professor R.P. Misra, is a senior geographer, Bhoogol Ratna, and Fellow of NAGI. Other geographers would also be involved to collaborate in this project as volume editors.

The series would cover practically every resources of India ranging from Natural to Human Resources, and all aspects thereof including Natural Resources and Development, Globalization and its Implications for Resource and Development, Biodiversity, Environmental Degradation, and Sustainable Management of Resources. This volume is the Third in the Series and focuses on Land Resources and related issues.

Land resources are one of the most important basic resources which form the main base for other geo- and bio-resources well culminated within them. Land refers not just to soil but to the combined resources of terrain, water, soil and biotic resources that provide the basis for land use. Land quality refers to the condition or health of land and specifically to its capacity for sustainable land use and environment management. Land quality indicators, such as nutrient balance, loss of organic matter, land use intensity and diversity and land cover are useful to task managers and decision-makers to monitor and improve the performance of projects with respect to their southeast and environment impact and to assess the trend towards or away from land use sustainability. Land quality indicator approach focuses on preventive maintenance rather than rehabilitation.

The land resources have also been the most ancient, most used and most impacted resources from a local to global level due to human factors. This may be due to (a) growing demand for food in terms of both quantity and quality for an expanding and more wealthy world population, (b) Competition for productive land for biofuel, urban expansion and other non-productive uses. (c) Decrease or lack of growth in productivity due to decline in soil health indicated by lower nutrient status and organic matter and other degradation process. Pressures also arise from natural factors such as climate variability, extreme weather events and wild fire. The pressures cited above are the making of extreme nature of interference of man with nature, which triggers of uncontrollable and unsustainable land management processes inclusive of land, water, soils and bio cover. Land with all its inherent physical attributes of lithology, structure, morphology, need to be understood in totality, so that their characters can be assessed regularly to involve them in planning perspective. It is a well known fact that every facet of land has an important resources potential and need to be understood in terms of their dynamic process amalgamation culminating into various land alterations and transformation on the earth surface.

Land degradation rate is high, and is mainly due to a range of natural and management pressures on land resources. It was always believed that normal degradation is a natural phenomena but accelerated degradation is an issue of contention. Land degradation is getting intensified due to land cut fills, slope instability, overgrazing along with climatic variability which enhances LD-soil health too is getting depleted due to nutrient decline, wind and water erosion, loss of soil bio diversity, soil sealing, contamination and salinization and compaction. It may extend further to include processes like pollution of aquifers and freshwater resources, water depletion, deforestation and biodiversity loss.

What is required to stop the menace of land degradation and maintain healthy land is to adopt Land resource planning in a sustainable way. The following practices are needed :

(a) Building soil organic matter for better soil moisture storage and nutrient status and improved soil structure,
(b) Integrated nutrient management
(c) Better crop selection targeting seed quality and locally adapted varieties combined with management for local conditions
(d) high efficiency of use of rain water to promote infiltration and minimize run off and erosion in both crop and grazing lands.
(e) management of surface erusting and compaction of soils through maintaining ground water.

Land degradation, climate change and food security are arguably the three greatest challenges facing the world today and Sustainable Land Management is fundamental to all three. As per Brundtland report, if human needs are to be met, the earths natural resource must be conserved and in particular, land use for agricultural and forestry must be based on sustainable assessment of both land capacity and the annual depletion of top soil. Addressing the problem of land degradation through ustainable Land Management approaches is truly multi-faceted with challenges on multiple levels of environment social and economical displines as well as human well being and socio-.cultural values. Understanding the consequences of unsustainable land management for key global issues of concern, including the compounding problems of food security and anthropogenic climate change, will help to prioritize these interventions. Land provides an environment for resource utilization in an amicable way. However, often it is misused. In this context, Sustainable Land Management comes into picture with the sole aim of harmonizing the complimentary goals of providing environment eco and social opportunities for the benefit of present and future generations, while maintaining and enhancing the quality of the land resource. Sustainable Land Management is the use of land to meet changing human needs, while ensuring long terms socio-economic and ecological functions of the land. Sustainable Land Management combines technologies, policies and activities aimed at integrating sustainable environment principles with environment concerns so as to simultaneously maintain or enchance prod, reduce the level of prod. Risk, protect the potential of natural resources and prevent soil and water degradation, be economically viable and be socially acceptable. These concepts span the scales of detail, application and levels of integration with sustainable environment data. Land quality requires integration of soil data with other biophysical information such as climate, geology and land use. Land quality is a condition of the landscape, i.e. it is a biophysical property, but includes the impacts of human interventions (Land Use) on the land space. Land quality should be linked to economic and social demands. The concepts differ in the kinds and scales of the processes being described, the data used for input and the amount and kinds of integration with other disciplines.

Land resources form the most important natural wealth of India and their utilization is a matter of utmost concern to its people The utilization of land according to its use capability ensures that this resource is utilized to the best advantage. Its improper use leads to wastage and can lead to progressive deterioration and loss of productivity of this vital resource. It is a moral obligation of the present generation to pass this valuable resource on to future generations as nearly unimpaired and over exploited as possible. In India during last 70 years agricultural land recorded a marginal increase, while the population increased by 70 percent. Out of total geographical area of 324.7 million hectares the land use pattern is available for only 305.69 million hectares. The land use has changed in India drastically since 1950-51. During 1950-51 the area under forest was only 40.48 million hectares (14.2%) in India. It has been increased for 80.20 million hectares (24.39%) in 2017-18. India has a net sown area of 46.2% of the total reporting land in India. There has been a phenomenal increase in the net sown area during last five decades. This increase is due to the reclamation of barren uncultivable land, pasture land etc. In contrast, the country is facing a major threat of desertification and land degradation. Contrasting figures on land degradation are available between 29% to 32% of the total geographical area of India. Certainly it is a very big area under going land degradation in the country. This issue needs to be addressed systematically. In this volume 19 articles have been included. The volume is divided into three sections. Section first includes general articles on Land use and Land Resources. In the second and third section the papers dealing with the problem of desertification/land degradation and models and techniques of land degradation mapping have been incorporated. The editors are grateful to the Executive Committee of the National Association of Geographers, India (NAGI) for giving them the opportunity to edit this very important volume. We express our profound gratitude to Prof. R. P. Misra, the Series Editor, who has kindly written a foreword to this volume. Without his inspiration and encouragement, it would have not been possible to bring out volume .

We express our sincere thank to the Concept Publishing Company Pvt. Ltd., New Delhi for bringing out the volume in a record time.

New Delhi
30th November, 2019

H.S. Sharma
S. Padmaja

Contents

List of Contributors

A. Dasgupta, SAC, Ahmedabad

Ajai, SAC, Ahmedabad

Md. Akther Ali, Assistant Professor, Department of Geography, University College of Science, Osmania University, Hyderabad.

Dr. (Mrs.) Anusuiya Baghel, Professor, S.O.S. in Geography, Pt. Ravishankar Shukla University, Raipur (Chhattisgarh).

Arunima Dasgupta, MOEFCC, Govt. of India, New Delhi.

Asoori Latha, Atomic Minerals Directorate for Exploration and Research, Hyderabad, India.

Baleshwar Thakur, Former Professor and Head, Department of Geography, University of Delhi, Delhi. **Email:** *bthakur21@gmail.com*

G. Praveen Kumar, Central Ground Water Board (C.G.W.B.) Hyderabad .

Email: *pavanaguru001@gmail.com*

Ganesh Sharma, Assistant Professor, Govt. College, Jaipur.

Guru Prasad Chattopadhyay, Former Professor of Geography, Visva-Bharati, Santiniketan. **E-mail:** *cguruprasad@rediffmail.com*

H.S. Sharma, Former Professor and Head, Department of Geography, University of Rajasthan, Jaipur. **E-mail:** *hssharma358@gmail.com.*

M.A. Paul, EOS/ISRO HQ, Bengaluru.

M.N. Kaul, Formerly Professor and Head, Department of Geography, University of Jammu, Jammu.

Madhavi Pore, Department of Geography, Pune University, Pune.

N.C. Gautam, Director, Centre For Land Use Management, 25-24/1, East Anandbagh, Malkajgiri, Hyderabad (Telangana).

Nandkumar Sawant, Principal and Associate Professor, Parvatibai Chowgule College, Margoa – Goa, India. **E-mail:** *nns001@chowgules.ac.in*

P.S. Dhinwa, TALEEM Research Foundation, Ahmedabad. **E-mail:** *psdhinwa@gmail.com*

Padmini Pani, Centre for the Study of Regional development, School of Social Sciences, Jawaharlal Nehru University, New Delhi. **Email:** *padminipani.jnu@gmail.com*

R. Pavanaguru, Department of Geography, Osmania University, Hyderabad. **Email:** *praveenkumar6@gmail.com*

S. Bandopadhya, EOS/ISRO HQ, Bengaluru.

S. Padmaja, Emeritus Professor, Department of Geography, University College of Science, Osmania University, Hyderabad. **Email:** *padmaja45@gmail.com*

S.K. Sharma, Retired Professor, Head and Director, Population Research Centre, Department of General and Applied Geography, Dr Harisingh Gour University, Sagar (M.P.).

E-mail: *shrikamal.sharma@rediffmail.com*

S.K. Nasib Ahamed, Research Scholar. Department of Geography, Pt. Ravishankar Shukla University, Raipur (Chhattisgarh). **E-mail:** *nasibahamed@gmail.com*

Saptarshi Mitra, Department of Geography and Disaster Management, Tripura University, Agartala (Tripura).

Srikumar Chattopadhyay and Mahamaya Chattopadhyay, Former Scientists National Centre for Earth Science Studies Thiruvananthapuram.

Subhajit Sen, Department of Geography and Disaster Management, Tripura University, Agartala, (Tripura).

Sunil Kumar De, Department of Geography, North Eastern Hill University, Shillong (Meghalaya).

T. Pratigna, Assistant Professor, Department of Geoinformatics, Telangana University, Nizamabad.

1
Land Use/Land Cover Studies for Sustainable Development

N.C. Gautam

Introduction

Land use/land cover changes assume significant importance in the study of global environmental change. The impacts of these dynamics coupled with the changing human dimensions have major implications on the sustainable development of land resources.

Land is the sole source of sustenance of mankind supporting the plants and animals on it providing food, fiber and shelter. The growing pressure and population coupled with an increasing variety of demand made on land resources, have brought extra pressure on the limited resources all over the country. Hence it is an obligatory pre-requisite to know the present or existing use of land in order to plan for its optimal use.

Timely and reliable information base is essential for effective land use management. In the past, land use/land cover mapping have been done by conventional ground methods which are labour intensive, time consuming and were being carried out relatively infrequently. While much of the information is being derived from existing paper maps which are not readily available in all parts of the country. With passage of time these maps soon become outdated, particularly in a rapidly changing environment. Therefore, acquisition of reliable, comprehensive, generalized and timely data on land use/land cover attains prime importance. At the same time, one cannot over-emphasize the importance of up to date and accurate data on land use for developmental purpose and management of resources.

Objectives

- To develop new definition of land use/land cover with land use system.
- To develop new land use/land cover classification system using modern technology (Remote Sensing) up to three levels and 1:5000 scale up to national planning for all agroclimatic regions.

Land Use Purposes

- Land use is captured by describing the land use purpose(s) and operation sequence. Operations are meant to have an impact on land so that the aimed land use purposes can be achieved. Resources used may include the soil, crop grown, natural vegetation etc.

- Land use always has one or more purposes defined by species/services. Product/benefit or combination etc.
- Grain
- Fodder
- Recreation – pleasure.

What is Land Use/Land Cover?

- A series of operations on land, carried out by humans, with the intention to obtain products and/or benefits through using land resources.
- The vegetation or man-made constructions which occur on the earth surface like water, ice, bare rock, sand and similar surface also count as land cover.
- Land use should not be confused with land cover.
- In land use human activities, whereas land cover natural activities.
- Land cover may be determined by direct observations.
- Land use requires an interview with land user.

What is Land Use System?

Combinations of one land units and one land utilization type constitutes a land use system.

- Land: Climate, Landform/soil/terrain, Flora (including crops), Fauna, Result of past land use (including infrastructure).
- Land use: Species, Services, Products, Benefits or combinations, Operation Sequence, Details of Operations.

Types of Land Use System

- Single Land Use System: This is the configuration whose performance is analyzed in assessment of land suitability.
- Multiple Land Use System: More than one crops on a field at one time.
- Compound Land Use System: i.e., single or multiple system in rotation.
- Farming System: Consists of one or more land uses system practiced by one household or management unit.

Major features of Land Use System are shown in Figure 1.1.

Present scenario of land use/cover information generation and usage
Presently land use information is available from the following sources:

- Revenue records compiled by the Directorate/Bureau of Economics and Statistics (DES/BES) which is mainly tabular. Further, the 9-fold classification does not encapsulate all the land use categories as is required.
- Topographical maps from Survey of India represent very broad land cover categories.

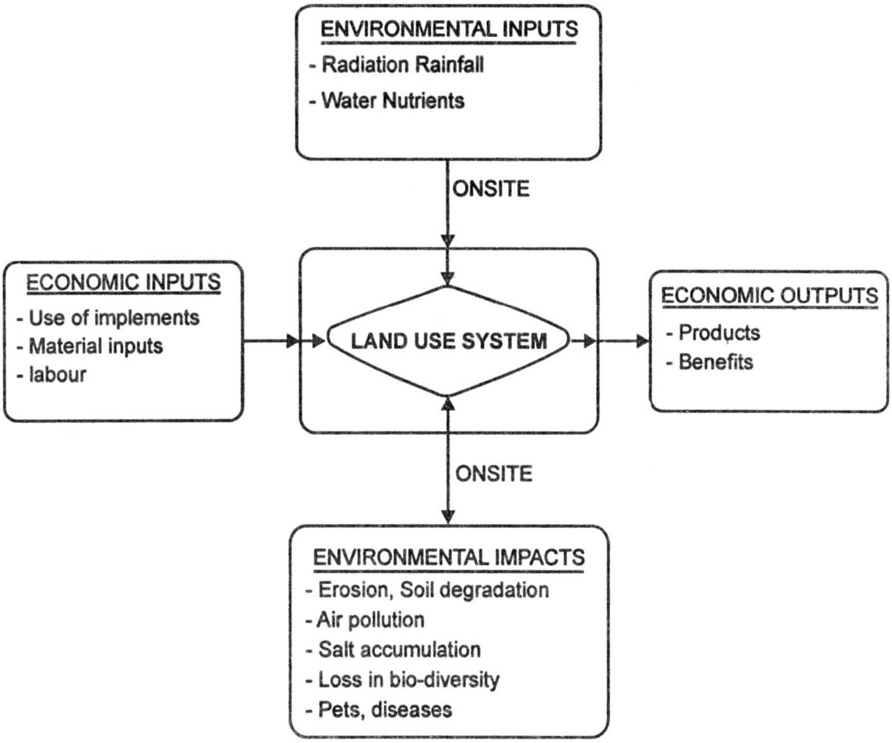

Fig. 1.1: Major Uses of Landuse Resources

These maps mainly provide topographical information.

- Land use Atlas from NATMO are mainly small scale and are secondary compilations.
- Soil Survey organizations (NBSS & LUP and AIS & LUS) generate soil and land capability maps for specific project/areas.

Why Classifying Land Use?

There is a countrywide a huge variation in land use to know the various aspects and states of land use, one has

- To map land use
- To report land use statistics
- To carry out land use planning
- To know the land capacity
- Common land use characteristics should be identified

Methodology for LU /LC

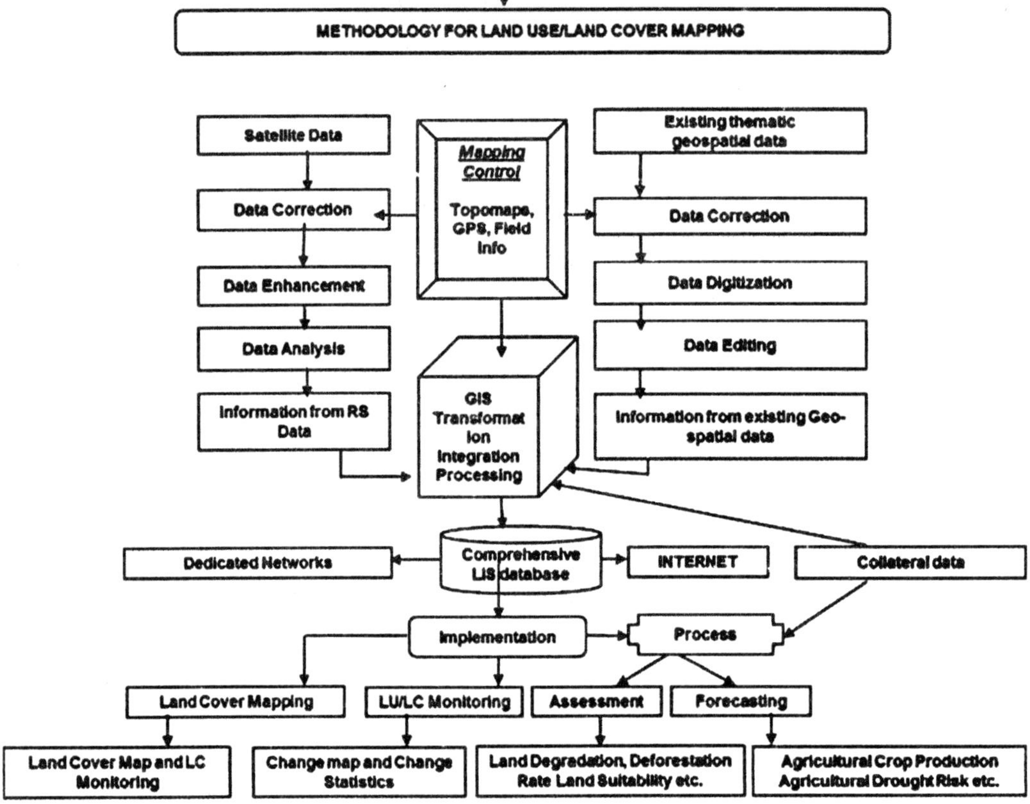

Comprehensive Land Use/Land Cover Classification System in India

LEVEL:	LEVEL-I	LEVEL-II	LEVEL-III
DETAIL:	MAJOR CLASS	TYPE	SUB-CLASS/STATUS
SCALE:	1:1 Million	1:250,000	1:50,000/25,000
01	Built Up	Built Up (Urban) Built Up (Rural)	Residential Industrial Commercial Recreational Public and Semi Public Public Utilities Transportation Mixed Built Up Area Open/Vacant Land Built Up area Transportation

02.	Agricultural Land	Crop Land Fallow Plantation/Orchards Tree Groves	Kharif Crop Rabi Crop Kharif + Rabi (Double Cropped) Zaid Crop Current Fallow Other Fallow Tea/Coffee Rubber Coconut/Arecanut/Cashew Mango/Citrus Horticulture (fruits and Vegetables) Floriculture/Nurseries Farm Forestry Social Forestry
03.	Forest	Evergreen/Semi Evergreen Deciduous (Dry/Moist) Scrub/Shrub Forest Forest Blank Grassland/Grazing Land Forest Plantations	Dense/Closed Open Degraded Dense/Closed Open Degraded Dense/Closed
		Mangrove Non-Forest activities	Open Forest Blank Dry/Moist Alpine Pastures Grassland with tree cover Sal/Teak Pine Bamboo Eucalyptus Casuarina Acacia etc. Dense Open Agriculture Mining Reservoir/Tank Built Up Sanctuary
04.	Natural/Semi-natural Grassland and Grazing land	Alpine/Sub-Alpine Temperate/Subtropical Tropical Grasslands	Moist/Dry alpine pastures Sub-alpine pastures Grasslands with tree cover Manmade Grasslands
05.	Wastelands	Salt-affected land Gullied/Ravinous land Scrub Land Sandy area	Saline/Alkaline Shallow Medium Deep

		Mine dumps/Industrial Waste or dumps Barren rocky/ Stony waste/Sheet rock	Land with scrub Land without scrub Desertic Coastal Riverine Mine dumps/Industrial Waste or dumps Barren rocky/Stony waste/Sheet rock
06.	Wetlands	Inland Wetland Coastal Wetland	Natural Ox-bow lakes/ Cut-off meanders Playas Marshy/Swampy Waterlogged Manmade
			Pisciculture ponds Natural Estuary Lagoon Creek Backwaters (Kayals) Tidal flat/Mud flat Marshy/Swampy Coral Reefs Algae/Weeds Manmade Aquaculture ponds Salt-pans/Salt Dumps
07.	Water Bodies	River/Stream Canal/Drain Lakes/ponds Reservoir/Tanks Ash pond/Cooling pond or reservoir Abandoned quarry pits with water	Higher order river/stream Lined/Unlined Canal and drain Perennial/Dry Reservoir/Tanks Ash pond/Cooling pond or Reservoir Abandoned quarry pits with or without water
08.	Others	Shifting cultivation Snow covered/Glacial area Mining Quarry Brick kilns Reclaimed land	Current Abandoned Fresh/Accumulated snow Active Abandoned Rock/Mud quarry Active Abandoned Active Abandoned

Fig. 1.2

Land use/land cover map using satellite data shown in Figure 1.3.

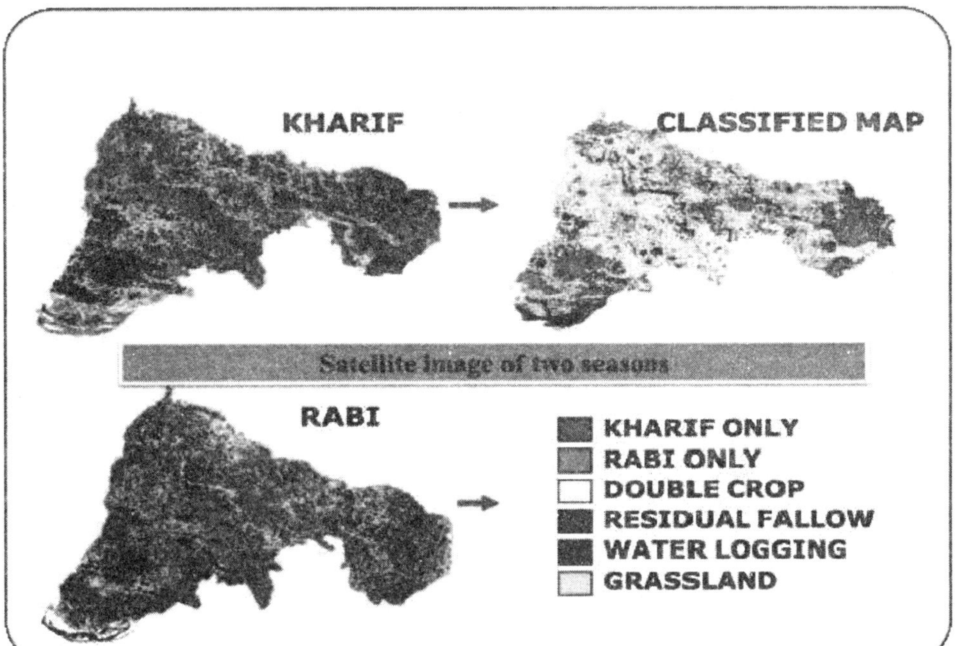

Fig. 1.3: Land Use/Land Cover Map

References

Anderson J.R. *et al.*, (1976). A land use and land cover classification system for use with remote sensing data. Professional paper 964 USGS, Reston, Virginia.

Clawson, M., (1965). *Land use information committee on Land use statistics, resources for the future,* the Johns Hopkins press, Baltimore, USA.

Directorate of Economics and Statistics, Ministry of Agriculture and Irrigation, New Delhi, 1978: Land Use Classification System.

FAO (1963). World Forest Inventory, Food and Agriculture Organisation of United Nations, Rome, Gautam, N.C., and Narayan L.R.A., (1982). Suggested Land use/Land cover classification for India using Remote sensing techniques, Pink Publishing House, Mathura.

NATMO (1986). National Atlas of India, Agricultural and Economic Maps, Calcutta. National Commission on Agriculture, 1976, Part-5, Resource Development, Ministry of Agriculture and Irrigation, New Delhi.

Gautam, N.C. (2015). National Land use/Land cover Classification System, CLUMA, Hyderabad, India.

2

Land Resource of Kerala

An Overview

Srikumar Chattopadhyay and *Mahamaya Chattopadhyay*

Introduction

Land (*Kshiti*) is one of the five fundamental elements of life as enshrined in our scripture. It is the upper most part of the lithosphere that performs several ecologic, economic and social functions to sustain life on earth. All human activities take place on land. Land is considered as prime resource. Access to land is a basic means for livelihood. By definition, land is a delineable area of earth's terrestrial surface and refers to a complex of surface and near surface attributes significant to human (FAO, 1995). Formation of land depends upon interplay of endogenic and exogenic processes. Climate, geology, geomorphology, surface and subsurface hydrology, topography, vegetation and soil characterise the land. Demand for land is escalating with growing population, development initiatives and increasing human activities. A land unit is integrated in nature, and distinctly differs from its surroundings. Each land unit has a production potential and multiple use. Often, these uses are determined through economic and political considerations with little attention to ecosystem services and sustainability issues resulting in land degradation and loss of this capital resource.

Kerala is rich in land resources. The State supports a wide range of crops from subsistence paddy, tapioca, oil seed, coconut to plantations and high value spice crop of cardamom. It is also one of the States in India where access to land resource has been ensured for a large section of people through land reforms. Arguably, land resource base and access to land resources have contributed to Kerala's high human development and poverty alleviation. However, Kerala faces serious challenges in land resource management. This paper provides an overview of land resources in Kerala covering land resource zones/units, present land use, and operational holdings. It also highlights some of the challenges that the State is facing to transit to sustainable land resource management. We attempt here a brief discussion on sustainable land resource management and challenges before dealing with the case of Kerala.

Concept of Sustainable Land Resource Management

Sustainable land management is a global challenge. The concept of sustainable development, which strives to promote a process of development that adheres to economic, social and ecological principles, stressed on land resource management for land use planning. The Chapter 10 of Agenda 21 addresses issues related to land resource planning and management, and called for an integrated approach through reorganising planning instruments and wherever necessary strengthen decision-making structures and policy initiatives. The issue has been deliberated at various levels across the world (FAO, 1995, Beek *et al.*, 1997). Subsequently, a series of guidelines have been issued by Food and Agriculture Organisation (FAO) with the objective of sustainable land use planning (SLUP). The SLUP aims to make optimal and informed choice of the future use of land on the basis of efficient and comprehensive data, a set of nodal institutions, and involving planners, stakeholders and decision-makers. FAO also suggested to document and scaling up of sustainable land management practices as noted in different parts of the world.

Three Basic Issues

Technically, two important questions warranting due attention for sustainable land management are: i) how to achieve land management in a dynamic environment satisfying the socio-economic needs and preserving production potential and (ii) how to plan and manage all uses of land aiming at sustainability and supporting human well-being. To accomplish the tasks indicated in these two questions three basic issues warrant due considerations. These are land evaluation, land use planning, and land governance. Land evaluation forms the foundation of land use planning as it spells out the ecological determinants, production potential of land, usability of land, crop suitability, and limitation to the type of use that a piece of land can be assigned. Identification and delineation of land units based on physio-biotic characteristics forms the first step in this direction. It helps to understand inherent quality of land.

Matching inherent quality of land and land use requirement forms the basic matrix to plan for sustainable use of land. All land units have their specific production potential. Land use may be devised to conform to this potential. Although, originally land use evolved according to natural potential there had been several changes with economic growth, technological innovation and progress of society. Land use changes with changing socio-economic requirements. Assessment of alternative use against needs and aspirations of all social and economic groups is necessary to take decisions to proceed with acceptable land use, and formulate alternative land use and non-use plan. There is renewed thrust on land use science, and integrated land use planning is now promoted as a tool for sustainable use and management of land resources. The objective of integrated land-use planning is to provide a guideline for decision makers and land users in selecting and putting into practice those land uses that will best meet the needs of people while safeguarding natural resources and ecosystem services for current and future generations. Tools and methods for integrated land-use planning at appropriate scales should encourage and assist the diverse and often competing users of land resources in selecting land-use and management options that increase their productivity, support sustainable agriculture and food systems, promote governance over land and water resources and meet the needs of society.

Ownership of land is vested with individuals, community and national and state governments. There are multiple stakeholders. Therefore, land governance plays important role to transit to sustainable land resource management practices and demand management It covers policy, strategy, land tenure, property rights, land market, institutional mechanism to implement land use policies and facilitate appropriate land use practices following change in societal aspirations. Besides, governance is expected to create ample space for public participation in land management **(Fig. 2.1)**.

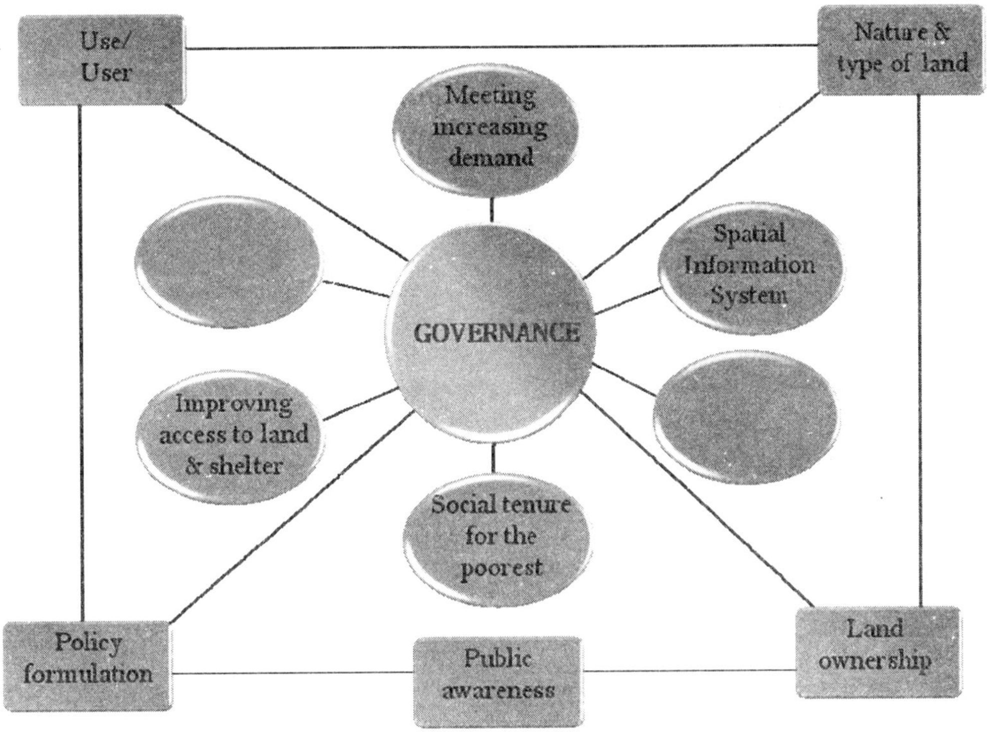

Fig. 2.1: Framework for sustanable land governance system

Land resource management plan may use a set of sustainability indicators to assess viability of the plan to achieve sustainable land resource development (FAO, 1995). These indicators may include both biophysical and socio-economic parameters. Useful indicators may be identified based on eco-regional characteristics of the area concerned.

Biophysical Setup of Kerala

Kerala located in the south-western part of the Indian Peninsula covers an area of 38,860 km^2 or 1.2 per cent of total geographical area of India and accommodates little over 33 million people or 3 per cent of India's population according to the 2011 Census data. Population density in Kerala is 860 persons/km^2, well above all India average of 382 persons/km^2. Sandwiched between the Lakshadweep Sea in the west and the Western Ghats in the east, the State is characterized by asymmetric area altitude ratio with as much as 62 per cent of the total geographical area lying below 100 m (Chattopadhyay and Chattopadhyay, 1995). Seventy per cent of landmass in Kerala fall in the slope category of >15 per cent. Abrupt rise of the Western Ghats from 100 m upward with precipitous slope is a characteristic feature of Kerala's topography that controls hydrology, climate, land use, infrastructural development and settlement distribution. The monolithic Western Ghats is broken by the 30 km wide Palghat Gap at the altitudinal level of 100 to 200 m. Geologically Kerala is a part of the south Indian shield dominated by Precambrian crystalline rocks (Geological Survey of India, 2005). The other formations include sedimentary rocks of Tertiary, and Recent and sub-recent sediments forming the low lying areas, coastal area and river valleys. Laterites occur as cap rock covering all three formations of crystalline, sedimentary and recent sediments. Rock types, their composition and degree of weathering, structure and geological and geomorphological processes influence land resource development.

Kerala falls under the climatic zone of tropical monsoon (Amw). Normal annual rainfall is around 282 cm (Krishnakumar *et al.*, 2009) precipitated throughout the year. Both south-west and north-east monsoons are active however, most of the rainfall occurs during south west monsoon. Spatially, rainfall increases from the coast to inland and the maximum precipitation (451 cms) is recorded along the foothills. It declines further east and the lowest rainfall (<100 cm) is marked in the rain shadow region along the leeward eastern slopes of the Western Ghats. Orographic influence on rainfall is well pronounced. Kerala does not experience winter season like other parts of the country. Average temperature is 28°C. Diurnal and seasonal variations of temperature are largely the same. The average minimum temperature in the winter months does not fall below 20°C in most part of the State. Fall in temperature is due to altitudinal variations. The elevated Western Ghat areas covering parts of Idukki and Wayanad districts record low temperatures during winter months. Kerala receives an average of 9 hours of sunshine per day. Due to high temperature and rainfall chemical weathering is intense as manifested through formation of laterites and deep to very deep soils. The state is well-known for tropical rain forest and luxurious growth of vegetation. Evergreen, semi-evergreen and moist deciduous vegetation are dominant types. Average biomass production is reported as 12 tonnes/ha. There are forty four rivers draining the state. Due to year round rainfall most of the rivers are perennial. The ground water reserve is also good. Biophysical setup of the State has endowed it with rich natural resource base of land, water and biomass.

Land Resources

Kerala is well endowed with highly productive and diverse land resources. Although, most of the lands in Kerala fall in the categories of Class III to Class VIII according to traditional land capability classification scheme, the land here is well suitable for high value cash crops. Land use is intense and multi-tier. Average agricultural income from one hectare of land in Kerala is around two times higher than that of the all India average. Kerala falls in a single agro-climatic region of the West Coast Plain and Ghat Region in all India perspective. The committee on Agroclimatic zones and cropping pattern, set up by the Government of Kerala at the behest of Planning Commission, Government of India identified 13 agro-climatic zones **(Table 2.1)** based on four principal parameters, namely, altitude, rainfall, soil type, and topography (Govt. of Kerala, 1974).

Land utilization has been worked out according to these zones. According to the Indian Council for Agricultural Research (ICAR) Kerala has been divided into five agro-climatic zones like, northern zone, central zone, high altitude zone, problem area zone and southern zone for conducting agro-climatic zone specific research (ICAR, 1991). There were subsequent improvements of this scheme of classification with additional data generated on soil. Following USDA scheme of three-tier land resource categories, like, Land Resource Regions (LRR), Major Land Resource Areas (MLRA) and Land Resource Units (LRU), the Soil Survey Department, Government of Kerala has identified 28 major land resource areas under 10 geographic units (Government of Kerala, 2007). MLRAs are identified based on physiography, geology, climate, water, soil, biological resources and land use practices. Major Land Resource Areas are geographically associated land resource units. Identification of MLRAs is important for state level agricultural planning and for regional and national planning.

Table 2.1: Agro-climatic Zones in Kerala

Name of Zone	*Area in %*	*Soil Type*	*Principal crops*
Highland	27.4	Laterite without B-horizon	Rubber, coconut, cashew nut, arecanut
High ranges	15.9	Red loam	Tea, coffee, cardamom
North midland (Malapuram type)	13.1	Laterite	Coconut, rice, cashew, arecanut, rubber
North midland	11.6	Laterite	Coconut, rice, cashew-nut, rubber, arecanut
Southern midland	10.0	Laterite with B-horizon	Rubber, coconut, rice, tapioca, arecanut
Central midland	8.2	Laterite	Rubber, rice, coconut, tapioca, arecanut, banana.
Coastal Sandy	4.8	Sandy loam	Coconut, rice,
Palakkad	3.9	Red loam	Rice, cotton, groundnut, millet
Chittoor Block	1.6	Black soil	Rice, sugarcane, cotton, groundnut, millet
Onattukara	1.6	Sandy loam	Rice, coconut, tapioca, arecanut
Red loam	1.0	Red loam	Coconut, tapioca, rubber, rice
Kuttanad	0.9	Peat	Rice, coconut
Riverbank alluvium	–	Alluvium	Rice, coconut, sugarcane, arecanut

Source: Report of the Committee on Agro-climatic Zones and Cropping Patterns, Department of Agriculture, Govt. of Kerala (1974).

Note: Total area covers 32,382 sq. km. The rest is under forest and not considered for this zoning.

Terrain Units/Land Resource Units

Although terrain is the base of all these classification schemes, there is pedological bias and process based data are not adequately used. Importance of incorporating process data in land resource management has been increasingly stressed during last couple of decades (Moss, Chattopadhyay and Chattopadhyay, 1995). Topographic grain of Kerala is longitudinal **(Fig. 2.2)**. Three major longitudinal terrain provinces in Kerala are: (i) Coastal areas and associated landscape mostly confined to <20 m dominated by marine/coastal and estuarine processes, (ii) the dissected pediplain mostly lateritic

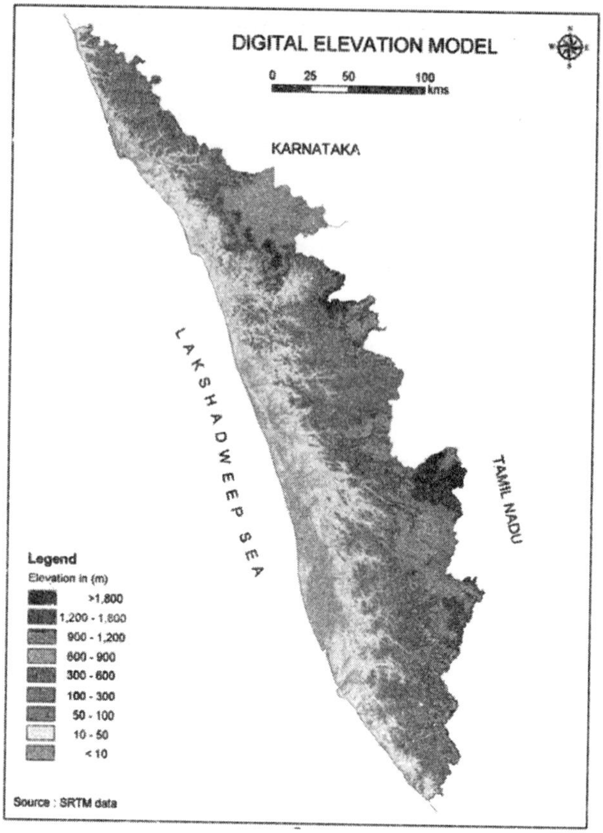

Fig. 2.2: Digital Terrain Model, Kerala

including foot slopes formed by interacting fluvial, denudational and chemical weathering processes, and (iii) the Western Ghats characterized by scarp slopes, plateaus and mountains sculptured by combined action of scarp retreat, slope wash, mass wasting and denudational processes. Considering process–response functions 22 terrain units have been identified under five main morphogenetic processes namely marine/coastal, estuarine, fluvial, denudational-cum-accumulational and denudational (Chattopadhyay and Chattopadhyay, 1995). There are specific agents like erosion and accretion operating under each of these main processes. A land unit under marine processes can be either erosional or accretional or a combination of both. We followed a hierarchical classification system to identify these 22 terrain units and characteristic features have been worked out. It is found that around 85 per cent of the total area of the State falls under the denudational category. The scarp face that is the near vertical slope of the Western Ghats accounts for 4 per cent area. Around 11 per cent of the area has an average slope of ≤10 per cent. A little over 30 per cent of the area is steeper than 35 per cent slope. The area lying below 100 meter altitude accounts for 58 per cent of the total geographical area of the state. Around 45 per cent of the total area is susceptible to erosion. The vulnerable area is about 11,000 sq.km. or 28 per cent of the total area. The isolated hills are mostly structurally controlled and consist of hard crystalline rocks. Brief description of each land units under the respective processes is provided here. A sample terrain map of a district of Kerala is provided in **Fig. 2.3**.

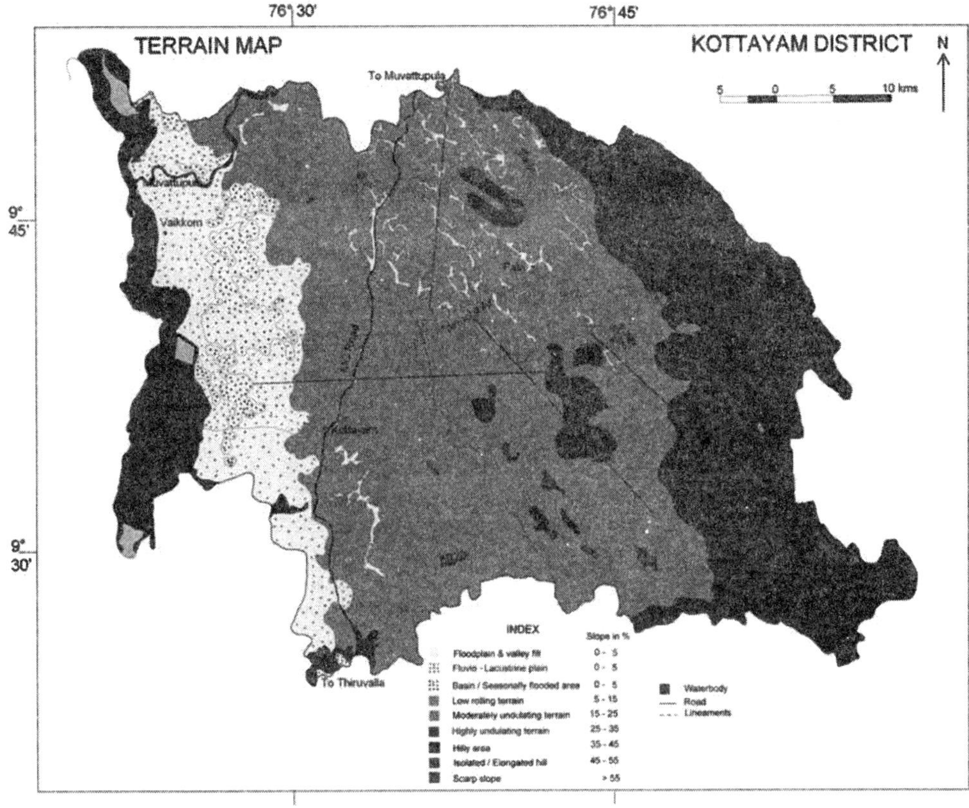

Fig. 2.3: Terrain map of Kottayam District, Kerala

Land Units under marine and fluvio-marine/estuarine Processes (C)

Kerala shows formation of six units under the processes of marine and fluvio-marine. These units can also be considered as have been formed under coastal processes. Four units have been developed under depositional environment and rest two units are of erosional origin. All these units combindly cover 5 per cent of total area of Kerala. Coastal plain cover the maximum area of 3.5 per cent of total area of the State.

Beach (C1): Beach is a landform along the land-ward margin of the sea and lake that lies between the lowest and highest tidal-water level, and whose land-ward extension is marked by permanent vegetation. This unit includes sand bar, barrier bar, spits and pocket beaches. This terrain unit is non-dissected and characterized by a slope of <0.5 per cent and relative relief of <10 m. It is mostly composed of bare sand deposits of various grain sizes. Characteristically, it is not uniform throughout the state and also has not developed everywhere. Width of beach records spatial variations. Beach formation is an indication of accretion process. The central Kerala coast is dominated by accretional process and thus exhibits the most well developed beach. There are newly accreted beach with unconsolidated materials and also eroded beach lying in close proximity. Parts of beaches in Kerala are eroding, which, in most places are protected by sea walls. The rocky coasts show pocket or sheltered beaches as is evident in the Kovalam area and in the Malabar Coast. Kerala beach is well known for placer deposits containing ilmenite, zircon, sillimanite, rutile, garnet, silica and monazite. Chavara coast shows high concentration of this deposit, which produces mineral of strategic importance and Indian Rare Earth industries have developed based on this deposits. Beach is also an important resource for tourism development in Kerala.

Coastal Cliff (C2): The coastal cliffs are mostly lateritic with few exceptions like Kovalam where it is composed of crystalline rocks. Lateritic cliffs at Varkallai and Quilon in the south and near Palur, Kadalur point north of Quilandi, Elathur cape, Azhikode, Ezimala, Cannanore, Pallikkere, Bekal etc. are notable features. These are mostly composed of Tertiary laterites. This unit is low-dissected and characterized by <15 per cent slope and relative relief of <30 m. Soils are mostly residuum and lateritic with land cover ranging from bare rock to natural vegetation. These are the sites of special geomorphic interest and potential areas of tourism development.

Coastal Plain (C3): In general, the coastal plain is narrow however in the tract extending from the north of the Thottapally spill way to the Ponnani mouth it is comparatively wide. The widest part is marked in the Sherthalai area, where it is around 10 km. This area is also noteworthy for glass sand deposits. Another important feature is that, in some places, average height of the sand dunes is more than 10 metre.

A series of dunes/beach ridges marking repeated regression and transgression events characterise this plain. The orientation of these beach ridges varies from 10° – 190° to 155° – 335°. The beach ridges appear to be diverging out from a point just north of the Thottapally spillway **(Plate 1).** This point, in all probability, was a spit during the time when Thottapally opening was hydrologically active in geologically recent times. This unit is non-dissected and is characterized by <05 per cent slope and relative relief of <10 m. Soils are mostly coastal alluvium dominated by sand. Land use consists of seasonal agriculture, tree crops mainly coconut and dense settlements with urban centres.

Plate 1: Satellite view of Parts of the Vembanad lake and surroundings

1. Vembanad lake, 2. Block reclamation, 3. Beach ridges, 4. Spit, C3-Coastal plain, C5. Mud flat, C6. Alluvial plain with paleo strand lines, P1-Old delta plain, P2-Young delta plain, P3-Transitional plain, P4-Fluvio-Lacustrine plain, D1-Low rolling terrain

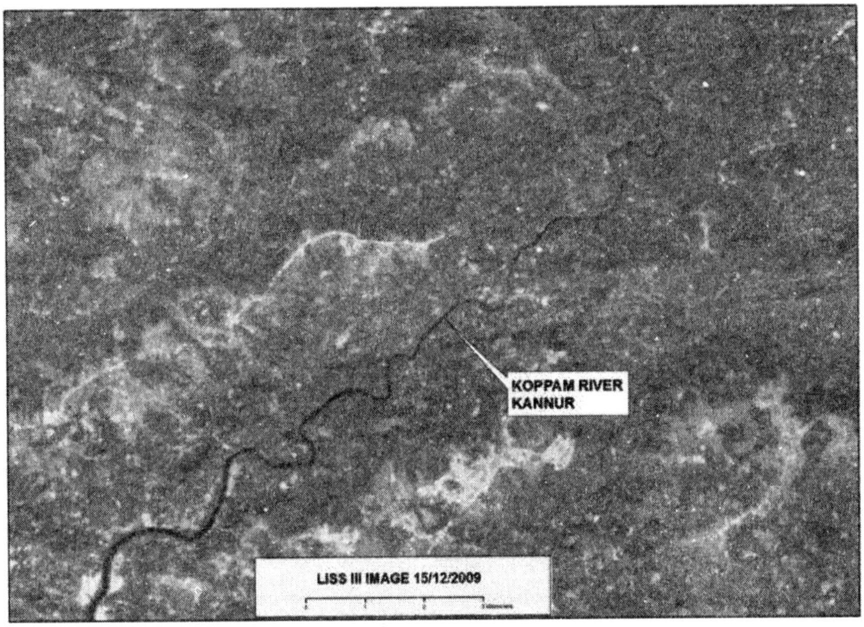

Plate 2: Satellite view of lateritic mesa, Kuppam river basin, Kannur district

1. Settlements with roads and human artefacts, 2. Soup plate in the lateritic mesa

Note the placement of the river in near straight line course. Roads have developed equidistant from the river indicating flatness of the mesa surface.

Coastal Plain-Lateritic (C 4): The lateritic coastal plain is marked mainly in the Malabar Coast and in some parts of Trivandrum and Kollam districts. Two patches, one around Varkallai and another in the Vizhinjam area in the district of Trivandrum are lateritic. In most cases this unit has developed on the Tertiary formations. In Kannur-Kasaragod area, the hard crust laterite is exposed in some places along the coast at 1-2 metre below laterite-derived materials as marine terrace, which is an important geomorphic feature. Teri sand deposits are found in parts of Trivandrum coast. The lateritic coastal plain represents a planation surface and in some areas it is eroding along the sea front. This unit is non-dissected and is characterized by <05 per cent slope and relative relief of <20 m. Soils are mostly lateritic and red loam. Land use covers tree crops mainly coconut and dense settlements with urban centres.

Tidal/Mud Flat (C5): The area which is flat with less than 2 per cent slope and is inundated during every high tide comes under this unit. These clay rich lands are highly fertile. Adjacent to the river mouth these are ideal sites for Mangrove vegetation. The district of Alappuzha has sizable area under this unit **(Plate 1).** These lands are mostly diverted to accommodate pisciculture and coconut plantation.

Alluvial Plain with Paleo-Strandlines (C 6): This unit is most prominently marked in the districts of Alleppey, Kollam and parts of Pathanamthitta. In Alleppey it covers about 13 per cent of total district area. The Haripad-Pallipad area is the type locality of this type of terrain. The eastern border of this unit perhaps indicates paleo coastline

(Plate 1). This unit is non-dissected and is characterized by <05 per cent slope and relative relief of <10 m. Coarse sand deposits of marine origin are marked as grayish Onattukara soil. The landuse is dominated by seasonal crops, tree crops mainly coconut and dense settlements. The clay rich Kuttanad plain (Old and young delta plains) appears to have been indented within this unit.

Coastal water bodies: The lagoons, estuaries (back waters), connecting tidal channels and canals are most conspicuous along Kerala coast **(Plate 1).** All the terrain units described under coastal above are associated with lagoon formation. Portion of these brackish water bodies are being reclaimed. In fact, water bodies are diminishing over the years. Apart from navigation coconut husk retting is an important activity in these water bodies.

Land Units Under Fluvial Processes (F1)

Running water is the most active agent in shaping Kerala's land form. It acts from the hill top to the sea front. The units identified under fluvial processes are those of mainly depositional and in the lower altitudes. Erosion dominated area has been separated under denudational processes. There are six units under fluvial processes. Specific origin covers deposition/accumulation, erosion, transition and complexes. These six units together cover 6 per cent of total area of the State. Each of these units is briefly described here. Flood plain is the most important land unit and covers 3.2 per cent of total area of the State.

Flood Plain (F1): Flood plain developed along the main river courses and its tributaries. This unit also includes valleys and valley fills. It is mostly flat bottomed and occasionally stepped and sloping in the upper part within the midland and high land region. Side slopes bounding these valleys fall under the slope category of moderate to steep. In most cases, these valleys are broad compared to the might of the stream. Apparently, it appears that all the rivers are misfit, however it must be borne in mind that the valley formation in Kerala dominated by lateritisation is principally governed by the process of parallel slope retreat, which results in widening/lateral expansion of valleys and also filling up of valley floors. This has significant impact on land management practices.

This unit is non-dissected and is characterized by <05 per cent slope and relative relief of <05 m. Soil of this unit is alluvial and also comprised of laterite outwash in the narrow parts. The productive potential of these units are very high as it receives sediments, water and nutrients from the vegetated slopes. Heavy monsoon rainfall generates huge amount of sediments, which move down with surface run-off of high velocity. Once it reaches the valley the velocity is lost due to abrupt fall in gradient. This results in deposition of loose materials at the valley bottom/lower part of the slope. In the midland-highland region the valleys are narrow and sloping. These are stepped to accommodate agriculture. These areas are mostly given for paddy cultivation, other seasonal crops and also coconut.

The natural landscape of Kerala is highly modified by human activities. In all these valleys the original channel has been pushed to the sides along the break-in-slope bordering the valley bottom, whereever possible. Flood plains are flooded during monsoon months and help flood cushioning and groundwater recharge. Due to high population pressure these areas are reclaimed particularly in and around the major settlement sites for building purposes and other allied activities. River sand mining from the river beds and flood plains is one of the major activities in the rural area. It generates revenue for the Panchayats. During monsoon, river bank erosion is one of the major problems. Due to incising nature of the rivers, the banks are steep and often experience failure during flood.

High Level Fluvial Terrace/River Terrace (F2): This unit is identifiable in patches at relatively higher altitudes. It is dominated by erosion. In the districts of Pathanamthitta and Kannur, this unit covers some sizable area. Some river terraces away from the main channel exhibit spring effect, due to which paddy cultivation is possible in non-monsoon months. This unit is low-dissected and is characterized by <10 per cent slope and relative relief of <40 m. Drainage density is >1 km/km^2. Soil of this unit is alluvial and also comprised of laterite outwash/coluvic materials in the narrow parts. Seasonal agriculture and mixed tree crops including arecanut dominate the land use practices. Many of these terraces have a pond in the valley head locally known as *Tala Kulam*.

Old Delta Plain (P1): High energy West coast does not favour development of deltaic landscape. However, there is an inland delta in the southern part of the Vembanad lake formed by the Pamba river **(Plate 1).** This unit includes levees, back swamps and derelict channels. This unit is known as the Upper Kuttanad. Similar deltaic landscapes of limited extent in the mouths of Periyar and Meenachil rivers also.

Young Delta Plain (P2): This unit is marked only in the district of Alappuzha. Around 17.4 per cent of the total area of Alappuzha district is covered by this unit. Characteristically, this unit resembles P1 with comparatively lower elevation. Part of this area has been reclaimed from lake bed as block reclamations surrounded by embankments all-around **(Plate 1).** These lands lie below mean sea level and are prone to flood in every monsoon. This area is commonly known as Lower Kuttanad.

The area falling under these two units presents a typical wetland ecosystem with numerous channels and waterlogged condition during monsoon. Both the land units are non-dissected and are characterized by <05 per cent slope and relative relief of < 5 m. Soils are riverine alluvium with high proportion of clay and organic matter. It is also acidic in reaction. Black clay soils are locally known as *Kari* soil. Large paddy fields and seasonal crops dominate land use practices. Young deltaic plain is more prone to flood. Conspicuous features of these two units are the embankments protecting paddy fields from water bodies. Large paddy fields locally known as 'Padasekharams' are given for agricultural practices on cooperative basis. Dewatering is an important activity for land preparation. Settlements have sprung up along the water courses over the embankments and relatively elevated areas. Upper Kuttanad is relatively densely populated.

Transitional Plain (P3): This plain is truly transitional between lateritic undulating terrain and fluvial/coastal plain. In the districts of Ernakulam, Alleppey and Pathanamthitta little over 2 per cent of total district area in each comes under this category. This unit is non-dissected and is characterized by <10 per cent slope and relative relief of <20 m. Drainage density is <1 km/km^2. Alluvium, colluvium and lateritic soils are found here. Coconut, mixed crops and settlements dominate land use practices. Due to alluvium-laterite interface the sub-surface water shows turbidity.

Fluvio-Lacustrine Plain (P4): This landform lies to the east of the Vembanad lake and covers little more than 16 per cent of total area of the Kottayam district. For the State as a whole coverage of this unit is around 1 per cent. Some lacustrine terraces are well identifiable. Patches of paleo sand dunes are also marked. Formation of this unit is genetically linked to the evolution of the Vembanad lake. This unit is non-dissected and is characterized by <5 per cent slope and relative relief of < 5 m. Drainage density is >2 km/km^2. Soil of this unit is alluvium. Part of the area remains waterlogged during monsoon months. Paddy fields, coconut and pisciculture dominate the land use practices.

Land Units Under Denudational Processes (D)

Weathering and erosion are two main denudational processes dominating land forms in Kerala. Weathering includes chemical processes like laterite formation and physical processes like mechanical disintegration of rocks, movement of material under gravity, water flow and sea waves and tides, and blowing winds. Erosion refers to sheet, rills, gullies, mass movement, land slides, land slip, soil piping, river and coastal erosion. Eight land units could be identified in Kerala under this process. They account for little less than 85 per cent of total area of the State. Brief description of each of these units is given here.

Low Rolling Terrain (D 1): Low rolling terrain covers around 17 per cent of the total area of the state. Alternate ridge and valley characterize this unit. Narrow valleys of alluvium and colluvic materials surround laterite mounds and ridges. Dissection index of this unit is low with <15 per cent slope and relative relief of < 20 m. Drainage density is <1 km/km^2. Laterite and alluvium dominate soil types. This type of land is distributed in all the districts of the State in varying proportions. However, its concentration is relatively high in the districts of Thiruvananthapuram, Kollam, Alappuzha, and Kottayam in south Kerala. These lands are intensely used for various purposes. Coconut is the main land use type associated with various other tree crops, rubber and tapioca. Settlements with mixed tree crops mostly along the lateritic ridges and slopes are distributed throughout this unit.

Moderately Undulating Terrain (D2): This type of land covers little over 21 per cent of the total state area. It has an average slope of 15 per cent to 25 per cent. Moderately dissected with relatively relief of <40 m and drainage density of >1 km/km^2 this unit of moderately undulating terrain is characterized by narrow valleys with steep side slopes. Except for the district of Alappuzha all other districts have area under this category of land in varying proportions. Each of the districts of Kannur, Kozhikode and Malappuram in north Kerala has >30 per cent of area under this type of land. The lateritic soil is intensely used for tree crops, coconut and rubber plantations. Settlements are distributed along with mixed tree crops.

Highly Undulating Terrain (D3): This land unit extending over the foot hill zone with high undulations covers 16 per cent of total area of the State. It is found in all districts except Alappuzha. More than 75 per cent of total area of Idukki district falls under this category of land. Although the average slope is less than 35 per cent, in some areas, the slope is considerably high. Relative relief is more than 250 m in some places. The land use is dominated by rubber plantation in south and central parts of Kerala, cashew in the Kollam and Kannur districts and forest patches in some areas. Pepper as the mixed crop is raised everywhere. A number of reservoirs are located within this unit.

The highest rainfall zones around Nariamangalam in south and Kuttiyadi in the north fall in this zone. Widespread deforestation coupled with soil erosion and mass movement are the major environmental problems here.

Hilly Area (D4): This unit represents the Western Ghat hills with more than 35 per cent slope and relative relief as high as 400 metre in some places. Lands are highly dissected with drainage density of >3 km/km^2. Parts of the plateaus are included in this category. Soils are shallow and of residual nature. It covers around 23 per cent of total state area. Forest and plantations of teak, rubber, tea and coffee are the major landuse. The virgin forest area is persevered in the slopes of this unit. All reservoirs are located in this land unit. Mass movement deforestation and soil erosion are immediate environmental problems.

Isolated/Residual Hills (D5): Residual hills are marked in all the districts except Alleppey. Idukki and Wayanad districts account for sizable area under this unit. The slope is more than 35 per cent and relative relief may be even more than 200 m as in the case of Pudalur mala in the Malappuram district. Kalpetta in the Wayanad district is one of the typical examples. These hills are structurally controlled and bounded by steep side slopes. This unit is of limited extent covering <1 per cent of total area of the State. Dissection index is high and drainage density is >3 km/km^2. The soil is shallow and is susceptible to erosion. Agricultural plantations, forest plantations, and grasslands are main land use types.

Scarp Slope/Rocky Cliff (D6): This unit represents the plateau bounding scarp slope. In some places like Pirmed-Periyar area this type of scarp face can be found in two levels. This unit is characterised by near vertical slope with bare rock faces. It covers about 4 per cent of the total area of the state. The prominent scarp faces are marked bounding Periyar plateau, the southern wall of the Palghat gap and western face of the Wayanad plateau. The height of the scarp face reduces towards north from the Wayanad scarp. This unit is characteristically rugged and has forest patches along the water courses. In some places like southern wall of the Palghat gap, it is composed of exposed rock without any trace of soil. The area with soil cover is prone to severe erosion and landslides.

Mesa Surface (M1): This unit represents the lateritic duri-crust widely developed in the northern districts of Kasaragod and Kannur. This type of landform has also well-developed in the districts of Kozhikode, Malappuram and Trichur. Angadipuram of Malappuram district is the type locality of laterite. There is a geological monument erected in Angadipuram. In all other districts of Kerala there are lateritic mesa distributed in patches. Characteristically, it is hard crust without any soil cover. The hard crust may be more than 10 m thick in places, however, in general it is 2-3 m thick. Below this hard crust, there lies soft clay, which, if exposed gets solidified. The CPCRI, Kasaragod has developed certain management practices to make this type of land usable. During monsoon, luxuriant growth of grasses is noteworthy. In some areas 'soup plate' has developed within the hard crust. The soup plates are naturally depressed area having soft materials surrounded by hard crust laterites. In many areas soup plates have been drained to raise crops. Slope of this type of land is <10 per cent with relative relief of <20 m. The surface is low dissected and has drainage density of <1 km/km^2. Hard crust laterite, grass and occasional settlements are important land use. Surangams, a unique traditional water harvesting structures are found in this land unit.

Mesa Side Slope (M2): All mesa surfaces are bound by side slopes, steepness of which may go up to 35 per cent. This land unit is well identifiable only in the districts of Kasaragod and Kannur. This unit is highly clay rich and dotted with laterite boulders (collapsed part of the hard crust) in places. Relative relief is of >20 m. Coconut, mixed tree crops and settlements are the major land use. It is erosion affected.

Unit Under Denudational-Cum-Accumulational Processes (H)

Hummocky Terrain/Units of Half-Orange Relief (H1): This type of landform is extensively marked in the Wayanad district. Other notable areas are Malappuram and Kozhikode districts. This type of land is associated with tropical rainforest. The slope is more than 25 per cent and relative relief is little above 40 m. This unit develops due to a balance between erosion and deposition. The slopes are near stable producing little sediment at present. This unit is extensively under coffee plantation in the Wayanad district. Mixed tree crops and settlements are also found. Erosion is a problem in certain patches.

Micro Level Diversity of Land Types

Diversity of land is also evident even at the micro level. In coastal area ridge runnel topography runs parallel to the coastline **(Fig. 2.3)**. Sandy ridges accommodate settlements, roads and tree crops and the runnels are given for seasonal agriculture-mostly rice and vegetables. Riverine alluvial tract adjoining the sandy coastal plain to the east shows large paddy fields. They are quite extensive in Kuttanad area of Alappuzha and Kottayam, kole lands of Ernakulam and Trissur and Bharatapuzha valley of Trissur, Malapuram and Palakkad. Towards east, the main valleys are in east-west direction indented within lateritic terrain and extending up to foot hills. Alluvial valleys/flood plains gradually narrow down from west to east direction. The land arrangement, often noticed in lower elevations of Kerala consists of a lateritic ridge/mound with plantations of rubber/tapioca surrounded by lateritic slopes with coconut, mixed tree crops and settlements which finally merges with alluvial valleys mostly given for rice cultivation bordered by banana and arecanut. All these arrangements are like ribbon pattern and recurring. Valleys are continuously enriched with sediment, other mineral nutrients, water and organic matter moving down from surrounding elevated areas through colluviation and alluviation. This diversity of land even at the micro level has contributed to diverse crop practices and enhanced productivity and contributed to livelihood security. Interdependence between valleys and ridges are well manifested in these spatial arrangements.

Land Use

Land in Kerala is intensely used. Hardly, there is any vacant or unused land in the State. Land use pattern of Kerala is quite different from the rest of India. Landuse statistics have been published for the State for the period beginning 1908-09. During this period, since the present Kerala state did not exist, different parts of the State had different administrative arrangement, with some parts under native kings and other parts under the Madras Presidency of the British rule, there are certain problems about the compatibility of data. Therefore only three broad categories—forest, non-forest and agriculture are discernable during this period. In these early years, land survey had not covered the entire State. Available data for 1908-09 showed that total area reported was 38,580 km^2, of which agriculture accounted for 53.5 per cent. The rest 46.5 per cent comprised of un-surveyed area (mostly forest) (25.6%), forest area (6.5%) and non-agricultural area (14.4%). In 1956-57 when the modern Kerala formed consequent upon reorganization of states in the country on linguistic line area under agriculture covered 61.2 per cent of total geographical area of the State. Most part of the State was surveyed by this time and area under agriculture increased by 7.7 per cent. By 2016-17, area under agriculture including fallow lands accounts for little over 55 per cent of total area of the State. Agriculture occupying more than 50 per cent of total area since the first decade of 20 th century indicates the extent of pressure on land experienced in Kerala even more than 100 years back. The high pressure on land is further evident from the fact that the population density was 165 persons/km^2 in 1901 which increased to 184 persons/km^2 in 1911. The corresponding figures for India were 77 and 82 respectively.

Nine-Fold Classifications

A nine-fold land use classification for the State is available since 1957. The Department of Economics and Statistics, Government of Kerala, continues to publish land use data following this standard classification scheme till 2004-05. Since 2005-06, four more categories, namely, marshy land, still water, water logged area and social forestry were added to the existing nine-fold classification, thereby increasing the number of classes to 13 **(Table 2.2)**. Brief description of all these land use classes is provided here.

Table 2.2: Distribution of Area under Different Land use Classes in 1960-61 and 2016-17

Land use Class	*1960-61*		*2016-17*		*Change*
	Area (in hectares)	%	Area (in hectares)	%	%
Forest	1,064,000*	27.4	1,081,509*	27.83	+ 1.64
Land put to non-agricultural use	206,000	5.3	441,934	11.37	+114.50
Barren and uncultivable land	155,000	4.0	11,780	0.30	– 92.40
Permanent pastures and grazing land	47,000	1.2	0	—	–100.00
Land under miscellaneous tree crops	206,000	5.3	2450	0.06	–98.8
Cultivable waste	144,000	3.7	101379	2.61	–29.60
Current fallow	67,000	1.7	72008	1.85	+13.80
Fallow other than current fallow	62,000	1.6	55530	1.43	–10.40
Marshy land$			106		
Still water$			98343	2.53	
Water logged area$			3210	0.08	
Social forestry$			2556	0.07	
Net area sown	1,934,000	49.8	2,015482	51.86	+4.21
State Total	3,885,000	100.00	3,886,287#	100.00	

Source: Department of Economics and Statistics Government of Kerala, Tivandrum

* There was some area readjustment in the forest over the years.

Area reported as per village records

$ Area under these categories enumerated since 2005-06

Note: 100 hectares = 1 square kilometer. Figures in various tables are in one or the other of these measurement units.

Forest: Before reorganization of State in 1956, the forests in Kerala were owned both by the government and individuals. Once the state was formed, all forestlands were vested to the government and was brought under the administrative control of the Department of Forests. In 1956-57 the forest area was 10,180 km^2 and constituted 26.39 per cent of the total geographical area of the State. There were some readjustments in the surveyed area over the years. Forest area was 10,640 km^2 in 1960-61 and since 1980-81 till date (2016-17) the reported area under forest was 10,815 km^2 or 27.83 per cent of total geographical area of the State. This much area is under administrative control of forest department. Besides natural vegetation, there are forest plantations, agriculture and other land uses practiced within the forest areas. Forest vegetation is mostly concentrated in the rugged Western Ghats slopes. Forest vegetation cover came down from 44 per cent in 1905 to 17 per cent in 1973 (Chattopadhyay, 1985). Littoral forests are limited to few patches along coast line and estuaries. There are six districts in Kerala with more than 30 per cent area under forest. Among those six one district namely, Pathanamthitta has as much as 59 per cent of its total geographical area under forest. Alappuzha district confined to coastal plain and lowlying areas does not have any appreciable natural forests.

Land put to Non-agricultural Uses: This category includes buildings and courtyards, parks, play grounds, industry, roads, railways and other non-agricultural uses like brick kilns, sites of clay and sand mining in the flood plains and water bodies used for pisciculture. Agricultural statistics of 2016-17 recorded 4420 km^2 or 11.5 per cent of total geographical area of the State under this category. Area under this category showed steady increase although four additional categories were identified to record water related uses and social forestry since 2005-06. There is positive correlation between population density and this category of land put to non-agricultural uses. Analysis of time series data indicates that the area under this class increased by 34 per cent between 1960-61 and 1970-71. However from 1970-71 to 1991-92, the rate of increase fell to 9.5 per cent, and again it rose to 42.9 per cent between 1991-92 and 2004-05. The later period (1991-92 to 2004-05) witnessed rapid expansion of construction activities, specially buildings and roads and clay mining from flood plains. In addition to these aquaculture is also spreading over. In the year 2006-07 when water related uses were separately accounted for, and this land use category of land put to non-agricultural uses represented more or less only the built up area it was found that urbanized and industrial tracts like Ernakulam-Trissur-Palakkad and Thirvananthapuram and surroundings were experiencing higher growth rates under this category. This trend continues.

Barren and uncultivable land: This includes rocky and sandy wastelands, marginal lands and degraded areas. By 2016-17 the State had 118 Km^2 or 0.30 per cent of total area under this category. Around 40 per cent of total barren and uncultivable land in the State is confined to Kasaragod district. These barren lands are mostly hard crust laterite exposures. The area coverage of this category declined from 4.02 per cent in 1960-61 to 1.42 per cent in 1991-92 and to 0.30 per cent in 2016-17. Idukki experienced the maximum reduction. Most of the usable and semi-usable lands are brought to some uses. Hard rocks are used as foundation for building construction in Kerala, therefore, there are rock quarries throughout the State. Environmental impact of rock quarrying is a serious issue drawing attention of the Earth Scientists from time to time. The barren hard crust lateritic wastelands in north Kerala are utilized for building purposes. Sandy patches are used for settlement and industry. Some of the barren areas are given for social forestry. The reduction of area in this category indicates increasing utilization of marginal lands.

Permanent pastures and grazing land: At present (2016-17) there is no land reported as permanent pasture and grazing land in Kerala. Land is so valuable and productive that all pasture lands have been put some other use. This has significant impact on livestock development. Stall feeding is the only means for raring livestock in the State today. Kerala has to import fodder from the neighbouring States to cater to the needs of livestock population. In 1960-61 the state had 470 km^2 or 1.17 per cent of total geographical area under this category. Reduction of area under this category indicates loss of village commons, which used to support livestocks and supplied fuelwood to the marginal sections of society. Most of these areas have been converted to agricultural use.

Land under miscellaneous tree crops: This land use class covered only 24.5 km^2 of area in Kerala in 2016-17. Palakkad is the only district which has 10 km^2 area under this category. State level area coverage of this category has come down from 2055 km^2 in 1960-61 to the present level. The average annual reduction is around 36.25 km^2. Between 1960-61 and 1991-92 the annual rate of decrease was around 56 km^2 and the period from 1991-92 to 2016-17 witnessed the annual reduction rate at around 19 km^2. This category falling under dry lands mostly covers garden lands. Varieties of tree crops were formerly grown in dry land areas. With expansion of agriculture these areas were transformed mainly for cash crop. The decline in this category indirectly indicates that village level biodiversity is being lost. With conversion to agriculture, the range of crops raised in a village is reduced. Loss of this type of land reduces the amount of free firewood available to the poorer sections of society. Idukki district which had around 113 km^2 area under this category is now left with only 2.3 km^2 area.

Cultivable waste: This category includes the type of land that can easily be brought to cultivation but has now been left as wasteland. It accounts for 2.6 per cent of total area in 2016-17. In 1960-61 the share of this category of land was 3.73 per cent indicating an overall reduction of 29.6 per cent by 2016-17. However, time series data indicate the trend was fluctuating. The decline trend continued till 2005-06 and subsequently it started increasing. There was 53 per cent increase from 2005-06 to 2016-17. As many as nine districts experienced increase in area under cultivable waste during this period. Substantial increase in cultivable wasteland in the districts of Alappuzha, Pathanamthitta, Kottayam, Ernakulam and Trissur is a matter of serious concern as most of the uncultivated paddy fields in these districts constitute bulk of this class of land. It would be expected that with an increase in population pressure and demand for land, the areas of cultivable waste would be substantially reduced. This was not the case however. One reason might be that the land is only cultivable for a certain type of crop that the cultivators do not favour. It is also possible that the size of holding is too small for profitable operation. In the case of highly urbanized districts like Ernakulam, Kannur etc. cultivable land might have been acquired/proposed to be acquired for some other use and for the intermittent period it was left as wasteland.

Current fallow: This category includes seed beds or fields which could not be put to cultivation for reasons related to agriculture. Over the seasons and years, spatial distribution of fallow land varies. The same lands are not kept as fallows year after year. This land use class accounts for 1.85 per cent of total area in the State in 2016-17. Ernakulam, Thrissur and Palakkad districts together recorded >50 per cent of all current fallows in the State. There was substantial reduction in area under this category from 670 sq. km. in 1960-61 to 230 sq. km. in 1970-71. This was a period of agricultural expansion, introduction of high yielding varieties and modernization of agricultural practices. The maximum area under paddy cultivation was reported in 1976-77, when the Government of Kerala was seriously thinking to make self-sufficient in foodgrain production. Subsequent years witnessed gradual increase in area under current fallow. By 1991-92 area under current fallow rose to 439 sq. km. and in 2010-11 it further went up to 760 km^2 and in 2016-17 it stood at 720 km^2. However compared to 2006-07 when current fallow was 816 km^2 there is a reduction of 96 km^2 by 2016-17. While there is increase in every district except Wayand and Kannur, the rate of increase is not the same. Increase in area under current fallow indicates that cultivable lands are not fully utilized. Most of the current fallow lands are uncultivated paddy fields. Recent reduction in current fallow is due to strong initiatives undertaken by the Government to promote paddy cultivation.

Fallow other than current fallow: This category includes lands which are kept fallow for a longer period. These lands are suitable for agriculture but are not put to use for various reasons that may or may not be directly related to agriculture. A little over 555 km^2 of area has been estimated under this category in 2016-17. Concentration of this category of land is mostly in the rice growing areas spreading over Alapuzha, Ernakulam, Thrissur, Palakkad and Malappuram districts. Both categories of fallow lands are common feature in rice growing districts. There are various socio-economic reasons underpinning this scenario. Time series trend indicates reduction of area under this land use class from 1.65 per cent in 1960-61 to 0.6 per cent in 1970-71. After 1970-71 the rate leveled off at 0.7 per cent till the 1992-93. Subsequently, area under this category started rising and by 2016-17 it accounted for 1.43 per cent of total area of the State.

Net sown area: Kerala has 20,154 km^2 area or 52 per cent of total geographical area estimated as net sown area in 2016-17 against an all India average figure of 41 per cent. Net sown area includes wetland agriculture, dry land agriculture, tree crops like coconut, and agricultural plantations. There is very little scope for further expansion of net sown area in most of the districts. Water bodies restrict the expansion of net sown area in the coastal area and there are forests in the Western Ghats to delimit the expansion. Net sown area in the State was 49.9 per cent in 1960-61, which rose to 57.86 per cent in 1991-92 and subsequently declined to 51.9 per cent in 2016-17. Net sown area varies from 30 per cent in Pathanamthitta district to 73 per cent in Kottayam district **(Fig. 2.4).**

Initial growth in net sown area had been at the expense of almost every other category of land. Expansion of rubber cultivation has contributed significantly in expansion of NSA and decline of land under other categories. The subsequent reduction since 1991-92 is a post-globalisation/liberalisation phenomena associated with growth of real estate business. Growth of area under non-agricultural use and keeping cultivable land fallow for longer period are two contributing factors of decreasing NSA. Compared to 1991-92, all districts, except Idukki, recorded decline in net sown area. However, rate of reduction varies spatially. Idukki shows more than 9 per cent growth in this category, which is at the expense of all other categories, particularly barren and uncultivable land, cultivable waste, and land under miscellaneous tree crops. Bringing land from cultivable waste to NSA is a positive trend. However, use of barren and uncultivable land is tantamount to expand agriculture to marginal lands, which may not be productive in future. Similarly, reduction in area under miscellaneous tree crops also indicates loss of biodiversity in non-forest areas.

Fig. 2.4: District-wise distribution of Net Sown Area (NSA) in Kerala

Additional Land Use Categories

Since 2005-06, four additional categories have been enumerated (Government of Kerala, 2006). These are marshy land, still water, waterlogged area and social forestry. All these categories were earlier counted under the class 'Land put to nonagricultural use'. Among these four categories only 'Still Water' covered 983 km^2 by 2016-17. Still water category includes lagoons, lakes, reservoirs and similar other large water bodies. The Vembanad lake is the main still water body in Alappuzha, Kottayam and Ernakulam districts and in Idukki district there are a series of impounded reservoirs like Idukki, Pallivasal, Sengulam and Periyar lake forming the still water body. Waterlogged area accounted for 32 km^2 followed by Social forestry (25.5 km^2) and Marshy land (1.0 km^2). Waterlogged area covers permanently waterlogged area, both natural like ox-bow lakes, left out channels, and shallow water bodies and man-made waterlogged areas mainly refer to the ditches caused by sand, clay and rock mining. These areas may be productively used through better management. Social forestry is a major organized social intervention to increase natural vegetation cover as a part of environmental management. Waste lands, vacant lands along roads and railways and degraded forest areas are usually given for social forestry. It is expected that once these vegetations are well established they will be included under forest cover.

Agricultural Land Use

Land use pattern of Kerala is quite different from that of the rest of India. Multi-tier cropping system and dominance of tree crops and plantations are unique features here. Based on duration of cultivation or life time of crops the common classification is (i) seasonal crops like paddy, tapioca, pulses, vegetables etc., (ii) annual crops like banana, plantain, sugarcane, betel leaves etc., and (ii) perennial crops like coconut, arecanut, jack, mango, cashew etc. and all plantation crops. Another common classification based on use of crops is food and non-food crops. Areas under seasonal crops, which are mostly food crops are used for cultivation more than once. Paddy fields are therefore used for two to three times in a year. Multi-tier cropping and inter-cropping help use of the same agricultural fields for a number of crops at a time. During 2016-17, around 5658 km² or 28.2 per cent of net sown area was sown more than once. The State had a gross cropped area of 25840 km² and crop intensity of 128.2 per cent. Both total cropped area and net sown area shows declining trend **(Fig. 2.5).** Compared to 2001-02 there was a fall of 13.6 per cent by 2016-17.

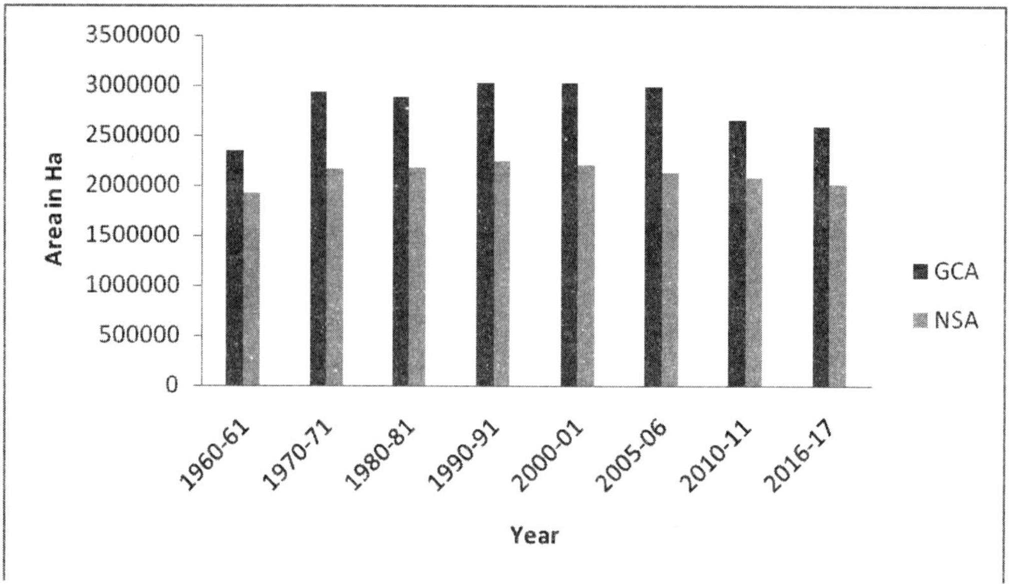

Fig. 2.5: Temporal variation of net sown area (NSA) and gross cropped area (GCA), Kerala

Food and Non-Food Crops

Crop husbandry witnessed a gradual change in favour of non-food crops. The ratio between area coverage of food crops and non-food crops was 67:33 in 1960-61, which turned to 50:50 by 1990-91 and now in 2016-17 it stood at 37: 63 **(Fig. 2.6).**

This is significant in Kerala as the State is deficit in foodgrain production. In case of rice there is a drastic change, from 40.5 per cent of total cultivated area in 1960-61 to 6.6 per cent of total cultivated area in 2016-17. India witnessed food shortage during 1960s and started importing food. There was all out efforts by the Government to produce food. This had culminated in Green Revolution and by 1970s India became self-sufficient in foodgrain production. Kerala was not in the belt of green revolution, however, it recorded introduction of high yielding varieties of paddy and steady increase in area under paddy till 1981. Subsequently, paddy acreage started declining. Another important crop, tapioca, also recorded decrease after reaching the maximum coverage in the year 1976-77. Compared to 1960-61 percentage share of food crops like paddy and tapioca

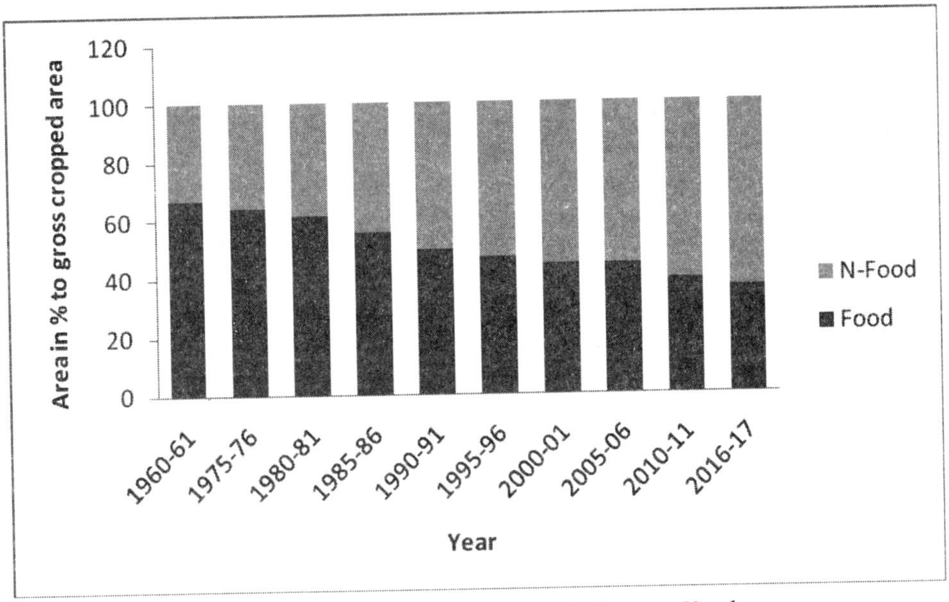

Fig. 2.6: Temporal variations in area under food and non-food crops, Kerala

decreased by 69 per cent. Area under nonfood crops like rubber, coffee, areca nut, and pepper recorded steady increase. Rubber and coffee gained the maximum by 328 per cent and 405 per cent respectively. Paddy was 1st ranking crop till 1980-81. By 1990-91, it was second ranking crop after coconut and by 2000-01 paddy ranked third below coconut and rubber. Coconut cultivation is expanding at the expanse of rice cultivation in coastal plain and low lands whereas rubber is replacing degraded forests, tree crop and tapioca. Land use change has serious implications on water quality and over-all hydrological conditions (Chattopadhyay *et al.*, 2005, Chattopadhyay and Chattopadhyay, 2014).

Challenges in Agricultural Sector

Decline in net sown area, fall in gross cropped area, increase in area under non-food crops and increase in fallow lands are challenging issues for agricultural land management. Urbanlands are encroaching over the productive agricultural fields. Bossurpian hypothesis (1965) postulates that crop intensity will increase with population density. District level data on crop intensity and population density do not show this relation **(Fig. 2.7)**. Crop intensity remains same and sometimes shows decline trends over the years. Contribution of primary sector in State Domestic Product is also diminishing. Agricultural sector is not remunerative and there is common apathy among younger generation to opt for agriculture. Revival of agricultural sector warrants total mechanization, land consolidation, group farming approach and integrated farming. The conventional approach may not be effective under these circumstances. Land consolidation is necessary to overcome this problem. Group farming as practiced in the waterlogged farm lands (*Padasekharams*) of Kuttanad area may provide certain lessons to devise the alternatives.

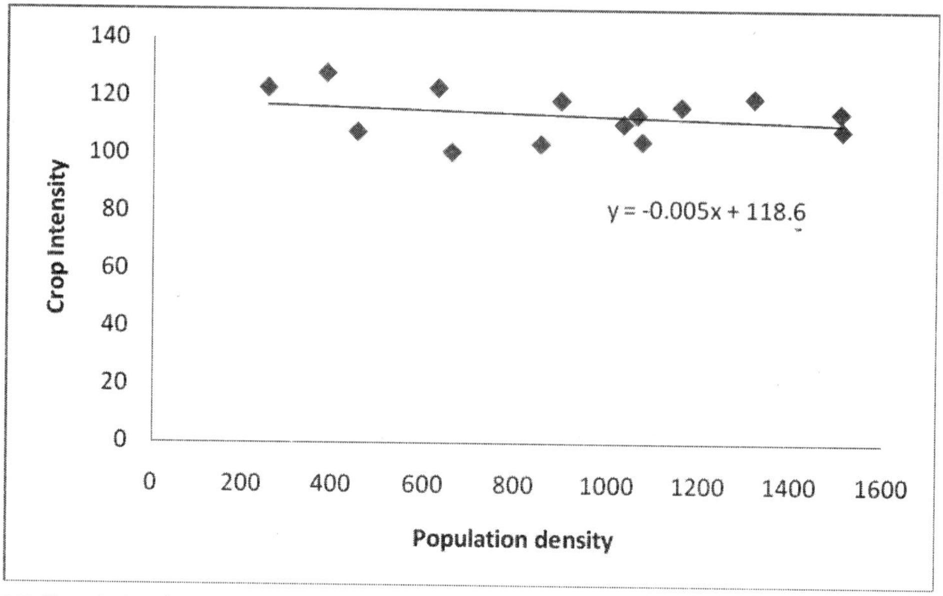

Fig. 2.7: Correlation between population density and crop intensity, 2011

Group Farming—A Topographic Necessity

Due to the undulating topography, which covers about 80 per cent of the landmass in the state, land parcels in the valleys/low lands receive sediment and water from the adjoining plots located up slope. Sub-surface water in any plot/area down slope is recharged through infiltration from the up slope plot. Under these conditions prudent land management may not succeed in individual plots. When the fragmentation level and number of holdings is so high it is quite natural that not an individual but a group of farmers should be involved in operating in a single topo-sequence, i.e., ridge to valley segment. Resource augmentation at the micro watershed level following this topo-sequence approach through group activity could possibly be a second phase of land reform (Chattopadhyay and Franke, 2006). Similar considerations may well apply to the house compound plots on which people grow coconuts, mangoes, papayas, teak, and other crops. When an individual owns only a small plot the preference might be to build a house. Many of the marginal landowners of plots in the rice fields have opted for conversion to residential use. This should also be an issue in any second phase of land reform (*ibid*).

Land Use and Landform

Relation between land use and landform is quite explicit in Kerala. Nature of land use varies with altitude **(Plate 3).** The coastal plain and the adjoining lowlands are dominated by coconut and rice. This segment of land is densely populated. All major urban centres are located in the coastal plain. The main transport lines, railways and roadways have developed in the coastal plain and adjoining lowlands. The undulated lateritic terrain

Plate 3: A panoramic view of land use across different landforms in central Kerala.

Note the sequence: Rice in the valley, banana in the terrace, coconut in the lower slope, Tree crops/rubber plantations over low lying ridges along foothills and natural vegetation in the upper hills.

hosts crops like coconut, varieties of fruit trees, rubber, cashew, tapioca, pepper etc. Coconut dominates the land up to 100 m altitude. Rubber plantation is an important component of land use up to 500 m altitude. Settlements are scattered amidst mixed tree crops. Further up, between 600 to 900 m altitude there are tea and coffee plantations. Cardamom is raised above 1000 m altitude. The rugged scarp slopes and high hills are forest clad. The boundaries between land use categories are not razor sharp. There are overlapping, nevertheless a broad zonation is well evident. Due to pressure on land and multiple usability crops areas are extended with little regard to and landscape ecology or land suitability. This impedes proper use of land resources.

Arable Land Availability and Demand

Kerala is a land scarce state with per capita availability of arable land standing at 0.10 ha against all India average of 0.11 ha by 2017. The average per capita requirement of arable land is 0.4 ha for sustaining a family as suggested by Mukherjee (1938) and recently adopted by Radha Devi and Ajit Kumar (2011). Field survey also corroborates this assumption. With technological improvement, use of high yielding varieties of seeds, fertilizer and water management, and having the provision of multi-tier cropping there may be contest about the requirement of amount of per capita arable land. However, given the improved standard of living and higher demand of non-food items the area requirement of 0.4 ha is reasonable. Even in case of North America and Western Europe, where food basket contains large proportion of meat an area of 0.5 ha of arable land is proposed as minimum requirement (Myers, 1999). 'It is realistic to suppose that the absolute minimum of arable land to support one person is mere 0.07 ha-and this assumes a largely vegetarian diet, no land degradation or water shortage, virtually no post-harvest waste and farmers who know precisely when and how to plant, fertilise, irrigate etc.' (Meyers, 1999; FAO, 1993). However, this refers to practicing precision land management, which is an ideal condition. A society may strive to attain this level of land and water management in due course. Considering per capita land requirement and availability it may be surmised that Kerala experiences four times pressure on arable land and therefore the State has to depend on external sources for sustenance of her people.

Access to Land

Access to land is an important factor for sustainable land management. Like rest of the country, Kerala inherited a feudal system of land tenure. Government of Kerala initiated the process of land reform immediately after formation of the State in 1956, and finally in 1970s land reform was introduced. It provided access to land resources to a large section of people, who were involved in land management but without any ownership right. Through a ceiling provision and a transfer of ownership process, 1.29 million tenants got title deeds to 1.97 million acres of rice land, about 1.5 acres, or 0.6 ha each. Another 340,000 households received house compound land of about 0.08 acres or 0.03 ha (Franke 1996). Land reforms have accorded ownership to people, who are not traditionally land owners. This has altered the land-based power structure, hitherto prevailing in Kerala. However, land reform is yet to be fully completed. There are issues like exclusion of plantation from the ambit of land reforms and non-implementation of land ceiling act. Besides, the task to accord entitlement to the underprivileged and tribal groups remained incomplete, which is causing social tension (Scaria, 2010) as evident from tribal movements from time to time. Government of Kerala launched Zero Landless Kerala programme. Kannur district was declared as the first district in India with Zero Landless Citizen in 2013 (Government of Kerala, 2013). Providing access to land for landless people and marginal groups is a challenge.

Operational Holdings

Size of operational holdings plays an important role in agricultural productivity. Land management becomes difficult when the size of the operational holding is too small. Small size farms are increasing in the State due to a variety of reasons like increasing population density and concomitant increase in number of house holds, breaking down of joint families, implementation of land reforms and distribution of excess lands.

Growing urbanization and increased investment in real estate also play their role in fragmentation of land holdings. This may lead to under-utilization of land resources, which is quite limited in the State given the high population density and intense use. Distribution of operational holdings bears close relation with physiography. Size of holdings increases towards highlands.

Growing Fragmentation of Holdings

Average size of land holdings in Kerala is extremely small and it is getting further reduced over the years. Average size of individual and joint holdings decreased from 0.37 ha in 1990-91 to 0.22 ha in 2010-11 (Government of Kerala, 2013). Fragmented nature of land holdings in Kerala is well-manifested in size class distribution. Number of operational holdings under different size class and their area coverage decrease with increase in size of holdings. Among the five size classes that the Department of Economics and Statistics recognized for presenting data on operational holdings the marginal group (up to 1.0 ha) accounted for 96.33 per cent of all operational holdings and 58.62 per cent of total area coverage in 2010-11. If the small group (1 to 2 ha) added to this class the number of operational holdings in Kerala will rise to 98.9 per cent and the area coverage will be little over 77 per cent **(Table 2.3)**.

Table 2.3: Operational holdings in Kerala (1966-67 to 2010-11)

Size class	*Number of holdings (%)*				*Area covered (%)*			
	1966-67	1985-86	2000-01	2010-11	1966-67	1985-86	2000-01	2010-11
Marginal (<1 ha)	81.79	91.53	95.17	96.32	31.07	46.02	56.23	58.64
Small (1-2 ha)	10.09	5.76	3.41	2.64	19.63	21.55	19.10	18.66
Rest (> 2 ha)	8.12	2.77	1.42	1.04	49.30	32.43	24.67	22.70

Source: Government of Kerala, 1975, 2013.

The number of operational holdings in the size class of ≤ 1 ha increased by 237 per cent in a span of 30 years from 1970-71 to 2000-01 against the overall growth of 189 per cent for all size classes together during the same period. All other categories showed decrease in the number of operational holdings. The maximum percentage decline was in the size class of >10 ha and the decline in number was in the size class of 2-4 ha. At the all India level number of operational holdings and area covered under the size class up to 1 ha was 46.02 per cent and 13.20 per cent respectively in 1985-86. Data available for 1995-96 showed that in Kerala the number of holdings below 0.02 ha was 7.48 lakh, which covered only 9741 ha of area. This indicates an average holding size of 0.013 ha or 3 cents.

Fragmentation—A Topographic and Social Reality

Fragmentation of holdings is another major issue in sustainable land management. Kerala's extreme landholding fragmentation threatens to undermine the beneficial effects of the land reform in terms of the distribution of wealth producing assets. Topographic control on fragmentation of land is well evident. Size of plots is small in the valleys and narrow flood plains, usually known as wetlands and given for paddy cultivation. Plot size increases along the side slopes and ridges, which are considered as garden lands or dry lands, mostly given for tree crops, settlements and plantations. Prior to land reform, almost all the land owners will possess some portion of wetlands and some portion of garden lands. With increase in population and consequent division of property both types of lands were divided among the heirs. Apart from population growth disintegration of joint family systems among upper castes contributed to fragmentation. Growth of number of operational holdings during 1990s and subsequent years can be attributed to liberalization, increased money flow, international economic melting down, problems in Gulf countries and growing tendency of people to invest in real estate. Age structure of population also plays an important role in fluctuating growth of operational holdings. As the valleys accommodating paddy fields are limited in Kerala intensity of fragmentation is relatively high in this type of lands compared to the lands in the slopes and ridges. With land redistribution following land reforms non-traditional land owning community benefitted however, size of holdings reduced through property division. Besides, the holdings of majority of working farmers are too small to provide adequate returns from farming and the small plots are fragmented in ways that reduce the effectiveness of technological and environmental interventions if managed separately.

Fall in Operated Area

Area under operational holdings shows a fluctuating trend. The maximum operational area reported since 1970-71 was 18.05 lakh hectares enumerated in 1980-81. Incidentally it may be noted that in 1966-67 operational area was still higher at 18.28 lakh hectares. Continuous diversion of agricultural land to accommodate other uses is a matter of serious concern at present, however, the trend was marked even before 1970. Growth in operational area has not only slowed down but also was negative in some periods. Operational area declined from 17.96 lakh ha in 1990-91 to 15.70 lakh ha in 2000-01 and further down to 15.10 lakh ha in 2010-11 (Government of Kerala, 2013). This indicates that 2.86 lakh ha area has gone out of operation in twenty years' period. The average declining rate was much higher (22,600 ha/year) during first ten years since 1990-91 compared to the next decade of 2000-01 to 2010-11 when the annual rate declined to 6000 ha. The higher declining rate is largely attributable to real estate boom and expansion of non-agricultural activities linked to economic liberalization/globalization initiated in early 1990s. This steady decline is a matter of serious concern.

In recent years agricultural land specifically paddy growing lands are being kept fallow for various reasons, as a result, operational area shows declining trend. There is a geographical limit to which operational area can be expanded and it has almost reached to a saturation level, whereas the number of holdings is growing primarily due to population increase and division of property. Therefore, the holding size is getting more and more fragmented. Since 1980-81 the number of operational holdings decreased under all size classes except the size class of below one hectare, and in the case of >10 ha size class the number increased for one time point between 1980-81 to 1985-86. By 2010-11 there are only 1854 holdings in the size class of >10 ha. Of these only 1261 holdings are in individual possession. Operational area under this large holding size class declined from 12.6 per cent in 1970-71 to 7.92 per cent in 2010-10.

Spatial Variations of Land Holdings

Among 14 districts in the State Wayanad has 1,74,416 operational holdings or 2.4 per cent of total holdings in the State which is the lowest and Thiruvananthapuram has 7,50,903 operational holdings or 12.4 per cent of total holdings in the State, which is the highest in the year 2010-11 **(Table 2.4).**

Table 2.4: District-wise distribution of number of operational holdings and area operated (2010-11)

Districts	*Operational holdings*		
	Number	Area (Ha)	Average size
Triruvananthapuram	750903	86352	0.114
Kollam	580297	90187	0.155
Pathanamtitta	296818	66771	0.225
Alappuzha	477343	67161	0.141
Kottayam	418736	123984	0.296
Idukki	259615	164859	0.635
Ernakulam	647395	120052	0.185
Thrissur	684740	99369	0.145
Palakkad	499524	145484	0.291
Malappuram	690472	132858	0.192
Kozhikkod	638351	110586	0.173
Wayanad	174416	90456	0.518
Kannur	485221	127115	0.262
Kasaragod	226958	85592	0.377
Kerala	6830789	1510816	0.221

Average holding size varies from 0.11 ha in Thiruvananthapuram district to 0.52 ha in Wayanad district. Spatial distribution of number of operational holdings broadly follows the trend of population density and relief character. It is found that the districts of Wayanad, Idukki and Pathanamthitta dominated by high and rugged relief and low population density, together, accounts for 10.7 per cent of total number of holdings and 21.32 per cent of total operated area in 2010-11. While the proportion of number of holdings reported by these three districts together hovered around 11 per cent to 12 per cent from 1985-86 onwards, proportion of operated area increased from 18.7 per cent in 1985-86 to 21.3 per cent in 2010-2011. This indicates that operated area in the hilly districts is relatively more compared to other districts and the rate of reduction in operated area is also relatively less in these districts. Nevertheless increase in proportion of number of holdings and decrease in proportion of area although within 1 per cent indicates the trend that fragmentation of holdings is taking place irrespective of geographical location of the districts.

In a ten years period from 1990-91 to 2000-01 number of operational holdings grew more than 25 per cent in Malappuram, Kasaragod, Thiruvananthapuram, and Ernakulam districts. In the case of Thiruvananthapuram and Ernakulam districts, number of holdings in the size class below 0.02 ha increased by 59 per cent and 91 per cent respectively in a ten years period from 1985-86 to 1995-96. Underlying reasons for this increase are not same for all the districts. In case of Thiruvananthapuram and Ernakulam districts which are traditionally urbanized, the growth of operational holdings is related to urban expansion and intra-district migration from rural to urban areas. In the case of Malappuram and Kasaragod districts the growth is mainly related to population growth, which is evident from the demographic data. Decadal (1991-2001) growth of population in Malappuram and Kasaragod districts were 17.2 per cent and 12.3 per cent respectively. However, number of operational holdings is decreasing in recent years. Data for the years 2005-06 and 2010-11 indicated that number of operational holdings in the State came down from 6.90 million to 6.83 million. Spatial trend shows that all districts in south Kerala barring Pathanamthitta, and Palakkad district in central Kerala recorded reduction in number of operational holdings, whereas all Malabar districts except Kannur experienced increase in number of operational holdings **(Fig. 2.8).**

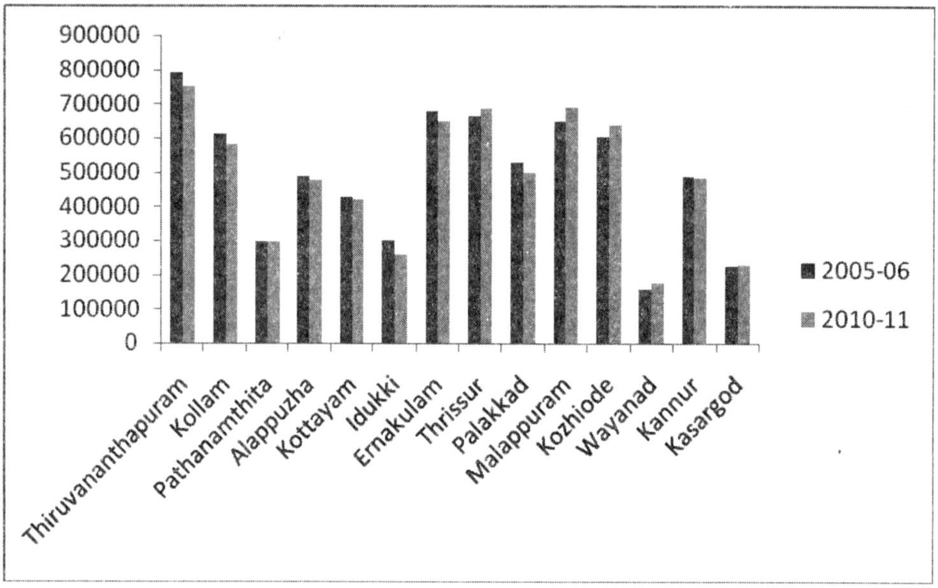

Fig. 2.8: Districtwise distribution of number of operational holdings, 2005-06 and 2010-11

Stakeholders

Land owners, forest dwellers and tribal groups are all stakeholders in land resource management. The 2010-11 agricultural Census indicated that 92 per cent area are operated by individuals followed by institutional holders (7.5%) and joint holders (0.5%). Considering the fact that individuals operate more than 90 per cent of agricultural land in Kerala it is a challenge to intervene in the private land and facilitate improvement of soil quality, strengthen land resource base and introduce conservation practices. Forest dwellers and tribal groups, original stakeholders of forest lands have lost their entitlement due to rules and regulation controlling forest areas. These rules were framed during colonial period and still continue with amendments. Other stakeholders are national and state governments for strategic issues like setting up defence establishments, resettlements to honour national and international treaties and establishing institutions of regional and national importance. State government, districts, municipalities and panchayats require land for building infrastructure, local institutions, community requirements, promotion of tourism, settlement, waste dumping and facilitate human welfare. NGOs spearheading conservation movements are also stakeholders for land resource management. All these issues deserve due attention while preparing land resource management plan for the State.

Land Market

Land market in Kerala is growing. Foreign remittance plays a very important role in demand and supply. A comprehensive assessment is necessary for land markets. It should include socio-economic characteristics of the buyers and sellers, spatial variations of market, purpose of buying (productive, speculation, real estate, hedge against inflation, residence) and selling (emergency, cash-in on investment, moving out, immediate survival) and cultural values. All these information along with data on land quality should be part of land bank information. This will help in proper valuation and decision-making in devising appropriate land use.

Land Resource Management in Kerala: Towards A Strategic Plan

High population density, corridor type settlement system, urban sprawl on agricultural land, landslides (4.7% highly prone) associated with high rainfall, soil erosion (9.5 lakh ha severe to moderately affected), reclamation of wetlands, occupation of flood plains, encroachment, increase in fallow land, intervention in water bodies, rock quarry and land mining, and land shaping are important issues impacting land management in Kerala. Besides, the actors like, commercialisation of agriculture, out-migration, labour shortage, spread of modern education, emergence of salaried class, disintegration of joint families, economic liberalisation and globalization are impinging factors in land resource management of Kerala. In one hand, lands are intensely used and there are growing demands for land to promote agriculture, sustain food production, establish growing settlements and to accommodate various development infrastructures like roads, air ports, community halls, hospitals, educational institutions and smart city programmes etc. On the other hand, there are problems of land fragmentation, land degradation, declining productivity, and loss of land due to landslides and erosion in addition to growing land requirement for waste management.

Kerala needs a strategic plan for sustainable land resource management. Integrated land use plan considering production and conservation components together is an instrument to attain this goal. It is also expected to balance between present demand for land and future developments. The FAO (1999) postulation for integrated land use plan can provide a guideline for Kerala. This plan proposes

(i) "to introduce mechanisms and incentives for changes in the allocation of land to uses for which it is suitable in biophysical and socio-economic terms, and prescribes appropriate management practices and options to ensure that land resources are conserved;

(ii) (it is) essentially a mechanism for decision support to guide the stakeholders in selecting the best sustainable land-use options which are consistent with their objectives. This approach is participatory and recognizes the rights of all stakeholders, including those such as indigenous groups or rural women who may be traditionally marginalized in some societies, to negotiate on the use and management of land resources;

(iii) (to use) negotiation as the basis for resolving land-use conflicts and agreeing mutually attainable objectives. Essentially, the approach embodies an interactive partnership between government and people to address their common concern to manage land sustainably for the benefit of present and future generations" (FAO, 1999).

A State land use policy is the most important instrument to introduce integrated land use planning. This policy should address land tenure, create conducive environment for sustainable land use practices, provide a fora for free and effective negotiation among stakeholders and link national policy with that of local initiatives. Key issues for strategic land use planning in Kerala should be knowledge based, customized and capable of meeting demand for future development of smart city, transport corridor and environment sensitive zones. Issues related to climate change and disaster management should also be taken into account. The present land use practices in Kerala can be broadly grouped into 10 land use zones **(Table 2.5)**. However, these zones are not spatially contiguous. Inventory of land for each of these categories across the State is necessary to draw a realistic picture. It may be prudent to identify these zones according to river basins and consolidated at the district level. Participation of stakeholders is an essential component.

Table 2.5: Land use Zones in Kerala

Zones	*Characteristics*
LUZ-1	Ecologically sensitive areas-mostly for conservation no development zones
LUZ-2	Areas for strategic development of national importance
LUZ-3	Areas of strategic mineral deposits, valuable minerals
LUZ-4	Urban centres and surrounding areas
LUZ5	Industrial areas and industrial parks
LUZ-6	Rural settlements and habitations
LUZ-7	Agricultural areas—food crops, cash crops, irrigation infrastructures, command areas
LUZ-8	Transport corridors and related infrastructures
LUZ-9	Areas of tourist interest, and heritage and religious sites
LUZ-10	Hazard prone areas: natural and manmade

Micro Level Land Resource Data

Kerala is perhaps the only state which has a wealth of data covering land resource, water resource and land use for all the panchayats collected under Panchayat Resource Mapping programme. Sample geomorphology/land resource maps for two panchayats, one in the coastal plain and the other in the mid-land sections and a land use map of part of a panchayat are provided here **(Figs. 2.9, 2.10, 2.11)**.

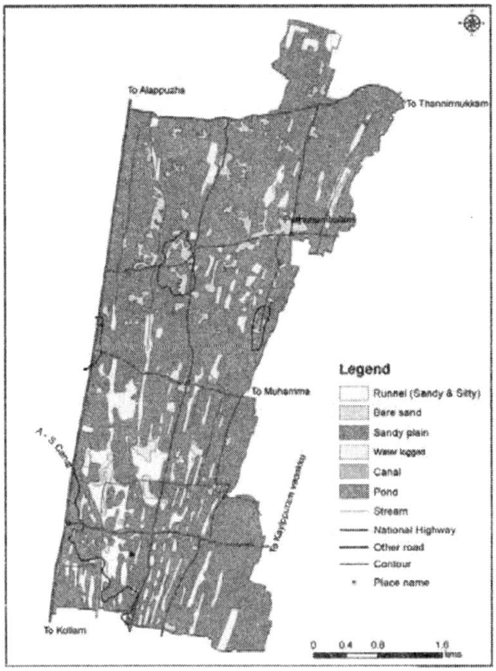

Fig. 2.9: Terrain map of a coastal panchayat, district, Kerala

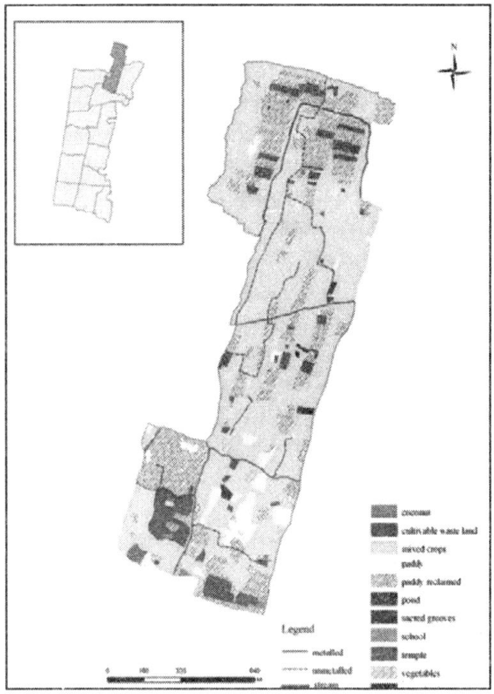

Fig. 2.10: Landuse map of a Ward of a coastal Alappuzha panchayat, Alappuzha district, Kerala

It is a participatory exercise for documenting data in cadastral scale for all panchayats in the State. This programme initiated by Centre for Earth Science Studies is now being coordinated by Kerala State Land Use Board. Spatial data are generated following participatory mode. These data are now being digitized and consolidated at watershed and district level. So far, these data are used only marginally. Plot level data generated under this programme can be consolidated at the micro watershed level, which can be further assimilated at the river basin level **(Fig. 2.12)**. A comprehensive programme may be devised to use these data, update and refine them and feed into data bank and finally to work out strategic plans for land resource and land use management.

Land Data Bank

Kerala needs a data bank on land resources. Data are being generated by various agencies. There are problems of compatibility and information are also partial. Problem of proper records in land resource assessment is an impediment noted even by the CAG, Government of India. It is important to initiate a programme to develop land resource data bank for the state covering all attributes of land and its spatial dimension. A comprehensive data base is essential for evolving strategic land resource management plan.

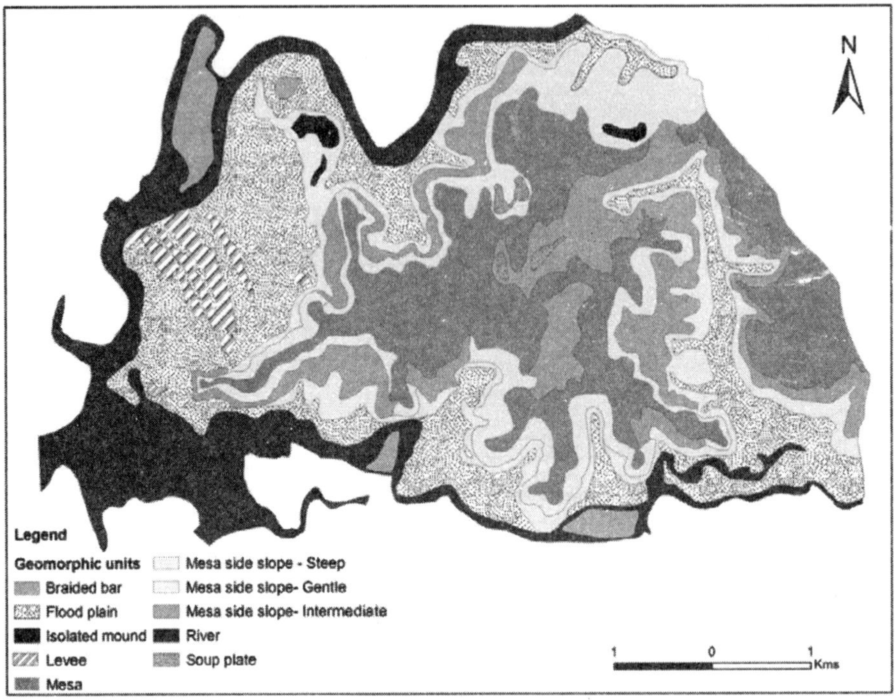

Fig. 2.11: Geomorphology of a midland panchayat, Kannur district, Kerala

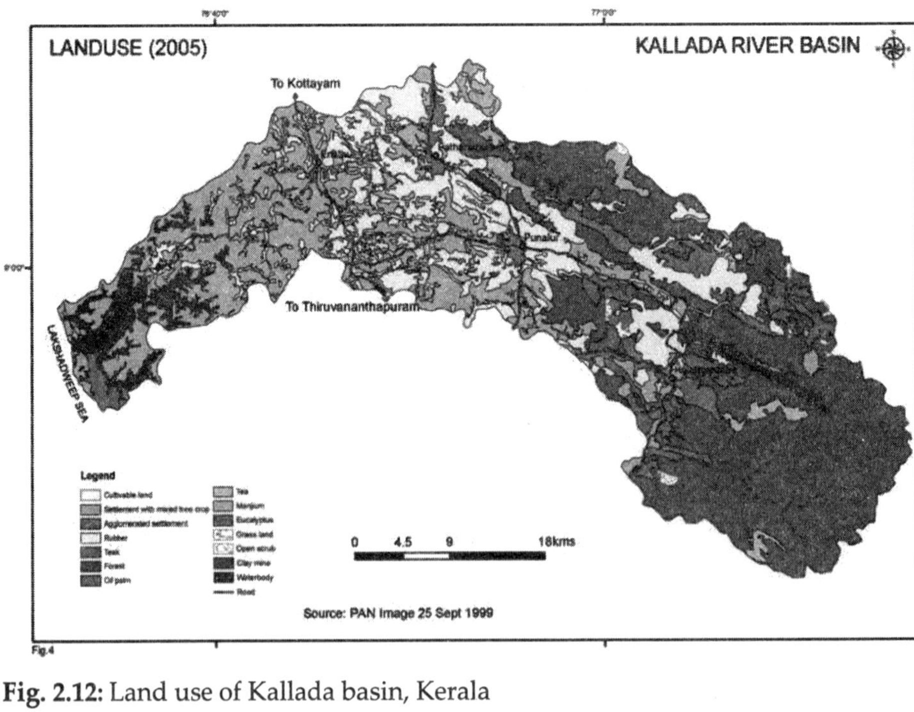

Fig. 2.12: Land use of Kallada basin, Kerala

Conclusion

Kerala is rich in land resources with high productivity and capacity to grow a wide range of crops. However, it is also a land scarce state due to high population pressure of >860 persons per km^2. Per capita availability of net sown area is only 0.06 ha. Economic growth, high development of infrastructure, globalization, and flow of capital all together has altered landman relations. Use and demand for lands are changing with a wide range of stakeholders from individual owners to national government. Land use decisions are often ex-situ, at present. Therefore, there are several challenges for land resource management in Kerala which can broadly be considered under three broad heads: characteristics of land, present land use and stakeholders. It is suggested that terrain approach may be followed for evaluating land condition. There are 22 terrain units under five dominant processes characterizing Kerala's land resource. More than 85 per cent land fall under the category of denudational process. Loss of land due to landslides in the hilly areas and is well evident. Coastal land also suffers erosion in several patches. So far as land use is concerned net sown area accounts for 52 per cent of total geographical area. NSA is declining over the years. Fallow lands are increasing. The ratio between food crop and non food crop is decreasing. There is also fall in operated area. Fragmentation of holdings is a major issue in Kerala. Another round of land reform is necessary to address several land management issues. Kerala is facing shortage of land to expand National Highways, State Highways and Railways. Land is also required for smart city programme, waste dumping sites and establishing industries/institutions. Due to occurrence of natural calamities land is required for erecting disaster shelters, rehabilitation of settlements and emergency operations. The State should devise a proper land resource management plan. It is necessary to revisit existing land governance and strive to work out land policy, strategy and land resource data base covering all aspects of land from quality to ownership and market-related information. A land data bank is necessary to accomplish this task. The state has taken unique initiative in generating land and water resource and land use data at the micro level in cadastral scale with people's participation. It is important to build up over this initiative and develop strategic plan for land resource development.

[**Acknowledgement:** Most of the data were generated while authors were is service. We thank all project fellows who have worked with us and helped in data analysis and preparation of map.]

References

Boserup, E. (1965). *The Conditions of Agricultural Growth: The Economics of Agrarian Change under Population Pressure,* London, G. Allen and Unwin.

Beek, K.J., Bie, K.D. and Driessen, P. (eds.) (1997). Proceedings, International Conference on Geo-information for Sustainable Land Management, ITC, The Netherlands.

Chattopadhyay, S. (1985). Deforestation in parts of Western Ghats region (Kerala) India. *Journal of Environmental Management,* London, 20 (3): 219-239.

Chattopadhyay, S. and Chattopadhyay, M. (1995). Terrain Analysis of Kerala: Concept, Method and Application. Technical Monograph No. 1/95, State Committee on Science, Technology and Environment, Government of Kerala, Thiruvananthapuram.

Chattopadhyay, S. and Chattopadhyay, M. (1997). Sustainable land management in Kerala, India—A biophysical approach. Proceedings (in CD-ROM) of the International Conference on Geo-information for Sustainable Land Management, ITC, The Netherlands. Eds. K.J. Beek, K.D. Bie and P. Driessen. pp. 165-174.

Chattopadhyay, S., Asa Rani L. and Sangeetha, P.V. (2005). Water quality variations as linked to landuse pattern: A case study in Chalakudy river basin, Kerala. *Current Science*, 89(12): 2163-2169.

Chattopadhyay, S. and Franke, R., (2006). *Striving for Sustainability-Environmental Stress and Democratic Initiatives in Kerala*. New Delhi: Concept.

Chattopadhyay, S. and Chattopadhyay, M. (2014). Rejuvenation of Kerala rivers: Geoenvironmental setting, potentials, problems and recent initiatives. In Vaidyanadhan, R. (ed.) Special Publication 3-*Rejuvenation of Surface Water Resources of India: Potential, Problems and Prospects*, Geological Society of India, Bangalore, pp. 1-12.

Food and Agriculture Organisation (FAO) (1995). Planning for Sustainable Use of Land Resources-Towards: A New Approach. *FAO Land and Water Bulletin*-2. FAO, UNO, Rome.

—— (1999). Future of our land-Facing the challenges. Land and Water Development Division, FAO in collaboration with UNEP, Rome

Franke, R.W. (1996). *Life is a little better: Redistribution as a development strategy in Nadur village, Kerala*. Delhi: Promilla.

George, P.S. and Chattopadhyay, S. (2001). 'Population and land use in Kerala' in *Growing Populations, Changing Landscapes: Studies from India, China, and the United States*. National Academy of Sciences, Washington D.C., National Academy Press.

Geological Survey of India (2005). Geology and Mineral resources of Kerala.

Government of Kerala (1974). Report of the Committee on Agro-Climatic Zones and Cropping Patterns. Department of Agriculture, Government of Kerala, Thiruvananthapuram.

Government of Kerala (1976). *Agricultural Statistics, 1975-76*. Department of Economics and Statistics, Thiruvananthapuram.

Government of Kerala (2006). *Agricultural Statistics, 2005-06*. Department of Economics and Statistics, Thiruvananthapuram.

Government of Kerala (2011). *Agricultural Statistics, 2010-11*. Department of Economics and Statistics, Thiruvananthapuram.

Government of Kerala (2013). *Facts and Figures of Agriculture in Kerala*. Department of Agriculture, Thiruvananthapuram.

Government of Kerala (2017). *Agricultural Statistics, 2016-17*. Department of Economics and Statistics, Thiruvananthapuram.

Indian Council for Agricultural Research (ICAR) (1991). Agroclimatic Zone Specific Research, National Agricultural Research Project, Government of India, ICAR, New Delhi

Radha Devi and Ajit Kumar (2011). Population Pressure on Land in Kerala. Working Paper No. 24. Centre for Socio-Economic and Environmental Studies, Kochi.

Mukerjee, R.K. (1938). *Food Planning for Four Hundred Millions*. London: Macmillan, p. 20.

Myers, Norman (1999). The next green revolution: Its environmental underpinnings. *Current Science*, 76 (4): 507-513.

Scaria, S. (2010). Changes in land relations: The political economy of land reforms in a Kerala village. *Economic and Political Weekly*, Vol. XLV (26, 27): 191-198.

Ziadat, F.; Bunning, S. and De Pauw, E. (2017). Land Resource Planning for Sustainable Land Management. FAO Land and Water Division Working Paper 14. FAO UNO, Rome.

3

Resource Management in Cryo-Conditioning Geomorphologic Landscape in Himalaya: A Case study of Suru-Zanskar Valley, Ladakh (J&K State), India

M.N. Kaul

Introduction

Geomorphologic resources include both raw material (related to geomorphologic processes) and landforms; both of which are useful to man or may become useful depending on economic, social, cultural and technological circumstances. Geomorphologic processes provides indepth investigation in determining the form and evolution of landscape starting with tectonic geomorphology, and then focusing on mountain and hill slopes, rivers, glaciers, coasts, and paleo-environment etc. Landforms are produced by erosion and deposition, as rock and sediment is worn away by these earth surface processes and transported as littoral deposit to different localities under different climatic environment. For instance, concerning raw material, littoral deposits can become important, economically valuable and considered as geomorphologic resource for sand quarrying. The natural stuff lying out at a place be measured, assessed and used by human for certain purpose ("development") or absolute "things" that are the foundation of material well-being of humans. Contemporary Geomorphologic resources has increasingly been characterizing by disjuncture with both the absolutist and positivist approach ingrained in geomorphologic processes in management of natural resources and by passage of time, the scope of geomorphology is becoming more and more widen and orienting towards human being. With the advent of technologies in remote sensing and GIS application, the scope of geomorphology has widened in natural resource inventories delineating the potential zones of environment and priority area identifying for planning and management of natural resources. Among the geomorphologic resources in mountain catchment areas are high relative relief that houses snow, glaciers, lakes, rivers and ground water territories rich in resource of water, energy, and, areas of recreation and store house of biological diversity. For instance the Indus river system contribute a lot to agrarian, industrial, as well the tourism potentialities of North India by providing perennial irrigation as well generating hydropower. Snow and glacial melt water contribute 400 to 800 cubic kilometre of water along with 100 tons hectare/yr of sediment (Bhadur 2000), the economically viable of geomorphologic resource. The Himalaya glaciers comprise nearly 30 per cent of glaciers outside the polar region and cover an area 20 times the glaciers of Alps (Weismann 1959).

Zanskar, the youngest mountain ranges of Northwest Himalaya, have a dramatic landscape with snow, glaciers, surging river system with limited tract of alluvial floor and verdant Zanskar upland that are judiciously cultivated by agrarian peasantry. The evolution of Himalaya, a result of geological events that goes back to almost 60 MYs, attained the present elevation 15 MYs that lead to initiation and establishment of monsoon system of India. Pleistocene has been the time period when pulsating upliftment resulted in pushing Himalaya and Zanskar further up that resulted precipitation to be in the form of snow around year giving rise to large glaciers (Valdiya, 1993), followed by Periglacial condition in higher Zanskar Himalayas. The present study has attempted a discussion of Glacial, Periglacial, Paraglacial and fluvial geomorphology of Zanskar valley and its impact on agriculture. Geomorphology and agriculture has synchronous association with each other and ostensibly the geography of agriculture begins with the morphology of concerned area. Land the carrier of ecosystem forms the basis of overall natural resources and is being increasingly exploited by man for enhancing production as far as possible. The impact of geomorphology is extremely complex as the land forms that support the cultivation of crops as well as the vegetation is synonymous with "Agriculture landforms" a surface deliberately fashioned by humans: venerable, ubiquitous but not well known and often ignored. In the era of agricultural modernization man is considered as most powerful geomorphologic agent globally and plays a pivotal role in shaping the agricultural landscape by taming the soil for creating a new landform managing environment. In the mountainous landscape like Zanskar human interaction with the agricultural landform is articulated variously now

Ladakh region has highest relative relief (2500 m over distance 5 km) in the world that constitutes a component of culture or scientific heritage of territory, having the same significance as historic monument (Gumpas) depicting work of arts a concrete support of architectural, by formulating component of tertiary cultural heritage with scenic and socio-economic values. Resources as constructed by cryo-condtioning processes of landscape development, both as driver of individual process but also by controlling interaction between surface processes, and developed concept of cryo-conditioning of landscape evolution. The climatic controls on Himalayan summit height resulted in steep fall in temperature profile that encouraged air mass to bring solid precipitation in the form of snowfall and development of glaciers resulting in glacial buzzsaw followed by permafrost environment and somewhat latter, per glacial buzzsaw. Enormous glaciers system originated at higher slopes latter by rapid advance and retreat of glaciers from the periglacial buzzsaw producing an irregular terrain of moraine complexes due to transfer of large sediments to the valley floor in the form of glacial fluvial terraces and flood plains with meadows at higher altitude leading to production of a typical geomorphic-site (a landform with particular and significant geomorphologic attributions) which qualify it as component of tertiary cultural heritage. The recent climate changes with regard to increase in temperature and rainfall supported by scientific technical measures have encouraged farmers to use littoral deposits in the form of glacial fluvial terraces and alluvial cones for cultivation of wheat instead of barley, where soil is conducive for its growth. Further in adjoining mountainous regions where crop production has least feasibility, livestock rearing is natural alternative supported by sub-arid climate during summer season. The substantial availability of green edible bio-mass has effectively routed through livestock link of food-chain through socioeconomic values.

Study Area

The Kargil region forms a vast mountainous region between the Great Himalaya Range in the South-southwest and Indus Valley in the north-east and occupies southern part of Ladakh. It has nearly 1796 glaciers, confined in Upper Indus basin, housed in Zanskar, Suru and Drass valley. Zanskar-Suru basin has thirty six sub-basins, out of which nineteen sub-basins are glaciered, housing nearly 268 glaciers encompassing an area of 1221.43 km^2 (17.44% of total area).

Suru-Zanskar sub-basin is the Vth order basin of IVth order Indus and it extends between the Suru-Zanskar (close to Pensila-La) in the west to Padam in the east. Panzila-La is gateway to Zanskar valley, situated on road connecting Kargil, with Padam (head quarter of Zanskar). The Panzila-Padam road remains closed to vehicular traffic during winter season due to closure of Panzila-La pass as a result of heavy snowfall. Zanskar valley is encircled by ridge crusts of higher peaks (3450-6270 m) of Himalaya studded with large glaciers, namely Haptal, Yaranchu, Shimling, Mulung, Kanthang, Denya, Durungdrung and others and their ridges descend precipitously.

The Zanskar sequence is composed of three tectonostratiagraphic units of Zanskar Crystalline complex consists of meta-sedimentary rocks (gneisses and migmatites), Zanskar super group (argilities and arenites with Permian volcanics) and Tsomorari Complex (schists, phyllit, gneisses, marble and amphibolites). Each group has a distinctive geomorphic setting produced by depositional environment in vertical and lateral sequence with variation in grain size. Conglomerate, quartzite, phyllite and slates characterize the basement of the region. The geomorphologic investigation indicates a possible location of Zanskar fault along Penzila resulting in up-thronging block on the west compelling some of the glaciers to turn a sinus course to north and northeast from its paleo position due west.

Suru-Zanskar valley has a distinct climatic characteristic due to its location in the shadow zone of Great Himalaya having an aerodynamic link with the air mass of westerly air flow and westerly disturbances (originating from Mediterranean and Caspian ocean) moving aloft the Pamir Range. The air mass develops cold high air pressure at higher altitude that ultimately sinks to lower altitude giving rise to cold anticyclone leading to production of thermal gradient during winter season (October to May) that is responsible for anchoring southerly jet. The study region has cold sub-arid type of climate. The winters are long chilly (mean minimum temperature -15°C to -35°C), lasting November to May. Summers are short (June to September) and mild (temperature varies between -8°C to 25°C). Nearly 72 per cent of its annual precipitation is received by western disturbances that is confined to November to May which sometimes prolongs to summers as well otherwise summers get scanty rains.

Methods

Most of the glaciers in Suru-Zanskar valley are located in remote and treacherous terrain that is inaccessible and not connected with motorable road. The monitoring of glaciers is difficult by direct field methods. Remote sensing has advantage of giving synoptic view of the region on regular basis. The IRS-IC-LISS III (October, 2001 and 2013) data was provided for Suru basin by Space Application Centre (SAC) at pixel resolution of 23.5 m for the detailed study. The base map of the area was prepared from Survey of India (SOI) 1965 topographic maps of the scale 1:50000. Only seven topographic sheets of Suru basin were available that covered 148 glaciers out of these 96 glaciers (>0.5 km²) selected for detailed study. All the satellite images were geo-referenced using SOI topographic maps. Images co-registered with each other resembled to same resolution. The glacier boundaries were delineated using topographic maps and the area was digitized using topographic Geographic Information system. The boundaries of glaciers delineated by visual interpretations and manual techniques using GIS False colour made from visible and near infrared satellite images could be used successfully to map various glacial features such as glacier boundary, accumulation area, ablation area, equilibrium line, moraines etc. The Normalized Difference Snow Index (NDSI) and TM band ratio has been used to demarcate glacier, peri-glacier, permafrost areas and also clean ice and debris ice along Ablation zone and Snout area of glacier. The streams emerging out of snout as well transverse crevasses developed near snout could be identified sharply in NDSI image. Shape file of the basin as well the glacier boundaries delineated from the satellite images was overlapped on topographic sheets and then it was digitized using Geographic Information System techniques. Benchmark glaciers like Kangrez and Machoi were monitored for long time fluctuation records. Thermal monitoring of permafrost and glacier areas was carried by opening pits by ice drill to monitor temperature at different depth by thermostats.

The location and sources of meteorological data used in this study chosen, for their proximity to glaciers length of their records. The climatological network in Ladakh is biased by predominant location of stations at Leh, Kargil and Drass on valley floor. To assess regionally representative trend, the meteorological data of Drass monitored by Indian Meteorology Department (IMD) and Snow and Avalanche Establishment (Government of India), using standard meteorological practices, is adopted as representative of the region. The data for record period of 28 years (1987-2013) is used to assess seasonal changes, if any in monthly mean maximum, mean minimum temperature and precipitation as tool to glacier stratigraphic system (Nov., Dec. previous year, and Jan. to Oct. current year). Hence, under this system length of season and duration of "mass balance year of glacier year" varies. The mass balance year is divided into winter (Nov-Mar), late winter (Mar-May), summer (Jun-Aug), late summer (Sep-Oct) is used in this study to isolate the inter-seasonal signals. The monthly mean maximum, minimum temperature and snowfall have been analyzed for each phase by fitting linear least square trend line (1988-2000, 2001-2008, and 2008-2013) to assess the impact of temperature and precipitation on the health of glaciers. The significance of trend line is shown by probability value (P value) for significance level $\acute{\alpha}=0.05$. The value of R-square (R^2) from regression is used to show correlation coeffient between glacier fluctuation and external climatic variables like temperature and precipitation (Haeberli and Beniston 1998, Kulkarni *et al.* 2002, Singh *et al.* 2005, Koul and Ganju 2010).

Results

Suru-Zanskar valley is veritable Shangri-La surrounded by high mountains on all sides. It shows greater relative relief and cuts through diverse structural and litho-logical elements of Zanskar Himalaya. The highest mountain (Mustages) consists of spectacular pyramidal peaks, horns, serrated ridges, and abrupt cols between adjacent ice filled basins. Nearly every where higher peaks and ridges (6478 meters) descend precipitously in snow and ice-covered avalanche slopes to glacier basin and bear rocks with avalanche cuties, moraines and massive scree slopes towards the low terraces and finally into the main valley floor (3340 meters). The region has diverse physical setup and is divided into four principal physical realms namely Basin floor (3340 meters to 3800 meters), Mountain rim (3800 meters to 4600 meters), Zanskar uplands (4600 meters to 5000 meters) and high Himalayan flank (5000 to 6478 meters) based on relief, lithology, topo-climate, geomorphology and vegetation, comprises the area of 293.56 km^2, 1795.65 km^2, 1568.43 km^2 and 3342.39 km^2 respectively (Fig. 3.2). Zanskar is the sub-district of Kargil it has 25 villages with sub-district headquarter at Padum. The geographical isolation with the esoteric nature of Buddhism practiced in the region has enabled it inhabitants to preserve and perpetuate their cultural identity. The entire area is physically remote and sparsely populated. This remoteness has resulted in concentration of 12,1,94 people only in 7000 Km^2, out of which 6230 (51.10%) are males and 5964 (48.90%) are females (Census, 2001). Padum has been important trade route to Central Asia and Zanskar is called Exotic land as Zanskari earn his livelihood by means, which do not allow pursuit of extra-ordinary leisure and traditional mode of living still, exists. In the traditional mode of living traditional domestic fuel system is a component that displays a strength and continuity of tradition. The domestic fuel pattern of linkages suggest that the physical isolation of the region has resulted in its behaving almost a closed system that has contributed significantly to preservation of its culture and tradition by its people. The entire demand for domestic fuel is fulfilled from the regions own resources: Livestock (cattle, sheep, goat and other specific Equines) and vegetation (bushes and shrubs). Livestock is a backbone of agrarian economy and necessary adjunct to mountain agriculture. It provides the Dung mass which is supplement to the total fuel resource. A study was conducted in the field to estimate the per day average consumption of grass and average excreta released by the animals (cow, yak, dzo, dozmo, sheep and goat) to find out the total production of dung in the region. The finding of the study reflects that the Zanskari cattle population can generate 32631.515 tones of dung annually to meet the demand of fuel in the region while consuming 205013.2 tones forages from grasses and shrubs of the region. This subsequently affects the fragile ecosystem leading to wide spread sign of slope failure and rill/gully erosion.

Impact of Climate Change on Cryo-Conditioning Landscape

The cryosphere is on one hand a very useful indicator to monitor changes in climate, on the other hand plays crucial role in many climate processes that directly affect human societies. Thus plays a major role in various dimensions of climate system: it is affected if climate changes, but it owns changes in turn affect the climate system. Monitoring these changes therefore provides crucial knowledge about climate changes.

The trends in annual, seasonal and monthly mean temperature and precipitation (snowfall) were investigated of Drass station (Representative of Study region) for Suru-Zanskar glacier basin from 1988 to 2013. The records have been analyzed by fitting linear least square trend line to assess the behaviour of temperature and precipitation. The analysis reveals that mean maximum temperature (1987-2000) shows no change in trend line during winter as well as late winter season temperature in relation to late winter (March-May) showing increasing trend in temperature). Similarly mean maximum temperature (2001-2013) trend line shows decreasing trend during summer (June-August) as well late summer (September-October) season, thereby indicating cooling, hence reducing degree-day melting of glaciers. The trend line of mean minimum temperature however shows marginal change in its behaviour.

Out of total precipitation, highest contribution (above 500 mm w.e.v.) of snowfall is, 984 mm (1995-96), 979.5 mm (2010-11), 565 mm, 562 mm, 589.1 mm. 495 mm and 587.3 mm w.e.v. (2007-2012), moderate, (300-450 mm w.e.v.) of snowfall recorded during the seasonal years of 2001-2007 and 2012-2014 and low snowfall of less than 300 cm during 1991-95, 1997-98, 1999-2000, and 2000-2001, respectively revealing 1991 to 2001 a dry decade with exceptional anomaly of 1995-96.

Analysis of rainfall records shows that study region experienced overall dry spells during the summer periods of 1995-2002. The rainfall scenario in summer period initiated a change in year 2002-03 latter as regular phenomenon since 2005-06. The total rainfall/snowfall record for the years 2011, 2012, 2013, 2014 from June to September is 42.0, 56.2, 90.1, 146.1 mm in water equivalent respectively. Whereas the analysis of precipitation recorded at the glacier, base camp and at higher ablation zone in vicinity of equilibrium line reveal that rainfall occurs at lower elevation near glacier snout and snowfall at higher elevation in ablation and accumulation zones. Further heavy snowfall in later part of winter and summer holds considerable significance in terms of health of glacier and helps in consolidating ice, reducing ambient temperature and degree-day melting that results positive impact on glacier stability and growth.

The role of temperature and precipitation have been examined in governing glacier cover extent in Suru-Zanskar valley and found that negative correlation (-0.471,-0.145) between mean minimum temperature and snowfall in 1995-2000, in comparison to weak positive correlation (0.191) during winter season of 2000-2013 time series, is insignificant as per "P" test values. The temperature and precipitation analysis and their trend shows an interesting shift of peak summers and winter season until late summer (August, September) and late winter (March to June), this shift has helped the overall health of glaciers and their stabilization process.

Geomorphology and Cryo-conditioning Landscape

The study suggests that the Zanskar valley was peripherally closed with limited detrital inputs during the pre-Pleistocene times. The Pleistocene uplift was responsible for the development of large valley glaciers, which originated at altitudes of 6000 meters and terminated at altitudes of 3500 meters and occupy all the basins of the Zanskar valley. They have deposited broad foot type terminal moraines, whereas avalanche fed present glaciers have deposited the material at higher elevations thereby indicating retreat of glaciers during recent past (Holocene period). As per the glacier inventory there are 268 glaciers in Zanskar valley out of which 131 glaciers are located in Greater Himalayas and 137 glaciers in Zanskar Mountains, majority of them are north facing (58.95%). There are eight large valley glaciers extending more than 10 Kms in length, largest being Durung Drung glacier (23.75 Km) a source of Stod river. The Suru-Zanskar valley contains nearly 30 per cent cryo-conditioning area in the form of seasonal snow, glaciers and permafrost grounds, glaciers (18%) occupy large area.

The field observation of different erosional and depositional features and their investigation reflects that the present landscape is polygenetic in nature produced by glacial, Periglacial and glacio-fluvial processes as a result of cryo-conditioning changes. The glacial processes were responsible for the deposition of polished, un-assorted sediments from the existing glacial snouts to Akshu (3860 ms) in northwest of Stod valley, Cha (4045 ms) southeast of Lungnak valley and at Padum (3505 ms) near the doab of Lungnak and Stod at northeast. Glaciers affect the landscape by erosion and deposition. Distinctive erosional features like striation markings, polished surfaces, and glacial groves characterize the glacial erosion landscape. Rochemoutonnees are observed at the different altitudinal levels of the region. The well-documented evidences in the shape of striations, groves and polished surfaces are preserved along the valley walls of Durung Drung, Haptal, Haskira, Tingzumche Nateo and Katkar glaciers basins. Rochemoutonnees are found in the areas where the ice velocity remained high enough and the effective normal pressure sufficiently low to allow cavities to open. The striations on stoss side and cavities on wind-ward side are preserved in Rochemoutonnees in Durng Drung; Akshu and Padum basins are suggestive of different glacier advances during the Holocene period.

A close examination of the glacial deposited landscape reveals a close relationship between the landscape orientation and ice flow direction. Moraines cover major part of glaciated section of the region, arranged in ridges approximately parallel to hanging valleys. The moraine ridges extend from an altitude of 5000 ms to the altitudes of 3800 ms in Lungnak and Stod basins. The broad foot moraine is attributed to broad trunk and tributary valley of Zanskar. In front of the snout of glaciers series of lateral and end moraines spreading in a huge Sandur plain up to an altitude of 4050 meters indicates that the glaciers were bigger in size and volume.

The Permafrost is thermally defined phenomenon describing ground or foot hills of mountains experiencing extreme cold arid climate and temperature during major part of the year is below freezing (-30°C to -35°C) leading to permafrost condition along glacier (ice) margin from snowline of Higher Himalaya flanks and Zanskar uplands to foot hills valley margins. The low temperature has led to freezing of supersaturated ice margins to permafrost conditions leading to production of ice wedges and rock glaciers on valley and solifluction is observed within basin slope and patterned ground with Pangolins in valley floor. This landscape category covered an area 249.58 km^2 as per Survey of India topographic map (1965) and as per satellite data 2001 and 2013 and field survey observation, the area has shrunk to 187.54 km^2 and 178.76 km^2 respectively. Permafrost line has shifted from the altitude from 3420 m altitude 3455 m from 1965 to 2001 as annual decadently in temperature has shot up from -0.226°C to +0.371°C decade. Further permafrost degradation at lower altitude has lowered active layer thickness 1-2 m which is thawing in summer and refreezing in winter leading to warm permafrost conditions since year 2001 particularly from 2005-2006. Fragmentation of permafrost line observed between Panikhar (3317 m) and Parkhachik (3634 m) stations along aerial distance 20 km.

The most important impact of climate change during last fifteen years is gradual increase in mean maximum and mean minimum during late winters (April-June) and precipitation pattern has resulted in degradation of valley slopes resulting in slope failures and slope gullies. The mass movement areas are confined to parts of Higher

Himalayan flanks and Zanskar uplands on the rocks phyllites, schists and quartzite that are devoid of ice bodies. During winter season the snowfall on the bare slopes is sometimes drifted to interior areas otherwise subjected to extreme freezing due to very low temperature (-45°C) for long duration of the winter (October-May). The freezing leads to large pressure and stress conditions on valley walls and solfluxion and navel fluxion in the inter-mountain basin floor and valleys. Along the valley walls there is mass wasting caused by supersaturated permafrost frozen wall that moves down slopes and produces scree cones and rock glaciers, leading to slides covering mountains along valley margins covering an area 68.95 km^2 out of which 88 per cent lies in Zanskar uplands having a slope of 32. The process of thawing in these areas leads mass movement process.

Soils

Soils and cryotic ground condition have significant influence on landforms and are inseparable and these have an intimate relationship with each other. In the study area, the Physio-chemical and morphological characteristics of the soil have been worked out according to scientific procedures. On the basis of morphological and textural characteristics the soils of the area have been categorised into two types: sandy loam and sandy clay loam. Soils of the study area are characterized by low water holding capacity, low organic capacity and high electric conductivity. Soil of the region also shows variability in the pH ranging from 6.40 and 8.30. The study also reveals that the soils are suffering from problem of soil erosion, and deficiency of nitrogen and phosphorous. They contain low to medium potassium content. Due to large variation in altitude and marked difference in temperature soils of the study region are classified into four categories. They are (a) Sandy clay loam to sandy loam, (b) Sandy skeletal to sandy sand, (c) Stony sand and (d) Fragmental stones. They are mainly Entisoles with suborder Orthents and great group Typic/ardic.

Impact of Cryotic Ground Condition on Land Use and Agriculture

Land use studies are of paramount importance when the resource base of any region is under evaluation. Rationalization, conservation and management of land resources play a crucial role in developing the agricultural economy of any region. The study region has been divided into five major land cover types on the basis of Satellite data, collateral data and on the basis of field evidences related to geomorphic processes. The delineated land cover types of the study region are (a) Area under agricultural activity and settlements (46.65 Km^2), (b) Grass land/grazing land (24.47 Km^2), (c) Water bodies (394.08 Km^2), (d) Barren rocky area (4220.8 Km^2), and (e) Area under Glacier and Periglacial activity (2314 Km^2). Being a mountainous region a very limited area is available for agricultural activity that too for monoculture on account of short growing season due to intense cold weather (Fig. 3.6). Under agricultural activity four land use types observed are: (i) Area not available for cultivation. (ii) Uncultivated land (iii) Fallow land, and (iv) Net sown area. During 1980 to 2005 the net sown area has increased by 24.75 per cent whereas area under other agriculture land use categories have decreased, thereby indicating that more area is being brought under net sown cultivation through rehabilitation of waste land as a result of various development programmes launched by Government of India since 1990.

The dynamics of agriculture studied is reflected not only in the extension of irrigation facilities and land use but also in the transformation of its cropping pattern in adverse climatic conditions. The area under major crops (grim and wheat) has increased (18.25% and 30%) significantly during 1980-2005. Maximum increase under wheat and pea cultivation (30%) is primarily attributed to introduction of mechanization and scientific application in agriculture practice. The application of mechanization has paid dividend as limited manpower is available. Due to introduction of modern techniques the yield per hectare in wheat and pea has increased by 30 per cent since 1980. At present yield per hectare of wheat and grim is 9.62 and 11.08 quintals per hectare respectively whereas yield per hectare of peas and vegetables is 9 quintals per hectare. Fodder crop has experienced marginal increase (1.43%) from 1980-2005. A comparative study of land use, cropping pattern and crop production indicate that the basin floor belt has experienced significant change in area under agriculture land use categories as well in cropping pattern and production. This is mainly because of extension of irrigation facilities, soil conservation, mechanization measures and better communication. The mountain rim recorded least change due to high relative relief, rugged terrain, adverse climatic conditions and traditional farming practice and excessive grazing by cattle leads to soil erosion. The pasture lands of mountain rim and Zanskar uplands have undergone change in last ten years due to summer rainfall that has regenerated dry scrub land to wet scrub land.

The significant feature of the present work has been the delineation of Zanskar valley into different land use zones as discussed in preceding paragraph in order to have an understanding of different land use practices that may be suitably adopted in such a region. Further the region is divided into broad Agro-ecological zones from parameters of soil texture, soil temperature, soil moisture adequacy index, water holding capacity, agricultural productivity and from the geomorphological evidences. Agricultural productivity has been worked out through Bhatia's (1967) method and moisture adequacy index has been estimated on the basis of Subramanium model (1963) while soil texture, soil temperature, soil water holding capacity and vegetation type has been observed and measured through field techniques. The land use zonal map has been superimposed on soil temperature regime, vegetation and agricultural productivity zones to carve out different agro-ecological boundaries. Consequently five micro-regions namely (a) Valley bottom isomesic, mixed cropping zone, (b) Rim land isomesic, gram dominated cropping zone (c) Upland bushy isocryic, meadow zone (d) High alpine isocryic, pasture and scrubby zone and (e) Great Himalayan perigelic, glacier and Periglacial zone (Fig. 3.7). The soils of the valley bottom are mainly Entisoles with suborder Orthents and great group Typic/ardic. In the valley bottom sandy loam and sandy clay loam soil supports cultivation of Wheat, Grim, Peas, Fodder and vegetable. In the rim land the shallow, sandy loam soil derived from weathered debris helps the cultivation of grim crop only. In the third zone soils with poor physical conditions like sandy land and thin soil profile development with high rate of infiltration are suitable for meadowland environment. Red clover and Lucerne grass species are doing well in the meadow zone. In the fourth zone of alpine pastures soils are cryic stony severely eroded and are classified as Arctic cryorthents. This zone is a domain of nutritious grass, herbs,

Scrubs and seasonal alpine flowers. The Great Himalayan perigelic, Glacier and Periglacier zone has predominance of fragmental stones. The zone experiences cold arctic type of climate in which practically all the precipitation is in the form of snow for more than nine months. Large number of glaciers/ice fields inhabits this zone with maximum concentration in the southwest part.

Discussion and Conclusion

Over record period of 28 years, there has been a small increase in annual mean temperature at Drass of -0.426°C decade prior to year 1995. However, since 1996 the rate of increase has accelerated to 0.275°C decade. The analysis of mean monthly temperature for period 28 years, lack of fit of lower portion of data (1988-2000) to upper portion of data (2001-2013) hence it is attributed to phase transition threshold. It indicates that winter is cooler, late winter warm humid, and summer cool and wet during time series 2001-2013 particularly from 2004 in comparison to 1988-2000. These changes were associated with change with inter-decadently of Pacific Oscillation and with increase in El Nino/southern Oscillation events that resulted lower ablation season during summers of 2004-2014 (Yasunari 1987). These trends in weather conditions have undoubtedly led to a favourable environment for decelerated retreat in glaciers to the extent of no change in 85 per cent glaciers (115 glaciers) of Drass and 81 per cent glaciers in Suru-Zanskar basin during last thirteen years (2001-2013). Further the 13 glaciers show gain in area and 28 glaciers show marginal loss in area including benchmark glaciers Drung Drung, Kangriz in Sur-Zanskar and Machoi glacier. Prior to the period 2001, the temperature was cold during early winter, warmer in late winter as well dry in summer and as such, the glaciers of the valley were under climatic stress leading to fragmentation of 19 glaciers and thawing of permafrost zone and fragmentation of permafrost line resulting in loss in area (Bahuguna, *et al.* 2014, Bolch, *et al.*, 2008, Fujila, K. 2008, Ye, *et al.*, 2006).

Geomorphologic landscape of Suru-Zanskar valley is built through tectonic upheavals followed by cryo-conditioning of landscape driven by temperature gradient for relevant periglacial, glacial, glacio-fluvial, mass wasting by permafrost processes and process interactions. These geomorphic processes are each influenced differently by cryo-conditioning of landscape by thermal variation with altitude, ecology and human activity and their interaction ultimately resulted in evolution of agricultural landform. Agricultural landforms are just taxonomy, then; they are also triumph over the soil brought out through the interaction processes that shape the landscape of the Zanskar valley. The agricultural activities in Zanskar valley depend on natural (climate, soil, availability of surface and sub-surface water for irrigation etc) as well as socio-economic factors including agricultural technology and irrigation facilities. The pedo-geomorphic features directly play a vital role in controlling the selection of specific cropping pattern and land utilization in a particular region as such are "treasurers" for understanding the various land use/land cover, cropping pattern and crop production and underground hydrological conditions. The recent climatic changes during last one decade has marginally shifted the winter and summer season that has led to increase in mean monthly maximum temperature during late winter and early summer season. This has caused the glacier surface to sublimate the air mass resulting in increase trend in summer rainfall in down valleys. These changes have encouraged the farmers to switch over to wheat cultivation from barley cultivation, in areas were soil is conducive for its growth. Further in adjoining mountainous region where crop production has least feasibility, livestock rearing is the natural alternative supported by more conducive sub-arid cold climate during the summer season. The substantial availability of green edible bio-mass has efficiently routed through livestock link of foodchain. The area under major crops (Gram and wheat) has increased (18.25% and 30%) significantly during 1980-2013. Maximum increase under Wheat and Pea cultivation (30%) is primarily attributed to organic farming and introduction of mechanization in agriculture practice. The application of mechanization has paid dividend as limited manpower is available. Due to introduction of modern techniques the yield per hectare in wheat and pea has increased by 30 per cent since 1980. At present yield per hectare of wheat and gram is 11.62 and 11.08 quintals per hectare respectively whereas yield per hectare of peas and vegetables is 10.5 quintals per hectare. The vegetable cultivation and pastoral activities are mainly practiced by women and government is also encouraging them as they are hard and dedicated working force. Due to their efforts fodder crop and vegetable production has experienced increase (3.43%) during 1980-2013. This has led substantial increase in Zanskari milking cattle leading to enhanced butter production meant for export to Indian and European market. This has helped the economic development of Suru-Zanskar.

References

Bahadur, J. (2001). Glaciations over Himalaya—The tallest water tower. *Proc. National Academy of Sciences,* India, Vol. LXXI A, SPL. ISSUE, Allahabad, August 25-26, 2000 pp. 38-49.

Bahuguna, I.M., Rathore, B.P., Brahmbhatt, R., Sharma, M., Dhar, S., Randhawa, S.S., Kumar, K., Romshoo, S., Shah, R.D., Ganjoo, R.K., and Ajai (2014). Are The Himalayan Glaciers Retreating ? *Current Science,* 106, 1008-1015.

Bhatia, S.S (1967). A new measure of agricultural efficiency in Uttar Pradesh, India. Economy Geography 43:238.

Bhutiyani M.R., Kale V.S., Pawar, N.J. (2007). Long-term trends in maximum, minimum and mean annual temperatures across the Northwestern Himalaya during twentieth century, *Climate Change* 89:159-177.

Bitz, C.M., and Battissi, D.S. (1999). Interannual to decadal variability of climate and glacier mass balance in Washington, Western Canada and Alaska. *Journal of Climate,* 12, p. 3181-3196.

Bloch, T.; Buchroithener, M.T, and Kunert, A. (2008). Plan metric and volumetric glacier changes in the Khumbu Himal, Nepal since 1962 using Corona, LANDSAT TM and ASTER data, *Journal of Glaciology,* 54, pp. 592-600

Bowling, S.A (1977). Relation between temperature and snowfall in interior Alaska, *Arctic* 62-64.

Crook, J., Osmaston, H., (1994); *Himalayan Buddhist villages; Environmental resources, Study and Religious life in Zangskar, Ladakh;* Bristol, University of Bristol U. K., 866 p.

Fowler, H.J, and Archer, D.R., (2006). Conflicting signals of climatic change in Upper Indus Basin, *Journal of Climate,* 19, 4276-4293

Ganjoo, R.K, Koul, M.N, Bahuguna, I.M, Ajai (2014). The complex Phenomenon of Glaciers of Nubra Valley, Karakorum (Ladakh), *India Natural Science,* 6, p. 733.

Haeberli, W. and Beniston M. (1998). Climate change and its impact on glaciers and permafrost in Alps. *Ambio* 27:258-265

Hewit, K. (2005). The Karakorum Anomaly? Glacier Expansion and the "Elevation Effect", Karakorum Himalaya. *Mountain Research and Development*, 25, 332-340.

Karl, T.R.; Groisman, P.Y., Knight and R.W., Heim Jr. R.H. (1993). Recent variations of snow cover and snowfall in North America and their relation to precipitation and temperature variations, *J. Climate* 6, p. 1327-1344.

Koul, M.N. (1990). *Glacial and Fluvial Geomorphology of Western Himalaya,* Concept Publishing Company, New Delhi, p. 1-332.

Koul, M.N, Ganjoo, R.K, Dhinwa, P.S, Pathan, S.K, and Ajai (2007). Use of Geo-Informatics for preparation of a Desertification Status Map of Stod Valley (1F4C2), Padam (Zanskar) District Kargil, J&K, Indian Society of Geomatics, Special Issue, Part-ii. pp. 27-35.

Koul, M.N. and Ganjoo, R.K. (2010). Impact of Inter-and Intra-Annual Variation in Weather Parameters on Mass Balance and Equilibrium Line of Naradu Glacier (Himachal Pradesh) NW Himalaya, India, *Climate Change,* 99,119-139.

Kuhn, M. (1984). Mass budget imbalances as a criterion for climatic classification of glaciers, *Geografiska Annaler* 66A. p. 229-238

Kulkarni, A.K.; Mathur P.; Rathore B.P.; Alex, S., Thakur, N. and Kumar, M. (2002). Effect of global warming on snow ablation pattern in Himalaya, *Current Sci* 88(11):1844-1850.

Kulkarni A.V (1992). Mass balance of Himalayan glaciers using AAR and ELA methods, *Journal of Glaciology* 38, p. 101-104.

Latouchi, T.H.D. (1910). Notes on certain Kashmiri glaciers, *Rec. Geol. Surv. Ind.,* Vol. 4, Pt. 4.

Lawrence, W.R. (1895). *The Valley of Kashmir,* London, Henry Froude Oxford University Press.

Oldham, R.D. (1904). Notes on the glaciations and history of the Sind Valley, Kashmir, *Rec. Geol. Surv. Ind.,* Vol. 31, pt. 3, pp. 142-161.

Raina, R.K. and Koul, M.N. (2011). Impact of Climatic Change on agro-ecological zones of the Suru-Zanskar valley, Ladakh (Jammu and Kashmir), India, *Journal of Ecology and Natural Environment,* Vol. 3(13), p. 424-440.

Raina, V.K. (1971). The snout of the Machoi glacier, Kashmir, *Rec. Geol. Surv. Ind.,* Vol., 96, Pt. 2, p51-54

Srikantia, S.V., Ganeshan. T.M., Rao. P.M., Sinha, P.K., and Tirkey, B. (1978), Geology of Zanskar area. *Himalayan Geol* 8.1009-33.

Subrahmanyam, V.P., Rao, S.B and Subramaniam A.R. (1963): Moisture adequacy in relation to the Distribution of some crops in India *I.A.S. H. Gem. Ass.,* Berkely Centre for Evaporation, Publication 62, p. 103.

Valdiya, K.S. (1993). *Dynamic Himalaya.* University Press (India) Pvt. Ltd.

Wissmann H. Von (1959). Die Heutige Vergletscherung and Schneegrenzc in Hoschasien. *Abh Wissenchaft* Lit. *Math-Nature,* K1 14, p1101-1407.

Yasunari T. (1987). A Global structure of EL-Nino Southern Oscillation pat 1: EL-Nino composites, J. Meteoros. Soc. Japan 65 pp. 67-80.

Yadev, R.R., Park, WK., Singh, J., Dubey, B. (2004). Do the western Himalaya defy global warming? *Geophys Res Lett 31*: L17201.

4

Quaternary Geomorphological Changes Around the Southeast-Facing Slopes of Kanchenjunga in Sikkim Himalaya

An Assessment Upon Field Observation

Guru Prasad Chattopadhyay

Introduction

The Sikkim Himalayan region, particularly the areas around Kanchenjunga summit complex, has drawn attention of many environmental scientists over the past centuries. Bose (*cf.* Mayewski & Jeschke, 1979), on his visit to this part in 1889, described the conditions of the glaciers nourished from the Kanchenjunga. Later Garwood (1902) managed to prepare a draft map of the position of Rathong, Onglaktang and other glaciers during his visit to this part of the Sikkim Himalaya in 1902. However, their reports were more of general documentations of the glacial environment of this part rather than pure geographical accounts. In the extensive tracts of high altitudes in this part, particularly above the timber-line, the variety of typical landforms bear with them certain geomorphological and geological mystery which is yet to be explored systematically. The wide variety of glacial, periglacial, nival and paraglacial features of this part reflect the pattern of typical Quaternary history. Such conditions and features have earlier been reported by a number of previous workers, *viz.*, from the Western Himalaya by de Terra (1936), from the Central (Nepal) Himalaya by Fuji and Higuchi (1976) and from the eastern Himalaya by Kar (1969). In these literatures, features have been reported and described from various parts of the Himalaya in much ambiguous and descriptive manner creating certain confusion. Therefore, a need has always been felt to recognise the Quaternary landforms of the Himalayan region under an organised classification. This author, through his extensive field investigation in this area, attempted two identify the significant Quaternary geomorphological features (Chattopadhyay, 1985, 1998, 2008, 2010, 2012) under systematic classification.

This paper is a discussion based on the investigation by the author over the area around the Dzongri Bugiyal and the Rathong and Onglaktang glaciers at 4,200 – 5,200 m (appx 14,000′ – 16,500′ ft) elevation on the southeast-facing slopes of the Kanchenjunga summit complex in the western part of the Sikkim Himalaya during the long period of 1984-2012.

The major difficulty for a systematic geomorphological study of this part of

Himalayas is the absence of any detailed geological and geographical report. As has been mentioned earlier the literatures published so far have only passing mention of the Quaternary environmental condition of this area which has further been complex and confusing. Another drawback is, for strategic reason, the Survey of India topographical sheets with detailed contour lines remain restricted and inaccessible.

Approach to the Area and Its Geological Environment

Approach to the area: Being situated in West Sikkim the Dzongri Bugiyal area, above the timber line (4,300 m), can be approached on foot from the last vehicle point of Yoksom village at a distance of about 30 km. The Dzongri Bugiyal is basically an alpine meadow of rugged terrain with occasional exposure of cliff walls and rocky ridges rising up to about 4,600 m. The HMI base camp is situated at the southeast-facing slope of Kanchenjunga summit complex in Sikkim Himalaya immediately below the Rathong Glacier valley, which can be approached at a distance of 8 km from the Dzongri La (4,300 m). The area around the HMI Base Camp (4,600 m) is a rugged mountainous tract. It is surrounded by Singalila ridges, Koktang glacier, Rathong glacier, several snowcapped peaks like Kabru Dom (6,600 m) and Forked Peak (6,108 m), and the stream channel of Churang Chhu. The Churang Chhu is the main stream of this area, fed by Rathong glacier. From the Dzongri Bugiyal to the east is the stream valley of Prek Chhu. This stream is fed by the Onglaktang Glacier lying on the eastern slope of the Kabru Dome summit of the Kanchenjunga.

Geological environment of the area: Geologically and structurally this part falls within the territory of "Darjeeling Gneiss". Gneissic rocks with occasional outcrops of

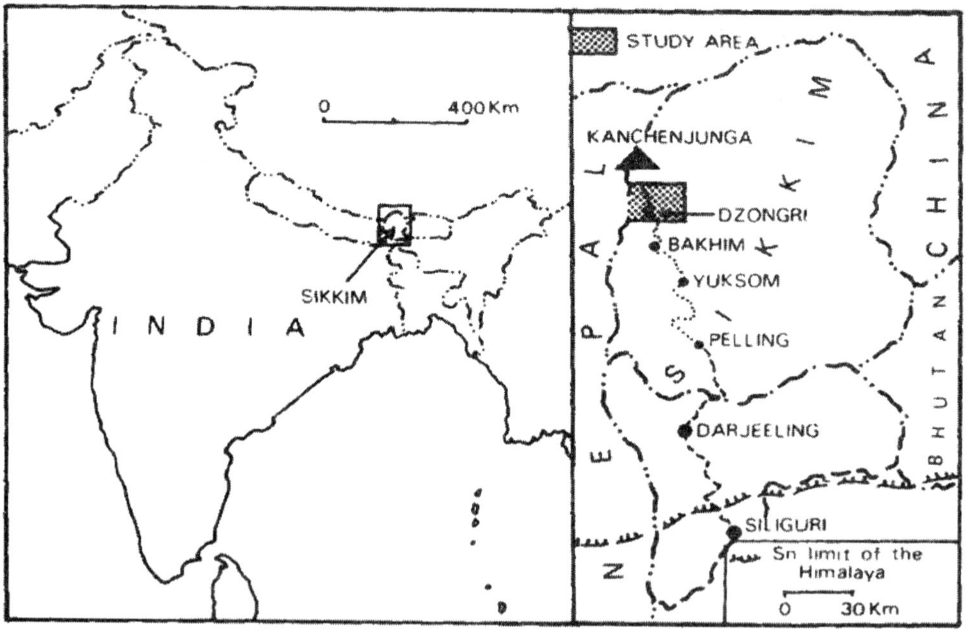

Fig. 4.1: Location of the study area in the Sikkim Himalaya

mica schist and quartz veins predominantly form the mountain ridges. The limits of pre-existing glaciers in the form of lateral moraine and terminal moraine are conspicuous down the valley and from the existing glacier. The extensive frost shattered debris slope, block slope, talus slope, comprising massive gneissic boulders provide the evidence of periglacial condition of the region. The study and field investigation of landforms were concentrated upon the areas in this region above the timberline (4,300 m).

INVESTIGATION OF GLACIAL ENVIRONMENT

The Physical Environment: From the timberline above, the first 200 m or so, the ground surface is characteristic of an Alpine meadow. On the bare mountain slopes frost-shattered (periglacial) detritus, comprising mostly massive gneissic boulders, occur extensively (Chattopadhyay, 1984, 1985). Numerous morainic ridges of pre-existing glaciers markedly traverse the main river valleys and the ridge-walls ascending up to the snowline. The two main valley glaciers, that exist today, are Onglaktang glacier (the source of Prek Chhu stream) and Rathong glacier (the source of Churang Chhu stream); both of these glaciers are fed by the ice from the main neveé field of the Kanchenjunga summit complex at its south face.

It has already been mentioned that the two main river valleys: Churang Chhu and Prek Chhu, and the ridge walls bounding them, are markedly traversed by numerous ridges of moraine loops of pre-existing glaciers. However, in order to explore the two main existing valley glaciers: Rathong and Onglaktang, of their maximum extension in the past, glacial and glacio-fluvial landforms, in the form of the successive loops of lateral and terminal moraines and other drift limits, were studied in detail. These morainic loops have been considered as the basis for reconstructing the glacial history of this region.

Characteristics of the moraine loops and their interpretation: In both the cases of the Onglaktang and Rathong glaciers three sets of moraine loops through the valley, each comprising lateral and terminal moraines, extending down the valley from the active zone of the glaciers, have been clearly identified. These three sets of moraine loops around the two valley glaciers have some distinguished characteristics are demonstrated in the following table prepared on the basis final field survey up to 2010.

Table 4.1: Characteristics of moraine loops and drift limits around the Ongloktang and Rathong glaciers

Moraine loops	*Average distance of the terminal moraine from the snout*	*Altitude at which terminal moraine occurs*	*Maximum height of the drift limit from the valley bottom*	*Morphology and surface characteristics of the terminal moraine*
The farthest loop (Stage I)	6.0 to 6.5 km	4,000 m	>500 m	Entirely vegetation-covered and traversed by melt-water channels
The intermediate loop (Stage II)	3.0 km	4,400-4,500 m	300 m	Partly vegetation-covered and traversed by melt-water channels
The nearest loop (Stage III)	1.5 km	4,550-4,600 m	200 m	Totally barren and composed of loose debris material

Source: Compiled by the author on the basis of field data 2010.

From the data assembled in the table above it is obvious that the two glaciers experienced three distinct stages of advance and retreat. It is assumed that the energy of transportation and the subsequent deposition of moraine debris were maximum during Stage I, when these glaciers extended downstream for farther 6.0 to 6.5 km from their present point of termination. The terminal moraine limits corresponding Stage II advance occurs at about 3 km from the existing snout position of both the glaciers. During Stages II and III advance the energy markedly diminished as is evidenced from the heights and extents of the terminal moraines and the associated drift limits. The fresh terminal moraine limits corresponding Stage III advances of the two glaciers occur at a distance of 1.0 km to 1.2 km from the existing snout position.

Rathong Glacier: Rathong glacier, having its source from the neveé field of the Kabru Dome (6,600 m) summit area, descends over the west-facing slope of this summit. The source of this glacier lies at an elevation of 6,200 m (27°36′N and 88°06′E). With a total length and average width of about 3.5 km and 700 m respectively its lower part extends southward through the main trough between Koktang (6,147 m) and Kabru Dome (6,600 m) down to about 4,800 m level from where the headwater of the Churang Chhu stream rises.

The fresh morainic ridge, showing the evidence of its recent retreat following the Stage III advance, occurs at an approximate distance of 1.5 km from the snout. A reconnaissance survey of the present condition of Rathong Glacier was conducted traversing the area on foot from the H.M.I. Base Camp (at 4,500 m) upward over the fresh (Stage III) morainic ridges and the glacier body itself up to the corrie lake of Dudh Pokhri. With close observation and measurements with reference to the base map the existing average length, width and depth of this glacier have been recorded as 3,500 m, 700 m and 50 m respectively. This gives the calculated existing volume of the glacier as (3,500 m×700 m×50 m =) 122,500,000 m^3.

The extent of fresh lateral moraine loop of the Rathong glacier has been used as the denominator for calculating the former maximum volume. Here the average height of the fresh lateral moraine, corresponding Stage III, from the glacier surface is about 70 m and the average distance between the highest parts of the lateral moraine ridge on both sides of the glacier is about 1,500 m. On the basis of this observation the reconstructed former depth, width and length of the glacier would be (50 m + 70 m =) 120 m, 1,500 m and 5,200 m respectively. Hence the calculated former volume of Rathong glacier would be (5,200 m × 1,500 m × 120 m =) 936,000,000 m^3, suggesting that with the recent retreat (following the Stage III advance) the volume of this glacier has reduced to about 13.09 per cent (less than one seventh) of its original size.

Most remarkable morphological feature around the Rathong Glacier is the accumulation of fresh hummocky moraines mingled with fragmented parts of the glacier-ice, which cover extensive part of the recently deglaciated valley area in front of this glacier for a distance of about one kilometre. A number of large and small melt-water lakes also occur through the recently deglaciated valley floor. This makes a certain indication of how rapid is the existing rate of wearing of this glacier. A close field observation of its upper part also revealed that after the complete disappearance of the original neveé field around the Rathong Pass this glacier now owes its supply only from the neveé field on the southerly slope of the Kabru Dome summit. Fresh morphological evidences are preserved to show that due to growing water pressure dam burst of these melt-water lakes has taken place, particularly over the last two decades, causing flash flood down the Churang Chhu valley. Also as the deglaciation continues at a remarkably faster rate in the present day, many such flood or even more devastating ones would occur in this part in the coming years (Chattopadhyay, 2010).

Onglaktang Glacier: Onglaktang glacier, being situated in the north-eastern part of the study area, is also nourished from the same neveé field as the other around the Kabru Dome (6,600 m) summit and extends to the south-southeast direction (to the opposite side of the Rathong), through the valley for a distance of about 4 km with an average width of 800 m. The lower part of this glacier for a distance of about 3 km is almost completely covered with debris of supra-glacial origin. The source of this glacier lies at an elevation of 6,000 m (27°36′N and 88°08′E). Average width of this glacier is about one kilometre and its snout part rests at about 4,700 m altitude from where the headwater of the Prek Chhu stream originates. Its two original tributary glaciers, one extending from the Forked Peak (6,108 m) to the west and the other from the Pandim (6,691 m) summit to the east, which used to feed its body, have now almost completely disappeared as a result of their apparent retreat.

The existence of fresh terminal moraine ridge, at a distance of about 3 km downstream from the present snout, clearly suggests that this glacier has assumed a significant recent retreat of about 3 km following the Stage III advance. In order to make a thorough investigation of the present condition and the stretch of this glacier the adjacent valley area beyond the Simiti Lake right up to Jemathang for a distance of about 6 km, was traversed on foot and a GPS Handset was used for recording altitudinal position of the successive parts of the glacier during the field investigation in 2010 and 2012. With close observation and measurements in the field, using the base map, average width and depth of this glacier have been recorded as 800 m and 60 m respectively. Hence considering the present approximate length (3,000 m), width (800 m) and depth (60 m) the approximate total volume of Onglaktang glacier would be (3,000 m × 800 m × 60 m =) 144,000 000 m^3.

The extent of fresh lateral moraine loop of the Onglaktang glacier has also been used as the denominator for calculating the former maximum volume. Here the average height of the fresh lateral moraine, corresponding Stage III, from the glacier surface is about 80 m and the average distance between the highest parts of the lateral moraine ridge on both sides of the glacier is about 1,600 m. On the basis of this observation the reconstructed former length width and depth of the glacier would be 6,000 m, 1,600 m and (60 m + 80 m =) 140 m respectively. Hence the calculated former volume of Onglaktang glacier would be (6,000 m × 1,600 m × 140 m =) 1,344,000,000 m^3, suggesting that with the recent retreat (following the Stage III advance) the volume of this glacier has reduced to about 10.71 per cent (less than one seventh) of its original size.

In common with the morphological features around the Rathong glacier this Onglaktang glacier also exhibits widespread accumulation of fresh hummocky moraine mingled with fragments of the glacier-ice, which cover extensive part of the recently deglaciated valley area in front of this glacier for a distance of about 2 km. A number of large and small melt-water lakes also occur through the recently deglaciated valley floor indicating rapid rate of wearing of this glacier. This glacier is considered to have shrunk significantly due to the withdrawal of supply of ice by the two tributary glaciers as have been mentioned earlier. In common with Rathong glacier valley there are also physical evidences and records that, on a number of occasions particularly over the last two decades, fragile moraine dams across the melt-water lakes occurring through the recently deglaciated valley floor in front of the glacier collapsed causing destructive flash floods down the Prek Chhu valley.

Plate 1: Segments of glacier ice of the rapidly retreating Rathong glacier (Photograph of 1995)

Plate 2: The exposed valley-floor after the disappearance of the segments of the Rathong glacier-ice (Photograph of 2009)

Plate 3: A large melt-water lake formed recently in front of the retreating Rathong glacier

Plate 4: The limit of the Stage III moraine across the Onglaktang glacier valley

Plate 5: Accumulated fresh hummocky moraine through the recently retreated Onglaktang glacier valley

Plate 6: Present form of the rapidly retreating Onglaktang glacier

However, the detailed field investigation also reveal the alarming rate of retreats of the major glaciers around the southeast facing slopes of the Kanchenjunga summit complex. The rapid retreat of the Rathong glacier indicates increase of the temperature due to environmental degradation and global warming. The location of the HMI base camp in the immediate vicinity of the Rathong glacier and rapid increase in the number of mountaineers, and trekkers each year cause marked environmental degradation in recent years. Rapid decay of glacier body often gives rise to debris-dammed lakes adjacent to the snout and at certain time intervals catastrophic dam bursts can cause flooding, and disturbing the equilibrium of the valley profile in the upper catchment areas of the glacier-fed river. The increasing pressure of cattle population on the alpine meadow area in recent years, who discharge higher amounts of methane gas in atmosphere through their dung and remnants, also cause rapid atmospheric warming.

Nowadays it is really a matter of concern that if such rapid rate of retreats of the glaciers continues around this part of Himalaya the most important rivers in this part of Eastern Himalaya and North Bengal will suffer from hydrological imbalance as well as scarcity of water in coming decades. There is enough reason to apprehend that in the absence of perennial snow fields and glaciers the hydrological character of the main rivers will be very much temperamental and be guided by the rainfall alone; they will be lacking water flow in the dryer months and be torrential in the rainy season.

Investigation of Periglacial Environment

The Himalayan periglacial environment is essentially identified as Alpine periglacial zone. The extent of available literature dealing with periglacial environment and processes in the Himalaya so far is not only very limited but also have, in most cases, incomplete as well as confusing owing to the absence of methodical observation and knowledge in this branch of geomorphology. Firstly, in many cases glacial drift deposits have mistakably been identified as the product of periglacial processes; secondly lumps of shattered debris actually resulted under paraglacial processes (a process which operate to collapse the deglaciated valley walls following rapid retreat of glaciers) often posed challenge to the Geomorphologists to separate them the periglacially derived ones particularly since they normally occur in the same cold environment, and thirdly, the presence of shattered blocks in the periglacial environment which have resulted entirely from activities of tensional and compression forces acting in the seismically sensitive rock structure in this young fold mountain system and they often look quite similar to that produced by frost shattering under periglacial process.

Outside the fresh moronic drift limits of the existing glaciers, as have been mentioned earlier periglacial features comprising frost-shattered debris and blocks occur on extensive areas of the mountain slopes. Through meticulous observations a wide variety of periglacial features have been identified in this region.

The identified periglacial features – their pattern of occurrence in terms of composition, gradients of slope upon which they occur, types of occurrence and position and characteristics have been listed below:

Table 4.2: Distinguished periglacial and nivational features and their pattern of distribution around Dzrongri

Periglacial features	*Characteristics and composition*	*Appx, slope angle*	*Type of occurrence*	*Position*	*Relict/Contemporary*
Talus Slopes	Accumulation of frost-shattered angular blocks of various sizes at the base of near-vertical cliffs.	30°	Extensive	Both inside and outside glacial drift limits	Both relict and contemporary.
Block Slopes	Mountain slopes covered with moderate to large size angular blocks	15°-30°	Extensive	Outside glacial drift limits	Relict (stones covered with lichens)
Debris Slopes	Mountain slopes covered with a near-continuous veneer of frost-shattered debris consisting of small to moderate size blocks with some fines	15°-30°	In patchy areas only	Both inside and outside glacial drift limits	Contemporary
Gelifluction/ Solifluction sheets	Lobate features on slope consisting of stones and fines	20°-30°	In patchy areas	Outside glacial drift limits	Relict (stones covered with lichens)
Pro-talus Rampart	An arcuate ridge with convex face outward at the foot of a cliff-wall	15°-30°	In many area	Outside glacial drift limits	Relict
Ploughing Blocks	Boulders on grassy moist slope showing a distinct line of furrow behind as the mark of past movement	5°-15°	In patchy areas	Outside glacial drift limits	Both relict and contemporary
Nivational hollows	Hollows on the mountain slope formed due to scouring by the semi-permanent ice patches	5°-25°	In patchy areas	On sheltered valley-walls above the timber-line	Both relict and contemporary

Source: Based on field survey by the author.

Plate 7: A relict Block Slope on the mountain slope near the Dzongri La

Plate 8: An extensive Solifluction (Gelifluction) sheet covering the mountain slope beyond the Dzongri La on the way to the Rathong glacier

Plate 9: A large Pro-tatus Rampart at the foot of a valley wall

Plate 10: Ploughing blocks on the solifluction/gelifluction slope

The principal characteristic features of the periglacial landforms in this area can be explained as follows:

(i) The ostensible periglacial phenomena in distinguished forms occur with widespread distribution above the timberline.

(ii) It can be inferred from the observation that the past climate of this region was sufficiently cold, with greater temperature range to permit extensive frost shattering. This is proved by the occurrence of massive periglacial features like debris slopes, talus slopes and block slopes with the component blocks covered with lichens.

(iii) The past periglacial climate in this region gave rise to permafrost in the ground, upon which massive features like rock glaciers, moved down slope. These occur in relict form in the present day.

(iv) All major periglacial landforms occur outside the moronic limits. This would suggest that a severe periglacial condition in the past that produced the massive features in this region during the later part of the Pleistocene. Through the valleys and on the ridge slopes, the snow line descended to lower levels nourishing glaciers of various sizes. The two main glaciers, Rathong and Onglaktang extended through the Churong Chhu and Prek Chhu valleys respectively to a considerably long distance. Successive limits of retreat (terminal) moraine are the evidence of such retreat. Large scale frost shattering and frost wedging under this harsh climate produced enormous amounts of detritus that subsequently moved down slope, giving rise to the major slope features. Present periglacial activity in this region operates to produce mainly rock fall features produced from the material derived from the cliff walls. Surface material of the old (relict) moraine landforms have also been found reworked by periglacial activity

Mapping of Quaternary Geomorphologic Features and Its Interpretation

Appropriate cartographic presentation of Quaternary geomorphologic features in this part of the Sikkim Himalaya is very rare. However, some attempts have been made in this respect in the previous decade (e.g., Chattopadhyay, 1998b). In this work Geomorphologic mapping of the distinguished and ostensible glacial and periglacial features has been done on the base map prepared upon 1:150,000 topo-sheet made available from the University of Bern, Switzerland on personal contact. The snow-covered area and the extension of glaciers were identified initially on the satellite Image (Google Earth Image) and placed accordingly on the map. This was further verified and necessary corrections have been made with ground truth verification during the field studies in 1995 and 2010. All periglacial landform features, however, were placed on the map entirely upon ground-truth verification. The Geomorphologic map, thus prepared, has been presented below.

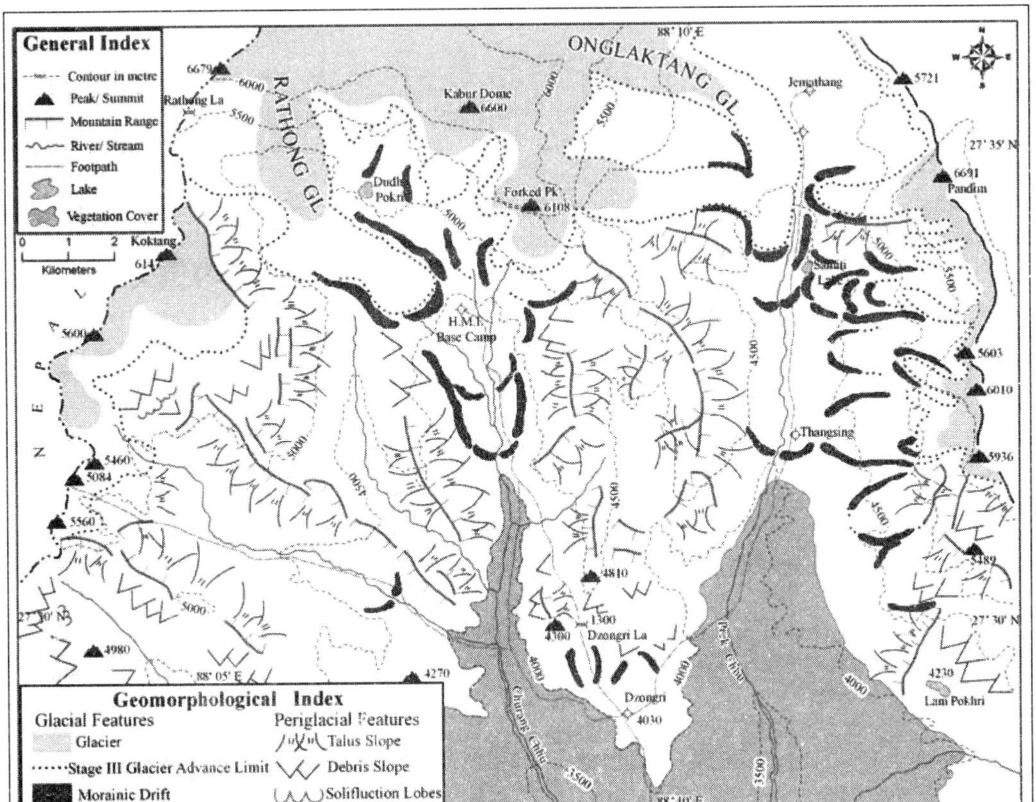

Fig. 4.2: Geomorphological map of the area around the southeast-facing slopes of Kanchenjunga

The prepared map exhibits two distinct valley glaciers, namely Rathong and Onglaktang descending from the neveé field of Kanchenjunga to the south-southeast. Three sets of terminal moraines were observed down through the glaciated valley before them, showing their retreats and halts during deglaciation. The amount of recent retreat following the Little Ice Age (16th to 19th century) can be clearly seen from the remains of fresh lateral and terminal moraines. The present snout point of the two glaciers rests at a distance of about 1.5 km from the limits of fresh terminal moraine, possibly dumped by the glaciers at the maximum of the Little Ice Age. Thus it can be referred that these two glaciers have retreated to the extent of 1.5 km from their advance limit attained in the Little Ice Age.

As has already been mentioned all the periglacial features resulted from frost wedging and shattering, occur outside the morainic limits. Among the large feature Talus Slopes, Block Slopes and Debris Slopes are ubiquitous. The component blocks are of large size (with average diameter >1.00 m) and spotted with lichens, thus showing their relict form. It can be assumed that the large features developed in the Late-Pleistocene period when periglacial climate in this part was much harsher. There are also several instances of Protalus Ramparts (a trough at the bottom of valley wall with an arcuate ridge of debris in front) as well. On moist grass and turf-covered slopes there are several instances of Solifluction (Gelifluction) lobes and lobate sheets. Ploughing blocks (individual blocks gliding down the slope with a line of furrow at the rear) on these sheets is also common. Nivational hollows are also common on the slopes where accumulated continues exist for a longer period during the summer. Occurrence of Solifluction (Gelifluction) lobes, lobate sheets and Ploughing blocks as micro-periglacial features indicate that limited condition periglacial climate, as well as its process continue to exist and operate in the present day in this region.

[**Acknowledgements:** This work is product of the author's Major Research Project sponsored and funded by the UGC, New Delhi for which the author remains grateful. Mr. Animesh Bose, Programme Coordinator, Himalayan Nature and Adventure Foundation, Siliguri, and other members of this organisation actively participated and extended generous co-operation during the field investigation at different times for which the author remains indebted.]

References

1. Chattopadhyay, G.P. (1984): High Altitude Slope Deposits in the Sikkim Himalaya and Their Periglacial Significance. *Himalayan Research and Development*, Vol. 3 (II): 9-13.
2. Chattopadhyay, G.P. (1985): Nature and extent of periglacial phenomena in the Kanchenjunga area, Sikkim Himalaya. *Abh.*, (Göttingen, Germany), Vol. 81: 145-52.
3. Chattopadhyay, G.P. (1998): Nomenclature and mapping of glacieal and periglacial phenomena in Arctic and Alpine Environments. *Göttinger Geographische Abhandlungen (Göttingen, Germany)*, 8-11: 145-152, 1986
4. Chattopadhyay, G.P. (2008). Recent Retreats of Glaciers on the Southeast-facing Slopes of the Kanchenjunga Summit Complex in the Sikkim Himalaya. *Himalayan Geology*, Vol. 29(2): 171-176.

5. Chattopadhyay, G.P. (2010): Glacial Lake Outbursts and Related Environmental Hazards: Some Observations in the Sikkim Himalaya. *Proceedings of International Conference on Global Climate Change – 2010, (19-20 Feb 2010)* In Balachandran, S., Padhy, P., Patra, P. and Chowdhury, S. (Editors) (Council for Scientific and Industrial Research and Visva-Bharati), pp. 166-171.
6. Chattopadhyay, G.P. (2012): Range of Quaternary Geomorphological Features Around The Dzongri-Hmi Base Camp Area on the Southeast-Facing Slopes of Kanchenjunga in Sikkim Himalaya. *Eastern Geographer,* Vol. 18 (1): 1-6.
7. De Terra, H. and Hutchinson, G.E. (1936): Data of Post-Glacial Climatic Change in Northwestern India. *Current Science,* Vol. 1: 5-10.
8. Fuji, Y and Higuchi, K. (1976): Ground temperature and its relation to permafrost occurrences in the Khumbu Himal and Hidden Valley, Nepal Himalayas. *Seppyo,* Vol. 38 (Special Issue): 125-28.
9. Garwood, E.J. (1902): Notes on a map of the glaciers of Kanchenjunga with remarks on some of the physical features of the district. *Geographical Journal,* 20(1): 13-24.
10. Kar, N.R. (1969): Some observations on periglacial features in the Sikkim and Darjeeling Himalaya. *Biuletyn Peryglacjalny,* (Lødz), Vol. 21 (2), 58-69.
11. Mayewski, P.A., Jeschke, P.A. (1979): Himalayan and Trans-Himalayan Glacier fluctuation since A.D. 1812. *Arctic and Alpine Research,* Vol. 11(3): 267-87.

5

Morphometric Characterization and Its Influence on Landuse/Landcover of Song River Basin, Uttarakhand

Sk Nasib Ahamed and *Dr. (Mrs) Anusuiya Baghel*

Introduction

Traditional geomorphological research is 'forward' problem-oriented where the researcher focuses on the evaluation of drainage pattern and system produced by known geology. On the other hand recently the 'inverse' approach has come into the light where geological information are derived by evaluating the drainage character (David Coblentz, Frank Pabian and Lakshman Prasad, 2014).

In this study, both the traditional as well as inverse approaches have been used for evaluating the land use/land cover in area of interest. Over the time drainage network develops by the local geological factors. Morphometry is the measurement of mathematical analysis of the configuration of earth's surface, shape and dimensions of its landforms (Ali Syed Ahmed, Khan Nazia, 2013). Morphometric analysis provides quantitative description of the basin geometry to understand initial slope or inequalities in the rock hardness, structural controls, recent diastrophism, geological and geomorphological history of drainage basin (Strahler, 1964, Rao G. Tamma *et al.* 2012). Morphometric characteristics help understand the morphology, hydrology and evolutionary history of any basin in a better way (Sharma and Sarma, 2013) that further influences the land use/land cover. The relationship between different drainage parameters and its underlying geology had been well established in the work of Strahler (1952).

Study Area

The area of interest is the Song river basin (Fig. 5.1) in Uttarakhand having the main urban center Dehradun is located between Siwalik in south and greater Himalaya in north. The geographical extension is 30° 00′ N to 30° 30′ N and 77° 45′ E to 78° 30′ E having an area of 1024.62 sq. km.

Objectives

The study has been made with the objectives of:

1. the drainage character produced by known geology.
2. Evaluating the geology in terms of permeability by analyzing the drainage character.
3. Understanding the spatial distribution of land use and land cover.

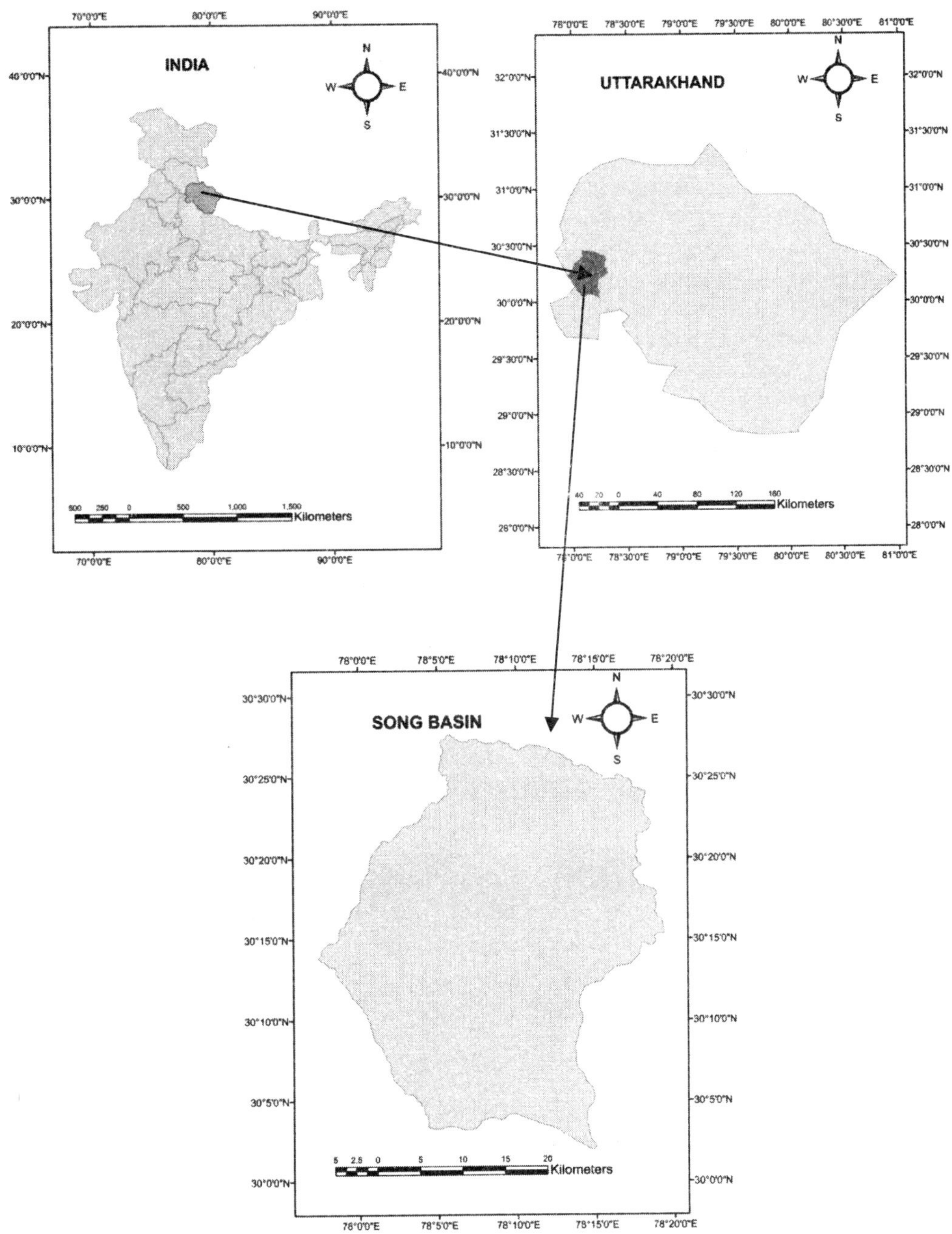

Fig. 5.1: Location of the Study area

Methodology

The drainage map has been prepared with the help of topographical maps, updated through LISS-IV imageries. The geological units are derived from geological maps made by G.S.I. with limited field check. Land use/land cover classifications have been done with the help of Landsat 8 imageries (having spatial resolution of 30 m) because it has 8 bands, orthorectified, projected and also freely available. Sub-basin demarcation has been done using SLUSI map. In this study, four sub-basins have been selected among all the sub-basins within the Song basin depending upon their lithological variations, drainage character as well as the land use pattern. These selected sub-basins are 2B5C8v, 2B5C8s, 2B5C8r, 2B5C8w. All these works have been done in ERDAS-11 and Arc GIS 9.3 environment. Details of all used data in this study are as follows:

- Topographical map – scale 1:50,000 from Survey of India.
- Geological map – scale 1:250,000 from Geological Survey of India.
- DEM is derived from contours.
- Landsat 8 – path (146), row (039), from USGS (2009).
- Sub basin map-from SLUSI, Dac (NRM Div) Min. of Agriculture.

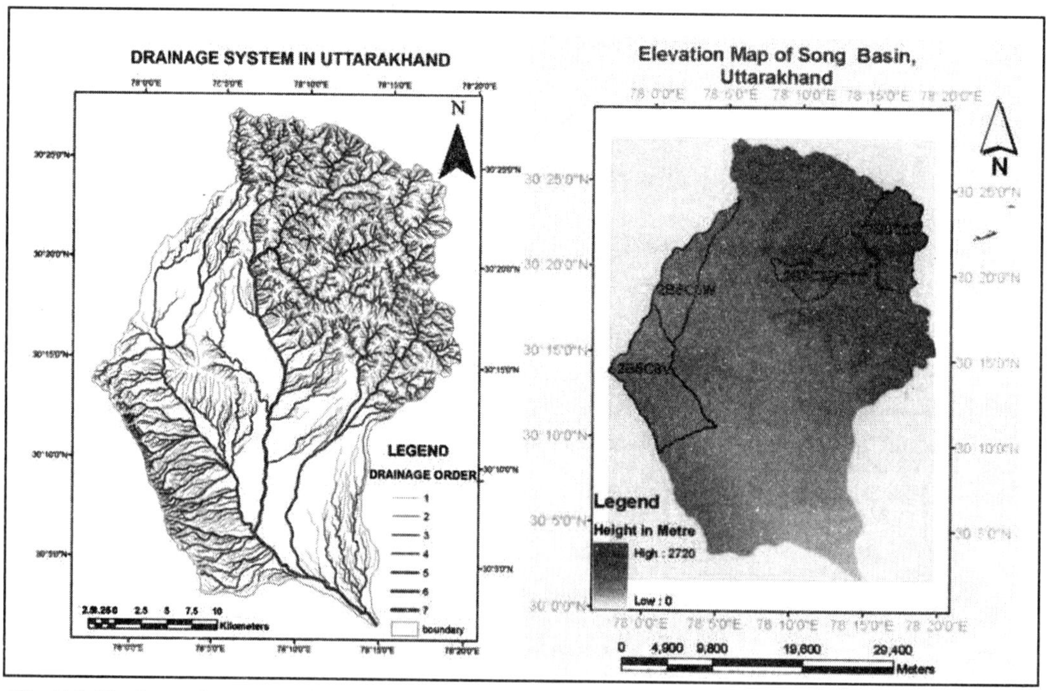

Fig. 5.2: Drainage System and Digital Elevation Model of Song basin

The formulas which are computed for calculating different morphomeric parameters are given below at a glance—

Morphometric Parameters	*Formula*	*Reference*
Stream Order (Nu)	Hierarchical rank	Strahler (1964)
Stream Length (km.)	Lu = L1+L2.Ln	Strahler (1964)
Bifurcation Ratio (Rb)	Rb= Nu/Nu+1 Where Nu= Total number of stream segment of order u Nu+1= Total number of stream segments in next higher order	Schumm (1956)
Drainage Density (Dd) km/km2	Dd =Lu/Au Where Lu =total length of stream segments cumulated for each stream order, Au= Basin area	Strahler (1964)
Elongation Ratio (Re)	Re=2 √(A/π)/Lb Where, Re=Elongation ratio Lb= Length of basin (km), A = Area of the basin (km2)	Schumm (1956)
Texture Ratio (Rt)	T = Nu/P Where Nu = Total number of streams of all orders, P = Perimeter (km)	Horton (1945)
Length of the overland flow (Lo)	Lo = 1/D*2 Where Lo= Length of the over land flow, D= Drainage Density	Horton (1945)

Results and Discussion

Forward Approach

Bifurcation Ratio (Rb)

Bifurcation ratio is a dimensionless property and it generally ranges from 1 to 5. Lower Rb value indicates that the area is structurally as well as drainage pattern both are less disturbed. The measured Rb values (Table 5.1) are – 1.95 (Song basin), 4.45 (2B5C8v), 2.4 (2B5C8s), 1.78 (2B5C8r), 1.56 (2B5C8w). After comparing all these Rb values of four sub-basins, it is clear that 2B5C8w is less disturbed than the other sub-basins where mostly disturbed basin is 2B5C8v. Rb should be decreased in the following higher order in a regular manner. But in 2B5C8v sub-basin, the Rb value is 2.32 in 5th order and 14 in 6th order which signifies the nature of disturbance having in the sub-basin (2B5C8v).

2B5C8v sub-basin is characterized by young folded mountain (Siwalik) where the top of the area is underlained by conglomerate (Fig. 5.3) indicates the high percolation but less than the unconsolidated material lies in middle and lower part of Siwalik Mountain and it helps to disappear the streams that lead to reduce the streams in number. Because of it there is a large variation in Rb values among the orders. For the whole basin there is a little bit of Rb variation, it means that some area of this basin is suffered from more structural disturbances. The correlation co-efficient is highly positive in 2B5C8w sub-basin (R^2 =0.75) where it is highly negative in 2B5C8s (R^2= 0.14) sub-basin.

Drainage Density

Drainage density (Dd) is concerned to the stream length per unit area. The Dds (Fig. 5.3 and Table 5.1) are 3.87 (Song basin), 0.28 (2B5C8v), 1.2 (2B5C8w), 6.64 (2B5C8r), 6.2 (2B5C8s). High Dd is the result of impermeable sub-surface material, sparse vegetation, low infiltration and high surface runoff. Low drainage density has been found in 2B5C8v and 2B5C8w sub-basin due to the soft rock with high permeability. Unconsolidated geological materials have been found in these two sub-basins and sediment by the eroded materials from the two mountains.

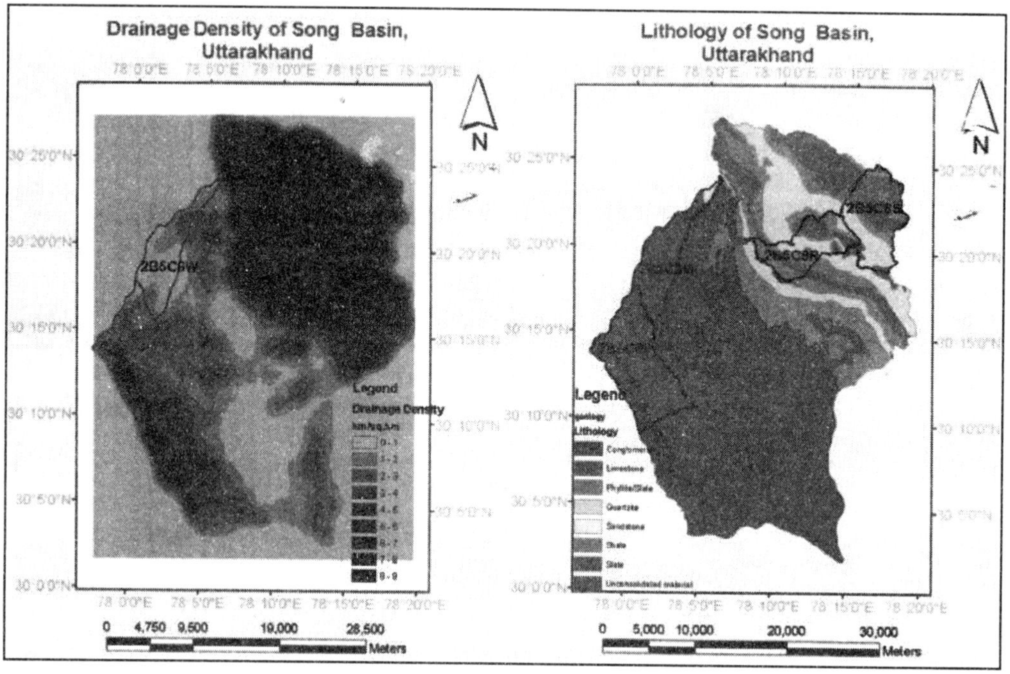

Fig. 5.3: Drainage Densities (DDs)

On the other side, another two sub-basins 2B5C8r and 2B5C8s both are characterized by high drainage density because they are characterized by limestone, conglomerate, phyllite, quartzite, shale, slate (Fig. 5.3) having moderate to low permeability which accelerates to flow through channel. The drainage density is very less in middle and lower middle part of the basin due to unconsolidated materials.

Elongation Ratio (Re)

Elongation ratio infers the shape of the basin. A circular basin is more efficient in runoff discharge. Strahler had classified the elongation ratio as circular (0.9-1.0), oval (0.8-0.9), less elongated (0.7-0.8), elongated (0.5-0.7), more elongated (< 0.5).

The elongation ratios (Table 5.1) are 0.72 (Song basin), 0.6 (2B5C8v), 0.43 (2B5C8w), 0.61 (2B5C8r), 0.69 (2B5C8s). 2B5C8w is more elongated and 2B5C8v, 2B5C8r sub-basins are elongated and the whole basin is less elongated as per the Strahler's classification. It means 2B5C8v is less in surface runoff due to permeable sub-surface geologic materials (unconsolidated) and the other sub-basins are less elongated than the 2B5C8w due to less permeable lithology such as limestone, conglomerate, quartzite, shale, slate etc. (Fig. 5.3). The whole basin is less elongated in shape because of moderate to low permeable geologic units are there in northern part of basin such as limestone, conglomerate, quartzite, phyllite, shale, slate.

We know that elongation ratio is 0.5 or less than that it must be elongated basin which linked with homogeneous geologic materials, highly permeable, low surface runoff and low soil erosion.

Texture Ratio (Rt)

Texture ratio (Rt) gives the information about the streams number (Fig. 5.2) which depends upon the underlying lithology, infiltration capacity and also relief aspects of the terrain. The Rts (Table 5.1) are 104.61 (Song basin), 22.69 (2B5C8v), 1.6 (2B5C8w), 44.81 (2B5C8r), 48.2 (2B5C8s). The texture ratio of Song basin is very high because underlying lithology has low percolation capacity except the middle part of the basin. 2B5C8r and 2B5C8s sub-basins are characterized by impermeable lithology and 2B5C8w sub-basin is characterized by maximum permeable lithological extent. The impermeable hard rock also gives the undulating terrain which give birth to more streams segment as like 2B5C8r and 2B5C8s sub-basins.

Table 5.1: Morphometric characterization

Morphometric Parameters	*Song basin*	*2B5C8v*	*2B5C8w*	*2B5C8r*	*2B5C8s*
Bifurcation ratio	1.95	4.45	1.56	1.78	2.4
Drainage density	3.85	0.28	1.2	6.64	6.2
Elongation ratio	0.72	0.6	0.43	0.61	0.69
Texture ratio	104.61	22.69	1.6	44.81	48.2
Length of overland flow	0.13	1.79	0.42	0.07	0.08
Stream length ratio	1.05	1.28	1.7	0.84	0.86

Source: Measured by researcher.

Length of Overland Flow (Lo)

Length of overland flow (Lo) gives the information about the distance between two streams. Large distance between two streams infers the less drainage density and texture ratio where the rainwater gets more time to percolate inside. On the other side, existence of sediment or unconsolidated materials (Fig. 5.3) reduces the stream number and length due to high infiltration. That's why the presence of (mostly) unconsolidated materials generates low number of streams in these two 2B5C8v and 2B5C8w sub-basins which increases the length of overland flow.

In 2B5C8r and 2B5C8s sub-basins and the whole basin, the length of overland flow is less due to the presence of impermeable lithological structure (Table 5.2).

Inverse Approach

In the inverse approach, it has tried to know the geological information analyzing the morphometric parameters.

Drainage density gives the information about the length of streams per unit area where the streams and its length both are high that indicates the area is less permeable.

Table 5.2: Lithological percentage shared by various sub-basins

Lithologies	*Song basin*		*2B5C8s*		*2B5C8r*		*2B5C8w*		*2B5C8v*	
	area	*%*	*Area*	*%*	*area*	*%*	*area*	*%*	*area*	*%*
Conglomerate	83.06	8.11	0	0	0	0	0	0	18.7	25.19
Limestone	90.34	8.82	19.38	37.49	5.36	13.06	0	0	0	0
Phyllite	61.3	5.98	0	0	1.6	3.9	0.05	0.11	0	0
Quartzite	116.36	11.36	15.55	30.08	10.27	25.02	0	0	0	0
Sandstone	1.86	0.18	0	0	0	0	1.82	4.22	0	0
Shale	100.41	9.8	8.65	16.74	14.92	36.34	0	0	0	0
Slate	32.37	3.16	8.11	15.68	8.05	19.6	0	0	0	0
Unconsolidated material	538.92	52.6	0	0	0.86	2.1	41.32	95.67	55.54	74.81
Total	**1024.63**	**100**	**51.69**	**100**	**41.06**	**100**	**43.19**	**100**	**74.24**	**100**

Source: Measured by researcher.

Due to impermeability of water it is resulted as streams and increased the number and length of streams which is nothing but increasing Rt. After comparing the Dd and Rt of each sub-basins (Table 5.1) it is clear that the permeable geologic materials are in 2B5C8v and 2B5C8w (unconsolidated materials—95.67% and 74.81%) sub-basins have been found where less impermeable materials have been observed in 2B5C8r and 2B5C8s (shale-36.345, limestone-37.49% dominant) sub-basins.

A circular basin is more efficient in surface runoff discharge than an elongation basin (Singh and Singh, 1997). Elongation ratio more than 0.5 tends to be circular. The elongation ratio (Table 5.1) of 2B5C8w (0.43) sub-basin refers that it is more elongated – means surface runoff discharge is very less which leads to more infiltration and it is possible in permeable geologic units. Geological Survey of India map validates it that the area is underlained by unconsolidated lithological materials. The elongation ratio in other sub-basins (Song basin-0.72, 2B5C8v-0.60, 2B5C8r-0.61, 2B5C8s-0.69) infers that they are tended to be circular which indicates that there must be hard and less permeable lithologies are available and it is proved by Geological Survey of India's map. In this area limestone, conglomerate, phyllite, quartzite, shale, slates are found.

Lithological structures can be understood through the length of overland flow technique. Very less length of overland flow indicates that the area is underlained by impermeable and hard geology. Due to the impermeability the water need to flow that give birth to stream. Large distance between two streams infers water percolatation inside the surface which reduces the number of streams. From the value (Table 5.1) of 2B5C8w (0.42) and 2B5C8v (1.79) sub-basins it is easily understood that the areas are underlained by unconsolidated (sediment) materials. On the other side, 2B5C8r (0.07) and 2B5C8s (0.08) sub-basins are underlained by cemented or consolidate materials (limestone, conglomerate, phyllite, quartzite, shale, slates have been found and the Geological Survey of India's map proves that).

Spatial Distribution of Landuse/Landcover

Eight land use/land cover classes (Fig. 5.4) have been identified in Song river basin where the highly dense forest is predominant – more than 1/3rd of the total area accounting 34.26 per cent (Table 5.3). High dense forest has been found in Siwalik mountain and

Dun regions, because these areas are fully covered by sedimentary rocks which is favourable for plant growth. Sedimentary rocks are naturally fertile and drainages are also less than the other area. That's why agricultural practices have been observed in only Dun areas (2.07%). In the northern part, agricultural land and settlement are not there due to the presence of hard rock (limestone, conglomerate, phyllite, quartzite,) and undulating terrain as well as high stream frequency.

Only 2B5C8w sub-basin is favourable for some anthropogenic activity (Fig. 5.4) rather than the other sub-basins. Settlement and agricultural practices have been found in this sub-basin only and Dehradun is located here.

Table 5.3: Land use/land cover distribution

Categories	Song basin		2B5C8r		2B5C8s		2B5C8w		2B5C8v	
	area (hect)	%	area (hect.)	%	area (hect.)	%	area (hect.)	%	area (hect)	%
High dense forest	35178.9	34.26	537.12	12.92	904.59	17.36	724.23	16.56	4863.51	65
Moderate forest	17156.3	16.71	1409.58	33.9	1847.07	35.44	318.33	7.28	332.01	4.44
Open forest	26941.6	26.24	1230.93	29.61	1323.27	25.39	1020.51	23.34	916.92	12.25
Scrubby area	3129.21	3.05	298.26	7.17	463.14	8.89	21.78	0.5	84.78	1.13
River sand	5801.49	5.65	50.94	1.23	0.09	0	117.72	2.69	718.83	9.61
Barren land	8315.37	8.1	602.72	14.5	654.66	12.56	334.26	7.64	224.01	2.99
Agricultural area	2130.03	2.07	1.62	0.04	0.45	0.01	277.29	6.34	99.27	1.33
Settlement	4019.04	3.91	26.46	0.64	18.72	0.36	1558.35	35.64	243.54	3.25
Total	**102671.94**	**100**	**4157.63**	**100**	**5211.99**	**100**	**4372.47**	**100**	**7482.87**	**100**

Source: Measured by researcher.

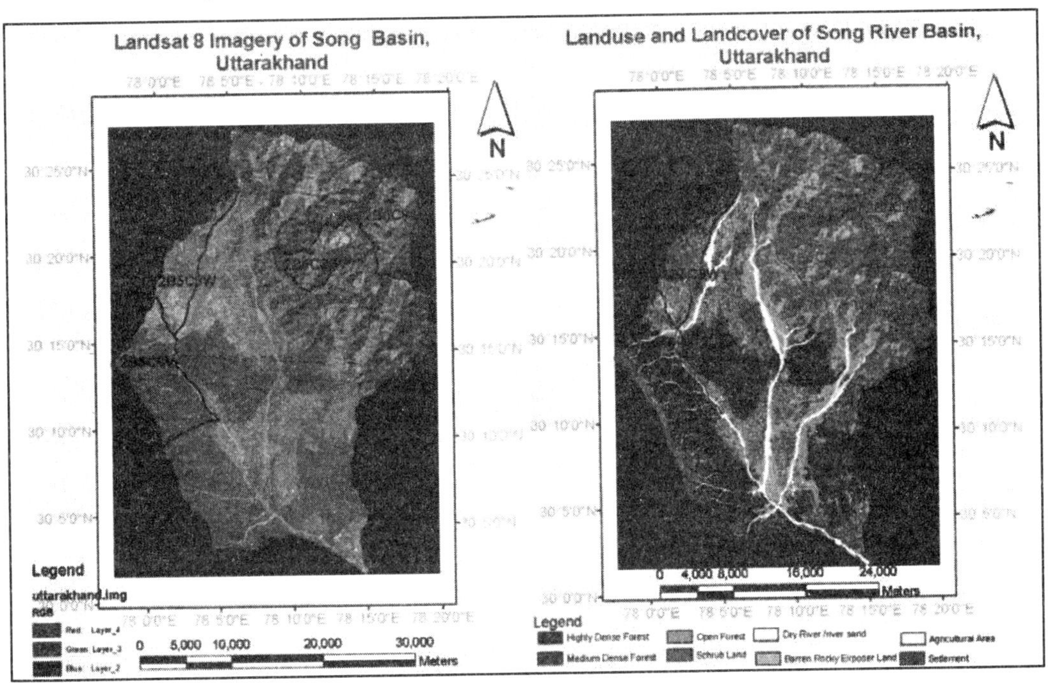

Fig. 5.4

Other predominant classes (Table 5.3) are open forest (26.24) and moderate dense forests (16.71) are mainly distributed in northern part (Himalaya) of the basin. 2B5C8r and 2B5C8s both the sub-basins are dominated by open forest and moderate dense forests due to high drainage frequency and impermeable hard rocks. The total forest cover is 76 per cent in Song river basin out of this 36 per cent is highly dense forest and remaining belongs to other forest types (Fig. 5.4).

Conclusion

Geospatial technology is an effective tool for morphometric analysis as well as land use/land cover classification. Morphometric characterization is an expression of geology. Bifurcation ratio indicates that the southern part of basin is more disturbed than the northern part of this basin due to the presence of conglomerate (unconsolidated) lithology which is more prone to landslide. High drainage density in northern and north-eastern sides of this basin refers to the impermeable hard lithological structure which is not favourable for any kind of anthropogenic activity. The value of elongation ratio suggests that Song basin is less elongated. The texture ratio in northern and north-eastern parts of the basin is very high due to the presence of impermeable sub-surface and the number of stream decreases to the south. Length of overland flow gives the information about the distance between two streams. Length of overland flow is very less in northern part and it increases to the south due to more ground penetration which leads to more anthropogenic activities such as agriculture, settlement etc. So, it is clear that the middle part of this basin mainly dun area is characterized by less number of drainages, less degree of slope, low surface runoff as well as sedimentary rock which encourages people to settle and practice the agriculture.

References

Ali Syed Ahmed, Khan Nazia (2013): "Evaluation of Morphometric Parameters—A Remote Sensing and GIS based Approach", *Open Journal of Modern Hydrology*, vol. 3, pp. 20-27.

Coblentz David, Pabian Frank and Prasad Lakshman (2014): "Quntitative Geomorphometrics for Terrain Characterization", *International Journal of Geosciences*, pp. 247-266.

earthexplorer.usgs.gov

Horton, R.E (1945): "Erosional Development of Streams and their Drainage Basins; Hydrological Approach to Quantitative Morphology", *Geol. Soc. Amer.Bull.*, Vol. 56, pp. 275-370

Kinthada Nooka Ratnam, Gurram Murali Krishna, Eedara Amminedu, Velaga Venkateswara (2013):" Remote sensing and GIS in the geomorphometric analysis of micro watersheds od hydrological scenario assement and characterization – A study of Sarada river basin, Visakhapatnam district, India", *International Journal of Geomatics and Geosciences*, Vol. 4, No. 1, pp. 195-212.

Pal Subodh Chandra, Debnath Gopal Chandra (2012): "Morphometric analysis and associated land use study of a part of Dwarkeswar watershed", *International Journal of Geomatics and Geosciences*, Vol. 3, No. 2, pp. 351-363.

Rao G. Tamma, Rao V.V.S. Gurunadha, Dakate Ratnakar, Rao S.T. Mallikharjuna, Rao B.M. Raja

(2012):" Remote Sensing and GIS-based Comparative Morphometric Study of Two Sub-Watershed of Different Physiographic Conditions, West Godavari District, A.P.", *Journal Geological Society of India,* Vol. 79, pp. 383-390.

Schumm, S.A. (1956): "Evolution of Drainage System and Slopes in Badlands at Perth Amboy, New Jersey". *Geol. Soc. Amer. Bull.,* Vol. 67, pp597-646.

Sharma, S. and Sarma, J.N. (2013): "Drainage analysis in a part of the Brahmaputra valley in Sivasagar district, Assam, India, to detect the role of nontectonic activity, *J. Ind. Soc. Remote Sens.,* Vol. 41(4), pp.-895-904.

Strahler, A.N. (1952): "Hypsometric analysis (area-altitude) of erosional topography". *Bull. Geol. Soc. Am.,* Vol. 63, pp. 117-142.

Strahler, A.N. (1964): "Quantitative Geomorphology of Drainage Basins AND Channel Networks", *In:* V.T. Chow (Ed.), *Handbook of Applied Hydrology.* MacGraw Hill Book Company, New York, Section 4-II

www.portal.gsi.gov.in

6

Identification of Landform Units of The Mej Basin, South-Eastern Rajasthan

Ganesh Sharma

Introduction

Geomorphology, as the study of the origin and evolution of earth's landforms and the processes that shape them, is clearly part of the disciplinary scientific basis of environmental management particularly land use/land cover studies. In India a number of studies on geomorphology have been done and new and new understanding of landforms are evolving. Land is a basic resource, hence understanding of land surface, evolution is necessary for better planning of land resources. The present chapter is an humble attempt to study landforms of the Mej Basin of Rajasthan.

The Study Region

The Mej river is a main left bank tributary of the Chambal. It forms on oval shaped basin over an area of 5,500 square kilometres. The basin is a transitional zone between the north-western lobe of the Great Vindhyan basin and north-eastern fringe of the Aravallis (Fig. 6.1).

Geographically, the Mej Basin extends from 25⁰0′ to 25⁰45′ North latitude and 75⁰30′ to 76⁰15′ East longitude. Politically, the basin covers the whole Bundi district along with marginal areas of Tonk, Bhilwara and Kota districts of Rajasthan.

Climatically, basin falls under semi-arid zone of India. According to Thornthwaite's (1948) moisture index, the basin has megathermal type of climate.

The Mej Watershed has developed over the matured topography of Aravallies and Vindhyan sedimentary basin. The area is rich in groundwater storage capacity but due to improper utilization of groundwater resource. Heavy soil erosion is getting degraded its natural resources as well as the productivity. The distribution of wasteland and degraded forest in the watershed is approx. 25 per cent of its total geographical area, whereas our country accounts to 20 per cent. The main geo-environmental problems of the area are:

(i) Heavy soil erosion resulting in development of gullies in some areas,
(ii) Poor productivity,
(iii) Depletion of groundwater,
(iv) Lack of surface storage of water, and
(v) Degradation of forests due to uncontrolled grazing and encroachment.

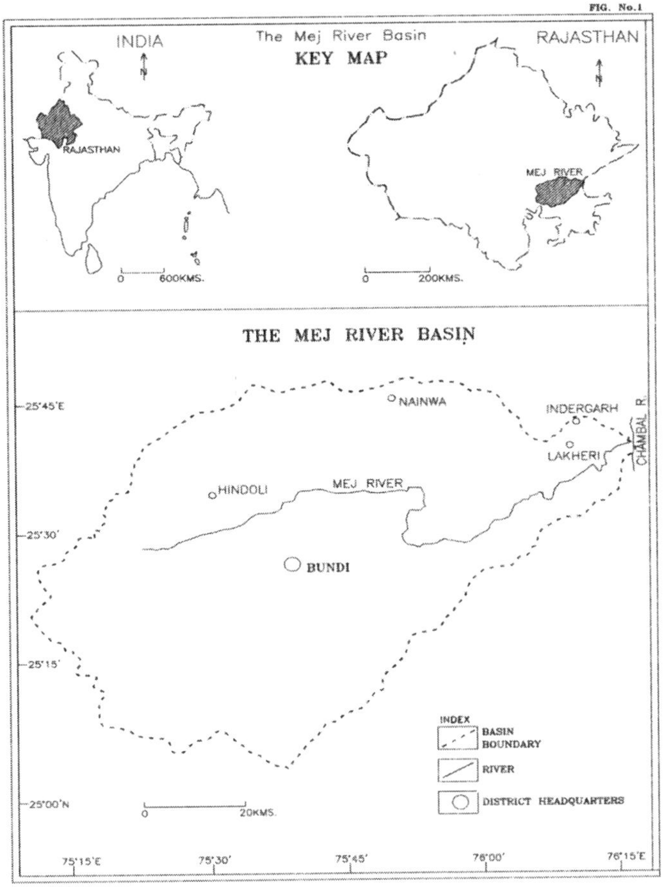

Fig. 6.1: Location of Study Area.

The area has a dominance of S.C., S.T. population, and is socio-economically backward and their main source of living is forest by-products and agriculture. However, the introduction of the Chambal canal system has transformed the economy of the people, but the problems of salinization and waterlogging have degraded the fertility of the soils. Earlier the area has been investigated by G. Padmaja (1976) and Sharma, H.S. and P.R. Binda (2007).

Objectives

Following are the objectives of this chapter:

1. to analyze the general physical character, geomorphology and geo-hydrology of the Mej Basin for better understanding the morphodynamics of the basin.
2. to apply IRS and GIS for generating Digital Elevation Model (DEM).

Methodology

Primary data has been generated through field work, analysis of Survey of India toposheets on 1: 50,000 scale and satellite image analysis. Data on geology, climate and hydrogeology were obtained from respective departments and survey organizations. Geomorphological maps has been compiled from the sources of Geological Survey of India. The compiled map depicts major landform units of the Mej Basin. GIS has been applied for data analysis and synthesis. Digital Elevation Model has also been made through Geographical Information System.

Geology

Geologically, the Mej Basin is a transitional zone in the sense that here the younger Vindhyans were folded and faulted against the older Gwaliors. The Great Boundary Fault of Rajasthan divides the Basin into two equal parts. This fault enters the Basin from north-east near Indergarh and passes through Bundi hills and further extends south-westward up to Chittorgarh. The central part of the Basin from south-west to north-east, along the Great Boundary Fault comprises of intensely folded Vindhyans. The area to the north-west of Great Boundary Fault is formed of Gwalior formations in which phyllites, shales, slates, and quartzites are common rocks. The Vindhyans generally lie to the south-east of the Great Boundary Fault and form hills which are inclined. The Vindhyans of the Bundi-Indergarh hills are inclined while the Vindhyans of south-western plateau are generally horizontal with a dip of 5°. The whole of the south-eastern plains is covered by alluvium which has been deposited on the Vindhyans by the streams of the region.

Geological History

The geological history of the Mej Basin is associated with the geomorphic history of the Aravallis and Vindhyans. The present topographical features of the area are an outcome of both the exogenetic and endogenetic forces. Unless those stages of evolution are not clarified, the present landform features can hardly be explained. Therefore, for the clear understanding of present geomorphic features it is must that we look back into the main incidents of its geological history. In the following pages, the geological history of the Aravallis and Vindhyans has been briefly sketched out separately.

Aravallis are the oldest folded mountains of pre-Cambrian Period. These ranges are formed of archaeans, which have been grouped into following groups:

- Delhi system,
- Railo series,
- Gwalior system,
- Arvalli system, and
- Bended gneiss complexes.

All the above listed formations are separated by well-marked unconformities. According to geologists like Heron, Coulson (1927-1928) and Fermor (1930), the materials of the Aravallis have been deposited by natural streams under a rather uniform climate. It has been further suggested by geologists that in the beginning of Protorosic era there was a vast and a long geosyncline which received vast deposits of Delhi sediments by various geomorphic agents. The Delhis are post-Archaean in age and post-Dharwar in formation. The Delhis are metamorphosed subsequently and this process took place maximum in the central part of the Aravallis. The deposition of Delhi sediments was being continued upto lower Vindhyan period.

The mountain building activity took place in the pre-Cambrian period and it was of great magnitude giving rise to a range even higher than Himalayas and extended from Gujarat in south-west to beyond Delhi in north-east. Throughout the Palaeozoic and Mesozoic periods, the Aravallis were constantly attached by exogenetic forces and were reduced to a Peneplain. But this Peneplain condition could not remain for a longer time and the ranges were rejuvenated in Cretaceous Period (Heron, 1953). This upheaval was not uniform throughout, but was like an arc being highest in the centre and sloping either way. It has been pointed out by Heron that at the close of the Eocene period, the Aravalli were once again cut down to show some signs of peneplanation in the eastern ranges, but regarding this, he himself was not clear. It is certain that the range was partially reduced to peneplanation.

The Vindhyans had been deposited simultaneously on the eastern and western sides of these mountains when the Aravallis were undergoing peneplanation in Palaeozoic and Mesozoic eras. As a matter of fact, the Vindhyans are the outcome of denudation of the Aravallis.

There is a controversy about the age of the Aravallis and deposition of the Vindhyans on the either side of it. Fermor (1930) considered Aravallis as a host having the Great Boundary Fault on the east and hypothetical fault of same magnitude in the west. He further contended that the Aravallis were brought to a peneplained condition in pre-Talchir times and the Vindhyans of western Rajasthan were deposited in continuation of those of eastern Rajasthan and Madhya Pradesh. On the contrary Heron (1938) says that it is not at all necessary to suppose a hypothetical fault along the western front of Aravallis. He further suggested that the Aravalli range was in existence throughout the period of deposition of Vindhyans and separated them into two disconnected areas. According to him the range was peneplained in Mesozoic period and was rejuvenated in Cretaceous period.

Structure

The Great Boundary Fault

The Great Boundary Fault of Rajasthan running in a direction from north-east to south-west direction is the most prominent structural feature of the Basin. This Boundary Fault has been mapped by Heron (1922) in north-east and by Coulson (1927-1928) in the Bundi district. Coulson has considered this fault as a reverse fault.

In the Mej Basin this fault enters near Indergarh and here it bifurcates into two lines. The north-western fault separates the Lower Bhander Sandstone from the Gwalior and runs about a displacement. From Satur, it continues without interruption to Khenia and then it passes out of this Basin, near Rooppura. Another southern branch of the Great Boundry Fault separates Upper Rewa Sandstones from shales and seems parallel to northern fault from Indergarh to Suraj Chatri from northeast to southwest. At Bundi it is clearly marked.

Other Fault

In the Mej Basin there are number of other secondary faults which are also a result of cretaceous faulting. About a kilometer of Bhairon, the Great Boundary Fault is joined by a small fault which commences near Ramgarh in the form of a horseshoe. Here, Lower Bhander Sandstone was faulted against Samria shales. Afterwards, the fault turns south-west and finally joins Great Boundary Fault near Bhaironpur.

Folds

Another structural feature in the area is the arrangement and nature of folds. Towards south, faulting is less important than folding. Here the rocks have been intensely folded against one another. Gently diping anticlines and synclines are common, and the Satur anticline is a good example of an eroded anticline.

On the whole, it can safely be concluded that the geological structure of the area under investigation is very complex. As a result of this intricate geological structure, complex and interesting geomorphic features have resulted in the Basin.

Geomorphology and Geohydrology

The existing topographic features of the Mej Basin are a combined result of underlying geological structures and fluvial erosion. The Great Boundary Fault movement has a significant effect on the evolution of relief features. The fluvial cycles which has been working since Mesozoic period has accommodated with the underlying structure. As a matter of fact, the present area is a good example of topographic features characterized by fault scarps, composite scarps and fault line scarps.

Broadly, the Mej Basin can be divided into three topographic units-Hills, Plateaus and Plains. The hills traverse from south-west to north-east and like a backbone divide the Basin into two equal parts (Fig. 6.2 and 6.3). The north-western area to the north of these hills is a Plateau having an average elevation of 330 metres. South-western portion of the Basin is an extension of the Kota plateau. On the other hand, the south-eastern part of the Basin is a plain overlain by recent and sub-recent sediments deposited by the Mej and its tributaries. The plain is generally below 300 metres, the average height being 210 metres. This area is also characterised by dissected ravines and gullies.

The Hills

Hills with an elevation of about 500 metres are found in continuous ranges from north-east to south-west. Those hills are composed of the Vindhyan Sandstone with intermittent layers of shales and limestones. The Kaimur, Bhander and Rewa—all sandstones form conspicuous ridges. The Kaimurs form extreme north-western hills extending parallel to the Great Boundary Fault. These hills are not continuous but are formed of disconnected ridges (Fig. 6.2). The Upper Rewa sandstone forms a continuous and elongated ridge right from Suraj Chatri to Indergarh. Those hills are at a lesser elevation than the Bhander sandstone range. The maximum elevation of those hills is 471 metres near Kaimla and have a minimum value of 368 metres. The boldest and most prominent among all these ranges is the Bhander sandstone ridge. This range extends from Khenia to Lakheri. The average elevation of this hill is 468 metres. The maximum elevation near Kankhra Dungar is 504 metres, while the minimum elevation near Bundi is 433 metres. All those hills are flat topped and form bevelled surfaces. The faces of all these hills are straight in their upper portion and give an impression of concavity in their lower portions.

The relative relief (Fig. 6.3) of these hilly sections is maximum, i.e., above 150 metres. It decreases on either side of the hills.

These hills are characterized by various scarps which are the result of faulting and later are modified by fluvial erosion.

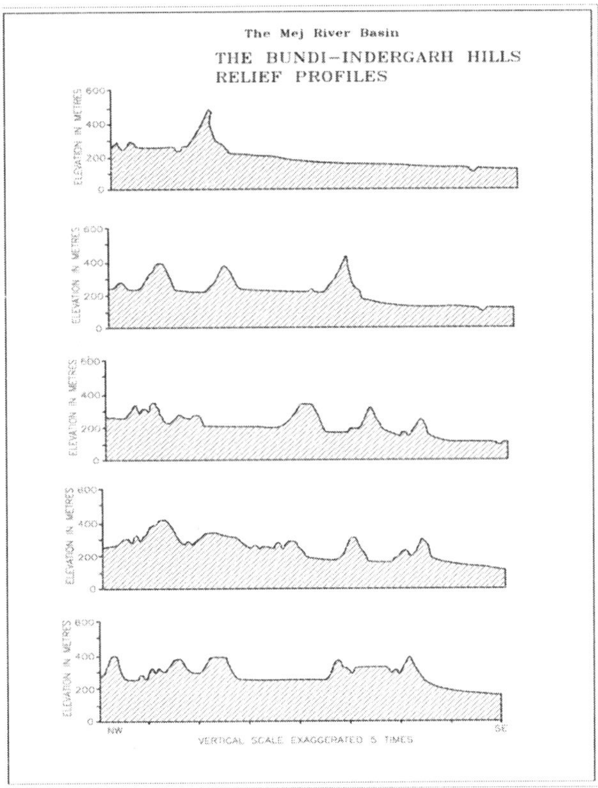

Fig. 6.2: The Bundi-Indergarh Hills Relief Profiles

Plateaus

In the Mej basin, there are two plateaus formed of different geological formations. The south-western plateau is formed of Bhander Sandstone and the beds dip towards northeast with 5° slope. This plateau merges into hills towards north-east and is terminated by an escarpment in the eastern portion of the Basin. Here, the plateau

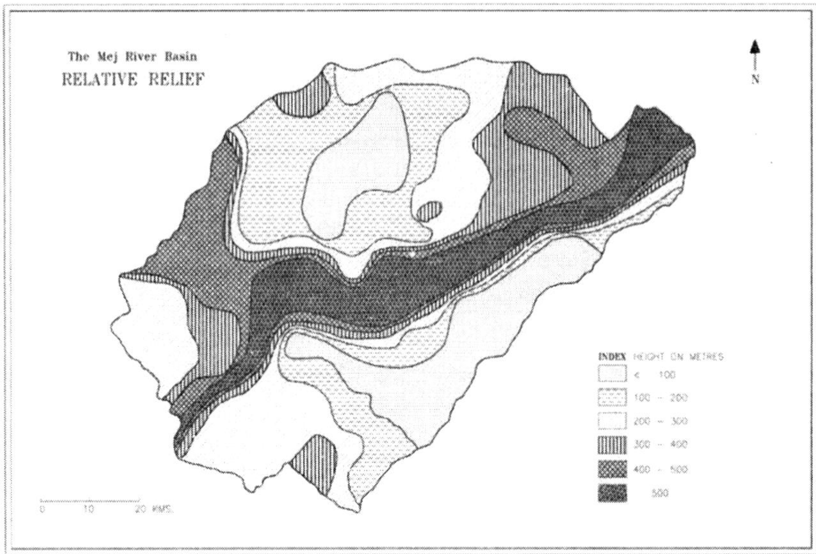

Fig. 6.3: The Mej River Basin Relative Relief

merges into extensive and vast riverine plain and the slope of this scarp is concave. Streams have greatly dissected this plateau, at present the dissection being more than 60 metres. The Bhimlst valley is a testimony of this fact. The general slope of the escarpment is usually concave and is directed towards north-east. The plateau towards the south is terminated by bold escarpment. The maximum height of this plateau is 546 metres north-west of Bijolia, whereas the general elevation ranges around 450 metres.

The relative relief of this plateau section is from 60 to 120 metres. In the central part it is upto 150 metres.

The whole north-western area north-west of the Great Boundary Fault is a part of Aravalli Mesozoic peneplain. This whole area is plateau—like in character with an elevation above 300 metres. It is composed of Gwalior quartzites and shales out of which Gwalior quartzites generally form rounded flat topped hills and give an impression of inselberge. Around Hindoli and the south-wetern portion of the plateau one can see numerous such hillocks with flat tops and gentle slopes. The whole surface of this plateau is composed of Gwalior shales which give an uniformity in its surface throughout. The shale bed are vertical and it is interesting to note that the streams of this senile plateau are not guided by the structure, which again leads us to conclude that it is definitely a part of Mesozoic peneplain with rounded flat topped hillocks amidst a vast stretch of land.

Plain

The whole south-eastern and north-eastern portions of the Basin are remarkably plain, and the later is covered with dark grey soil. The soils around Nainwa are generally fine textured and become sticky during rains. The average height of the plain is 270 metres. The southern part of the plain is dissected into ravines and the whole southern and eastern portion of the Basin is a plain and is covered with younger alluvium brought by the Mej and its tributaries. The chief characteristics of this plain is that the stream banks are attended by gullies and ravines. The maximum height of the plain towards south-west is 309 metres and minimum is 210 metres near the confluence of the Mej with the Chambal.

The relative relief of the lowland is mostly less than 150 metres, but in some parts, it is upto 260-290 metres.

Geomorphology

In the earlier description, the general physiographic characteristics of the basin have been discussed. In view of this, the following geomorphic/landform units have been identified in the Mej Basin (Fig. 6.4):

1. Fluvial origin
   - Alluvial Fill
   - Alluvial Fan
   - Ravine
2. Denudational origin
   - Pediment
   - Buried Pediment
3. Structural origin
   - Plateau
4. Hills
   - Structural Hill.

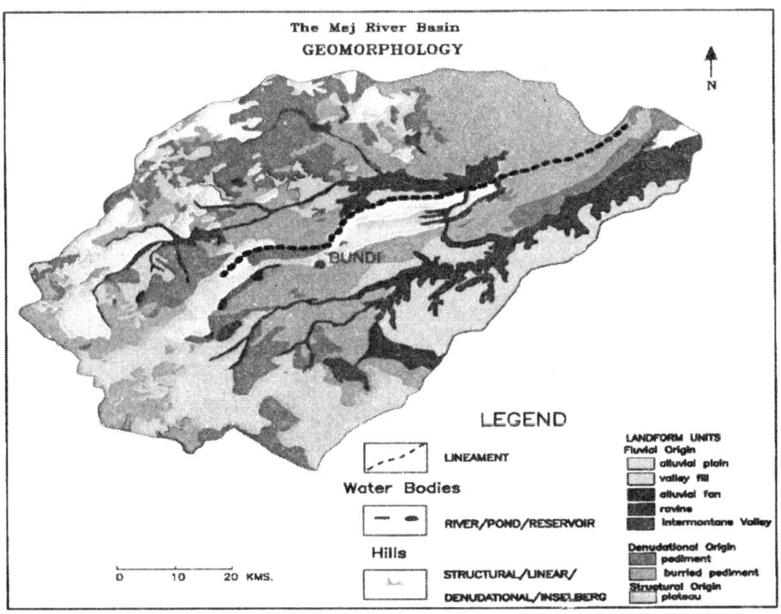

Fig. 6.4: The Mej River Basin: Geomorphology

The further description of each unit of landforms has been given in Table 6.1.

Table 6.1: The Mej Basin Geomorphology

Landform	*Symbol*	*Lithology/Material/ Description*	*Occurrence in district*	*Land use/Land cover*
Fluvial Origin Alluvial Plain	AP	Mainly undulating landscape formed due to fluvial activity, consisting of gravels, sand, silt and clay. Terrain mainly undulating, produced by extensive deposition of alluvium.	In between river Mej and river Chambal	Double crop, single crop (Rabi/Kharif), fallow, open scrub.
Valley Fill	VF	Formed by fluvial activity, usually at lower topographic locations, comprising of boulders, cobbles, pebbles, ravels, sand, silt and clay. The unit has consolidated sedmient deposit.	Scattered in upper half of district.	Marginal double crop, single crop (Rabi)
Alluvial Fan	AF	A fan shaped mass of sediment deposited at a point along a Nallah river where there is a decrease in gradient.	West of district.	Gravely wasteland.
Ravine	RV	Small, narrow, deep, depression, smaller than gorges, larger than gulley, usually carved by running water.	Along river Chambal and river Mej.	Open scrub.
Denudational Origin Pediment	P	Broad gently sloping rock flooring, erosional surface of low relief between hill and plain, comprised of varied lithology, criss crossed by fractures and faults.	Scattered in north-west, west, south-west and north-east.	Marginal single crop (Kharif), fallow and open scrub.
Buried Pediment	BP	Pediment covered essentially with relatively thicker alluvial, colluvial or weathered materials.	Covers northern and central part.	Marginal double crop, single crop (Kharif), fallow, open scrub.
Structural Origin Plateau	PT	Formed over varying lithology with extensive, flat, landscapes, bordered by escarpment on all sides. Essentially formed over horizontally layered rocky marked by extensive flat top and steep slopes. It may be cris-crossed by lineament.	South-west of district.	Marginal single crop, open scrub.
Hills Structural Hill	SH	Linear to arcuate hills showing definite trend-line with varying lithology associated with folding, faulting etc.	Trending south-west to north-east.	Marginal single crop (Kharif), fallow and open scrub.

After having discussed the landform units of the Mej Basin, it would be worthwhile to mention the salient features of Digital Elevation Model of the Mej Basin. This model represents square grid arrays of terrain height, has become an integral part of digital geomorphometric analysis for characterization of both specific landforms and continuous landscapes. This model provides basic elevation data for digital geomorphometric analysis. SRTM-3 data set pertaining to the Mej river tributaries has been imported into Rives Tools Software as irregular XYZ ASC II files and placed into different folders. The DEM of the Mej Basin depicts drainage lines, plateau of south-west, northern peneplain, Vindhyan hills and southeastern plain area in 3 dimensional form (Fig. 6.5) The DEM is extremely useful in studying hydrology of the basin.

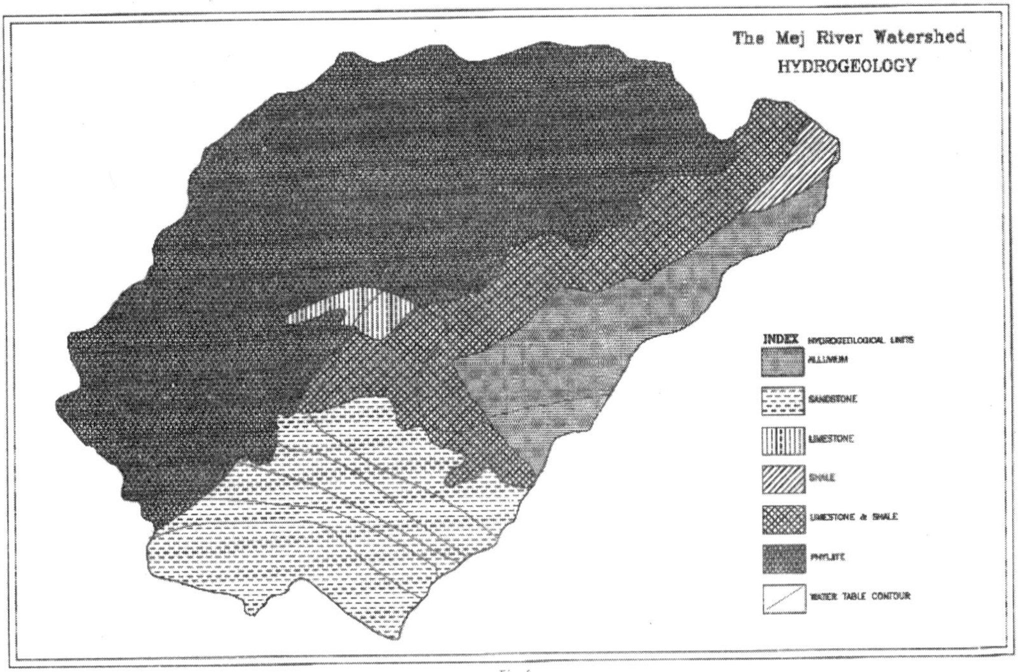

Fig. 6.5: The Mej River Watershed: Hydrogeology

Hydrogeology

The hydrogeological map (Fig. 6.6) indicates that occurrence of groundwater is dependent on various hydrogeological units given rise on the basis of lithological character. The Table 6.2 gives a detailed picture of hydrogeological condition of the basin.

Fig. 6.6: The Mej Basin Digital Elevation Model

Table 6.2: The Mej Basin Hydrogeology

Hydrogeological Units	*Description of the unit/Geological section*	*Occurrence*	*Groundwater flow*
Alluvium (Quaternary)	The litho unit comprises unconsolidated to semiconsolidated fine grained sand, silt, clay and kankar. Thickness of alluvium varies upto 38 m. Very thick alluvial sequences have been encountered in area between Chambal and Mej rivers.	Alluvium is mainly confined to central and south-eastern part of the area where occupies drainage courses and flood plains of Chambal, Kural and Mej rivers. Alluvium encampasses nearly 27% potential area.	The general direction of groundwater flow at inferred from water table contour is from southwest to north-east. Hydraulic gradient is steep in the southern part (2.0 m/km) moderate in the northern part (1.2 m/km) and gentle in the eastern part (1.0 m/km.)
Sandstone (Vindhyan Super Group)	The litho unit belonging to Kaimur and Bhander groups are fine grained, grey to red in colour, very hard and compact, at places appearing like quartzite.	It occupies southern part within the limit of Talera block. Sandstone covers nearby 9% potential area.	
Limestone (Vindhyan Super Group)	The litho mainly belongs to Bhander groups. It is fine to medium grained, grey, red, yellowish, pink or buff coloured; Red and brick coloured varieties are non-siliceous whereas grey bands being siliceous are very hard. In general Bhander limestone is cavernous.	Litho unit occupy small area in part of Hindoli and Talera blocks. It covers nearly 6% potential area.	
Shale and Limestone with Shale (Vindhyan Super Group)	Shales are grey, light green and purple in colour. It is mostly splintery in nature.	These litho units cover extensive area in Talera, Keshoraipatan and Nainwa blocks, shale and limestone with shale cover nearly 15% potential area.	
Phyllite (Bhilwara Super Group)	The super group in order of super position includes Jahazpur group, Hindoli group and Mangalwar complex. Jhahazpur group comprises dolomite, phyllites, carbonaceous phyllites, quartzite and basic metaolcanics. These rocks are intruded by sills and dyke of dolomite. Hindoli group of rocks are represented by shales, slates, phyllites, metagreywack, limestone, dolomitic, marble, meta basic volcanics, quartzites and mica schists. Sills and dykes of dolomite often occurs. Mangalwar complex comprises megnitites	These litho units belonging to Bhilwara Super Group occupies north-western part of the area and share about 39% of the potential area.	

	genisses, feldspathic mica schist, garnetiferrous mica schist, quartzite, impure marble and amphibolite.		

Conclusion

The present study has brought out different geomorphic units in the Mej Basin. Major units of fluvial, denudational and structural origin. The Mej river is a major left bank tributary of the Chambal. It forms an oval shaped basin covering an area of about 5,500 square kilometres. It is of 9th order basin. The ultimate objectives of the study is to generate Digital Elevation Model for watershed planning. The study has also demonstrated the utility DEM in hydrological study.

References

1. Coulson, A.L. (1927-1928), Geology of Bundi State, *Rajputana Record,* Geological Survey of India, Vol. 60,167-186.
2. Formor, L.L. (1930), On the Age of the Aravalli Range, Record Geological Survey of India, Vol.. 62, Part 4, 391-409.
3. Heron, A.M. (1922), The Gwalior and Vindhyan Systems of South Eastern Rajputana, *Memoirs Geological Survey of India,* Vol. 45, Part 2, 129-179.
4. Heron, A.M. (1938), The Physiography of Rajputana, *Proceedings of 25th Indian Sciences Congress,* Presidential Address, Vol. 25, Part 2, 119-132.
5. Thernthwaite, C.W. (1948), An Approach Towards a Rational Classification of Climate, *Geographical Review,* Vol. 38, 85-94.
6. Padmaja, G. (1976): Geomorphology of the Mej Basin, Unpublished Ph. D., Thesis, University of Rajasthan, Jaipur.
7. Sharma, H.S. (2007): Report on An Integrated Approach for Watershed Management: of South Eastern Rajasthan. Project Sponsored by the UGC.

7

Land Resource under Stress caused by Issues and Problems of Water Resources

S.K Sharma

Introduction

Land is an essential natural resource, both for the survival and prosperity of humanity, and for the maintenance of all terrestrial ecosystems. It is a vital resource especially for food production and also is a vital component of many natural systems. But the increasing human population and its increased expectations and demands are putting pressure on land resources, leading to land degradation and causing severe stress on land. Most often, increasing pressure of population compels farmers to ignore the natural limits and suitability of land for uses; and to expand farming to physically unsuitable and forested lands, creating several environmental crises. Land degradation, soil erosion, land and water pollution, deforestation are some of the points of stress. These problems not only reduce the productivity but suitability of land for different uses also. For alleviating this situation, therefore, an integrated approach of land use planning, accommodating the qualities and limitations of each component including water resource, is required.

Water as Integral Component of Land Resource

While the attributes of the physical environment have their own individual significance in conditioning and development of other natural resources, it is the combined effect of all and the interaction between them which are important. The term 'land' has been used to denote this complex (Christian and Stewart, 1953) and refers to land surface including all its characteristics of importance to man's existence and success. It is an integration of all these factors. Christian and Stewart further elaborate that the 'land comprises the physical environment including climate, relief, soils, hydrology and vegetation to the extent these influence potential for land uses' (Christian and Stewart, 1968, 238). In this context, it is pertinent to recall that land has several dimensions and components. But it was a common practice to equate land with soil. Major deviation from this concept emerged in '*Framework for Land Evaluation*', a manual of land evaluation (FAO in 1976, 9). It specifically mentioned that land, a basis for agriculture and other rural land use activities, includes also the climate, vegetation, slope conditions, and other natural resources. Hence the Framework defined land as an area of the earth's surface, the characteristics of which embrace all reasonably stable, or predictably cyclic, attributes of the biosphere vertically above and below this area, including those of the atmosphere, the soil and underlying geology, the hydrology, the plant and animal populations, and the results of past and present human activity, to the extent that these attributes exert a significant influence on present and future uses of the land by humans. Later, UN in its document on desertification (1994) further elaborated that:

> "Land is a delineable area of the earth's terrestrial surface, encompassing all attributes of the biosphere immediately above or below this surface, including those of the near-surface climate, the soil and terrain forms, the surface hydrology (including shallow lakes, rivers, marshes, and swamps), the near-surface sedimentary layers and associated groundwater reserve, the plant and animal populations, the human settlement pattern and physical results of past and present human activity (terracing, water storage or drainage structures, roads, buildings, etc.)."

It is pertinent to note that land means different things to different people (Whyte, 1976, 1-7). Land, in its broad sense as a factor of economic production, is synonym of natural resources complex (Zobler, 1962, 80). It represents the fundamental resources to which other resources are related to and are part of it. Graham (1967, 170) says that land expresses all various natural factors existing in the area. A more complete concept of land is that it is 'the sum total of the natural and manmade resources over which possession of the earth's surface gives control' (Barlowe, 1960, 7-8). Most often, land is deemed synonym of the physical environment (Young, 1976, 383).

Included in the land resources along with other things, thus, are surface and near-surface freshwater resources. Water adds value to the land resources. The linkages between water and land are so intimate that the water element cannot be excluded from land. Land is intermixed with water, depends on access to that water for its use, and at the same time, affects the quality and quantity of water. Land regulates the storage and flow of surface and groundwater resources, and influences their quality. Excessive and improper land use reduces both quality and quantity of this valuable resource. The suitability of the land for different uses depends, to a great extent, on the availability of water. Because of this fact, methods of land evaluation has been given due importance to the moisture content of the soil. Direct and high correlation between cropped area and production and precipitation, the principal source of water resources, can be seen on all scales, ranging from local to national and international. Evaluation and development of land in conjunction with water resources has been earmarked as essential preliminary to development, particularly to the agricultural development planning (Robertson 1970, 148). In the present attempt this view of water resource as an integral component of the land resource has been behind the current.

Concern of Water Resources

While availability, accessibility, utilization and renewability of water resources depend, to a great extent, on nature of precipitation, terrain and characteristics of soil, their temporal and spatial distribution and utilization also pose serious problems to the land resources, such as land degradation, salinization, waterlogging and lowering of the water table and depletion of groundwater resources. Large scale submergence of valuable agricultural land and displacement of people and water pollution are other serious problems (Sharma, 2016, p. 78). Some of these issues of water resource development in perspective of land resources in wider sense are examined below, citing Madhya Pradesh state as a case.

With the advancement of the society, awareness and requirements of water resources have been increasing phenomenally, resulting in expansion of the list of uses of water from traditional uses for drinking, domestic and irrigation purposes to generation of power, industrial ingredients, and media of transport, waste remover and purifying agent as well as a recreational agent. Contrary to expanding demand, its supply is limited and is being depleted or made unusable by excessive utilization, pollution or careless management. Uneven distribution is another characteristics of water resources. There is ample surplus water at one place and highly deficit on the other. Such situations manifest in uses of land. Under such conditions, careful examination of different problems and issues of water resources utilization is imperative.

Potential and Availability of Water Resources: Uneven Distribution

In context of water resources, Madhya Pradesh has certain unique physical features. Dependent primarily on monsoon rainfall, the geological structure and physiography determine the condition of the water table, run-off and aquifers etc. which ultimately determine the feasibility of water resources. Structurally this state is predominantly composed of hard rocks, ranging in age from the Archaean to the Recent period with limited potentiality of groundwater. Barring a few areas, most of the state is hilly and dissected plateau with sloppy land, increasing the velocity of running water and inhibiting percolation and therefore facing water scarcity. Climatically, western half of the state is semiarid region undergoing water strain throughout the year.

The primary source of water supply on earth is precipitation. Much of it is lost as evaporation; a good deal of it goes as run-off and is called 'surface water' and a small amount penetrates into ground and is known as 'groundwater'. Both availability of surface water and potential of groundwater depend on amount of rainfall. Precipitation decreases from southeast to northwest in this state (Figure 7.1). Average annual rainfall is 62 cm at Gohad in Bhind district in the northwest and 78 cm at Jhabua in the west to 212 cm at Pachmarhi in Hoshangabad district in south. South of the Satpura-Kaimur ranges, running diagonally from southwest to northest, annual rainfall is more than 130 cm but it is less than this amount north of this line. Along with low rainfall, it is highly variable in the Madhya Bharat, western Malwa and middle Narmada valley. Because of these characteristics, Indian Metrological Department has identified 27 drought-prone districts of this part of the state. This pattern is manifested in available potential and utilization of both surface and groundwater.

Surface Water

The surface water is feasible through ponds, tanks, rivers, streams and reservoirs. Rivers are major source of surface water in the state. Rivers of this state constitute part of the

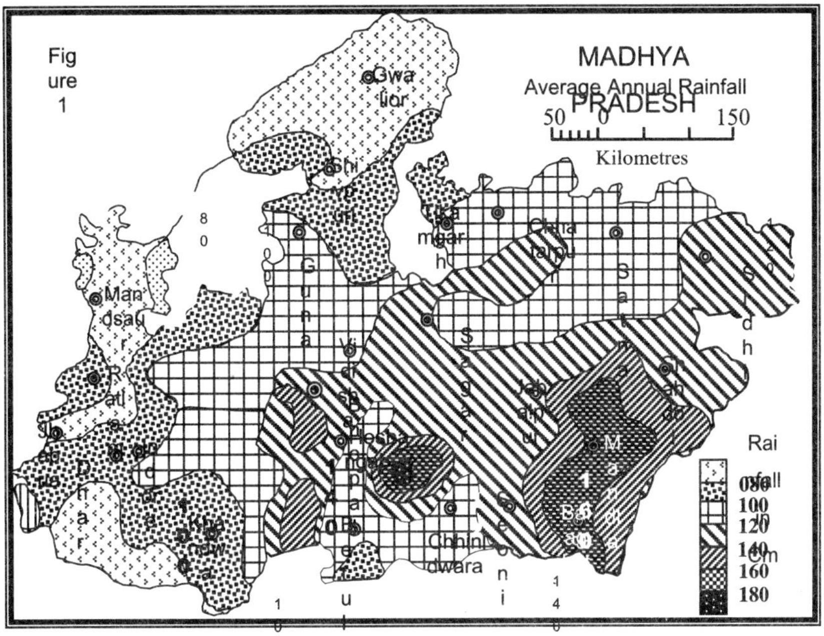

Fig. 7.1: Madhya Pradesh: Average Annual Rainfall

Ganga, Narmada, Tapi, Mahi and Godavari drainage basins. The Ganga basin, comprising the sub-basins of Chambal and its tributaries, Betwa, Ken (all part of Yamuna basin), Tons and Son, drains the largest area of the state. The northern part of the state is drained by these rivers and their tributaries, which flow northerly through Bundelkhand region and ultimately join river Yamuna. The Son river, flowing east–north-east, joins Ganga near Patna. The Narmada, flowing in a westerly direction, is a major river between Satpura and Vindhayachal ranges draining to Arabian sea. Tapti is also westerly flowing river debouching its water in Arabian Sea. Wainganga and Pench are tributaries of Godavari river draining in south-eastern part of the state. Mahi drains small area of Dhar and Jhabua districts in the west.

The water resource department of the Madhya Pradesh state has estimated the annual flow of rivers of the state at 81.7 billion cubic metres at 75 per cent reliability. Out of this, 57.1 billion cubic metres can be utilized in this state and rest 24.6 billion cubic metres are allocated in share of bordering states. Basin-wise distribution of this potential is very uneven (Figure 7.2). Largest availability, more than two-fifths (41.4% of the state) is in the Yamuna sub-basin followed by the Narmada basin (39.5%). The Ganga sub-basin is at third place (10.9%). Godavari basin possesses only 4.7 per cent of potential surface water of the state. The combine potential of Ganga-Yamuna basins (52.4%) is largest in the state. However, this share in potential water resources is lower than their share in total catchment area with in this state.

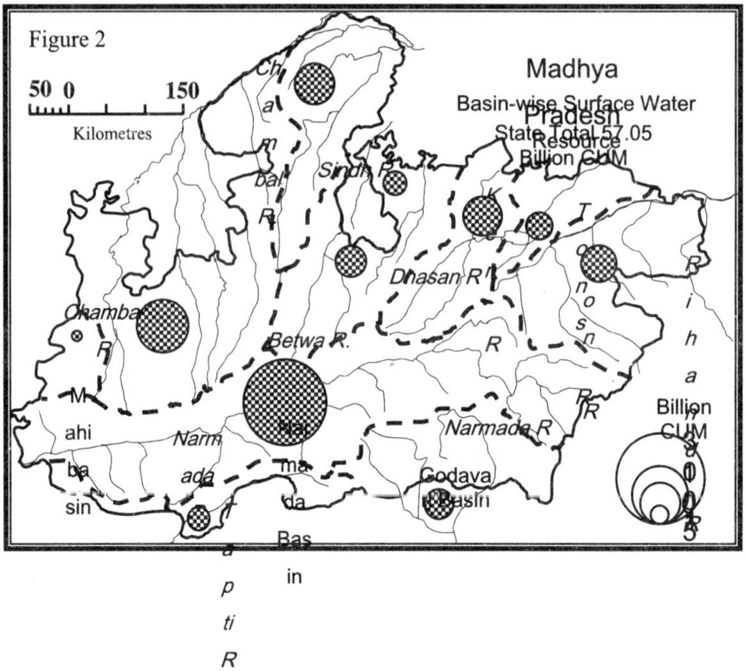

Fig. 7.2

Besides rivers, there are a large number of tanks, ponds, lakes and reservoirs preserving surface water. Estimated area under these water bodies is about 60 thousand hectares. Along with this, about 2.27 lakh hectares are under large man-made reservoirs.

Groundwater

Groundwater constitutes a major and widely used resource in the state. However, its distribution is very uneven because of varied hydro-geological characteristics. Central Groundwater Board, Government of India (2013) has divided this state into 12 Principal Aquifer Systems and 25 major aquifers based on hydro-geological characteristics. The replenishable groundwater potential is associated with unconsolidated formations, while semi-consolidated and consolidated geological formations have very limited potential. Areas of unconsolidated rocks and soils are localized in the Chambal valley in the northern and the Narmada valley in the southern part which form excellent aquifers and constitute only 14.4 per cent of total geographical area. The northern, eastern and southern parts are covered by hard rocks. These hard rock areas show wide variations and complexities in nature and composition of rocks, geological structures, geomorphologic setup and hydrological conditions. The crystalline rocks of Archaean age like granite, gneiss, granulites, schist, quartzite and granitoids occupy about 14.7 per cent of geographical area of the state. In these areas groundwater occurs in upper weathered mantle and fractured zones underlying them. The consolidated sedimentary rocks of Vindhyan and Cuddapah Super Groups occupy about 19.1 per cent of total geographical area and the semi-consolidated Gondwana formation occupies about 6.7 per cent. Gondwana formations are granular where tubewells are capable of yielding 100-500 cum/day for drawdown of 6 to 10 m. The basaltic rocks of Deccan lava flows are the predominant formations and occupy nearly 44.5 per cent of western part of the state, separable into vesicular and massive units. The vesicular units, at various depths below ground, are moderately productive with wells capable of yielding 250 – 750 cu.m/day for drawdown of 3-6 metres.

As per 2015 re-assessments by the Water Resource Department of the State and Central Groundwater Board (2017), total recharge from rainfall in the state is 28.99 BCM, whereas from other sources is 7.05 BCM. Thus, total recharge from all sources is 36.04 BCM (Table 7.1). The natural discharge during non-monsoon season is 1.83 BCM. The net ground water availability in the state is 34.21 BCM.

Table 7.1: Madhya Pradesh: Availability and Utilization of Groundwater (BCM) 2015

Recharge from rains	28.99
Recharge from other sources	7.05
Gross Recharge	**36.04**
Natural discharge	1.83
Net annual availability	**34.21**
Present draft for irrigation,	18.18
Draft of domestic and industrial uses	1.34
Gross draft at present	19.52
Allocation for domestic and industrial uses	2.41
Future availability for irrigation	**13.62**
Groundwater development in %	**56.07**

Source: Central Ground Water Board, Dynamic Ground Water Resources of India, (as on 31st March, 2015)

Distribution of groundwater potential is very uneven in the state. It is as low as 0.198 BCM (19834 ham) in Alirajpur and 0.241 BCM (24,065 ham) in Jhabua districts and as high as 2.62 BCM (2,16,173 ham) in Hoshangabad district. District-wise potential of groundwater is depicted diagrammatically in **Figure 7.3**. Out of 51 districts of the state, only 7 districts, viz. Hoshangabad, Narsinghpur, Sagar, Chhindwara, Dhar, Betul and Ujjain have groundwater potential above 1.0 billion cubic metres. These districts are located in south-central part of the state and have developed canal irrigation resulting in higher recharge from other sources than rainfall. Other 44 districts have availability below 1.0 BCM per annum and 16 out of them have potential even below 0.5 BCM. These are non-command areas and some of them, such as Dindori, Jhabua, Barwani, Sidhi and Singrauli, are dominated by the hilly and dissected terrain unsuitable for rapid recharge of groundwater and others are small in size such as Harda, Ashoknagar, Datia, Bhopal and Neemuch. As such, uneven distribution of groundwater is mainly due to geological structure, characteristics of soil, nature of terrain and slope, amount of rainfall and its characteristics, vegetative cover and size of the district.

The effect of size of the district can be eliminated by calculating the replenisable groundwater per unit of area. In calculating total recharge of groundwater, hilly area of the district/block is deducted from the total area and hence per unit replenishable groundwater is per unit of replenishable area only. Areal unit used is hectare. The annual replenishable groundwater resource in metre per hectare is highest in Hoshangabad (0.38 m/ha) district followed by Narsinghpur (0.26 m/ha). It is as low as 0.07 m/ha in Alirajpur followed by Sheopur, Damoh and Singrauli, all 0.08 m/ha. In 15 districts, the annual replenishable groundwater resource is between 0.07 to 0.10 m/ha, whereas in 25 districts, it is between 0.10 to 0.15 m/ha and 8 districts have 0.16-0.20 m/ha. The average annual replenishable groundwater resource of the state is 0.13 m/ha. Replenishable of groundwater is above 0.15 m/ha in districts of eastern Malwa and middle Narmada valley (Figure 7.3). Contrary to it, districts with replenishable rate lower than 0.01 are concentrated in three areas. First major tract extends from Chhatarpur in the Bundelkhand region to Rewa and Singrauli in the Baghelkhand region in the northeastern part to Mandla and Dindori districts in southeast. Second tract extends from Sheopur to Ashoknagar in the Madhya Bharat plateau region in the northwest and third tract encompasses Alirajpur and Jhabua districts in the southwest.

Utilization of Water Resources

Major economic use of water in the state is for irrigation. Irrigated area increased from 4.7 lakh hectares in 1950-51 to more than 103.0 lakh hectares in 2014-15, recording average annual growth rate of 4.9 per cent. Consequently proportion of gross irrigated area also rose from 4.4 per cent to 42.3 per cent during the same period. Spatial and temporal variation in growth of irrigated area has been very wide (Sharma & Jain 1994).

Gross irrigated area ranges from only 1.26 per cent of gross cropped area in Dindori district to 75.64 per cent in Datia district in 2014-15. However, net irrigated area varies from 1.8 per cent in Dindori to 99.3 per cent in Raisen district in the same year (Figure 7.4). Areas with higher intensity of irrigation than the state average are (i) the Madhya Bharat plateau including upper and lower Chambal basin, (ii) Narmada valley-Nimar plain-Dhar upland, (iii) eastern Malwa, (iv) Bundelkhand upland, and (v) Wainganga valley. Contrary to them, proportion of irrigated area is low and very low in the north-western Malwa, Satpura region, Rewa plateau and Baghelkhand plateau. Potential of groundwater is limited in districts of these regions. Reasons behind low development of irrigation in these areas are dissected and hilly terrain, shallow and comparatively low fertile soils, dense forest, dominance of tribal population, marginal and small landholdings and predominance of coarse grains in crop structure. These regions are deprived of such basic facility essential for improving farming. In certain districts such as Dindori (1.3%), Anuppur (3.3%) and Singrauli (9.0%), Mandla (14.8%), Shahdol (15.0%) and Alirajpur (17.8%) irrigation could not be started in real sense. Intensive use of land and modernization of agriculture in these areas, thus, is prohibited by the limited use of water for irrigation.

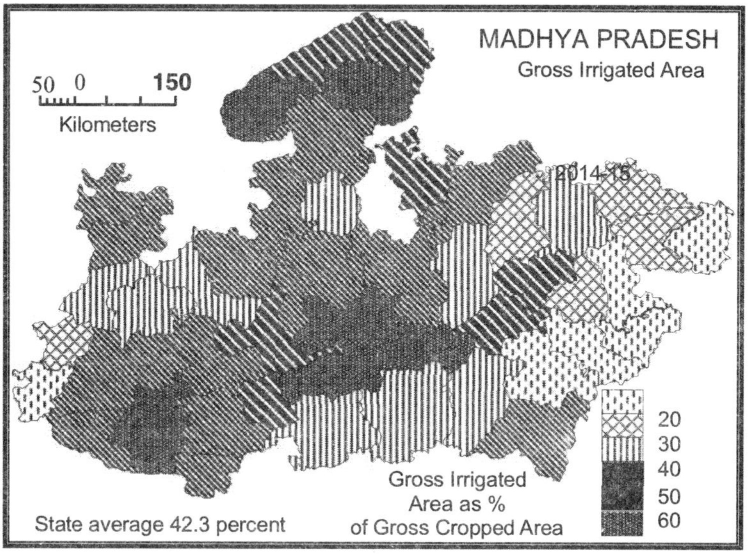

Fig. 7.4: MP: Gross Irrigated Area

Sources of Irrigation

There are three major sources of irrigation in this state. They are canals, wells and tubewells and tanks. Besides them, some other sources are used for irrigation. Changes in relative share of these sources itself is one of the major cause of anxiety. For example, canal irrigated area declined from 26.0 per cent of net irrigated area in 1950-51 to 17.2 per cent in 2014-15. On the other hand, significance of wells and tubewells increased enormously with the introduction of diesel and electric pumping-sets (Figure 7.5). Net irrigated area by them increased from 241 thousand hectares in 1950-51 to 6403 thousand hectares in 2014-15, recording more than twenty-six fold growth. Consequently, proportion of well and tubewell irrigated area rose from 53.2 per cent to 66.8 during this

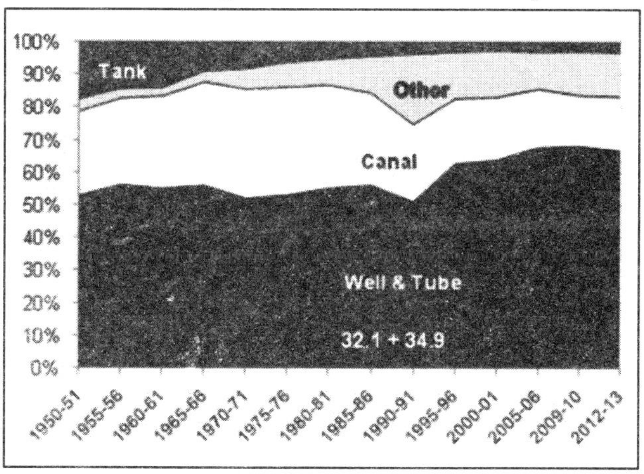

Fig. 7.5: Madhya Pradesh: Trend of Percent share of Net Irrigated Area by Different Means

period. Out of this, more than half (34.2%) is irrigated by tubewells and remaining 32.6 per cent by dug wells. There are 1623.1 thousand irrigation wells and 665.6 thousand tubewells in the state in 2014-15. Out of them only 2.45 per cent wells and 0.58 per cent tubewells are under government control and majority of them are owned by farmers. It means most of the area is irrigated by non-governmental sources in the state. It involves the issues of the ownership and management of water resources and over-exploitation and lowering of water table. Other sources also gained significance, by raising their share in net irrigated area from only 3.9 per cent in 1950-51 to 13.2 per cent in 2014-15. They are also putting extra stress of fresh water sources. Problem of even drinking and domestic water has emerged very fast in this region.

Issues of Water Resources

Utilization of water resources for different purposes has created certain problems (Sharma 2016, p. 78). Irrigation has largest share in water utilization. The spatial variation in availability irrigation facilities has given birth to conflicts of interests in several forms. At the same time, several problems such as land degradation, salinization, waterlogging have crept in the irrigated areas. The rapid development of groundwater based irrigation in most parts of the state has caused lowering of the water table and depletion of groundwater resources. To meet the irrigational necessity inter-basin transfer of water is proposed and practiced, causing disputes. There has been emphasis on the construction of big dams. They have caused large scale submergence and displacement of people from their original habitat. Low efficiency of water use and water pollution are other serious problems. The major sources of water pollution are domestic and industrial waste and agricultural runoff. Some of these issues are reviewed below.

Declining Trend of Rainfall

The base of the water resources itself is dwindling in the state. Rathore, Attri and Jaswal have recently (2013, p. 142) analysed the trends of average annual rainfall for the period 1961 – 2002. Though there is an inter-annual variability of average monsoon rainfall in the 41 year period, there is trend of declining rainfall also. Further a study carried out by Goswami *et al.* (2006), for observations spanning 50 years in the Central Indian region including Madhya Pradesh, indicates that the extreme precipitation events which are above 100 mm are increasing in terms of their intensity and frequency, with low and moderate events becoming more and more infrequent. On the basis of rainfall data for 1951-2013 Mishra and others (Mishra *et al.* 2016, pp. 1-94) have concluded that most part of the state of Madhya Pradesh experienced a significant decline in the monsoon season precipitation during this period. The monsoon season precipitation is projected to decline in the near future (2016-2045). Further, they also indicated that the frequency of severe, extreme, and exceptional droughts has increased in the state. Droughts in the recent years were severe and wide-spread. The number of hot days has increased significantly in the state. It will increase the need of water. The trend analysis of rainfall from 1970 to 2014 for rain gauge stations of district headquarters of Madhya Pradesh, done for the present study and presented in Figure 7.6, does not support uniform spatial trend of rainfall; however, majority of the districts present declining trend. This is just an indication that the present potential of water resources is likely to decline in future which must be taken into account while planning for their development as well as of land resources in the state.

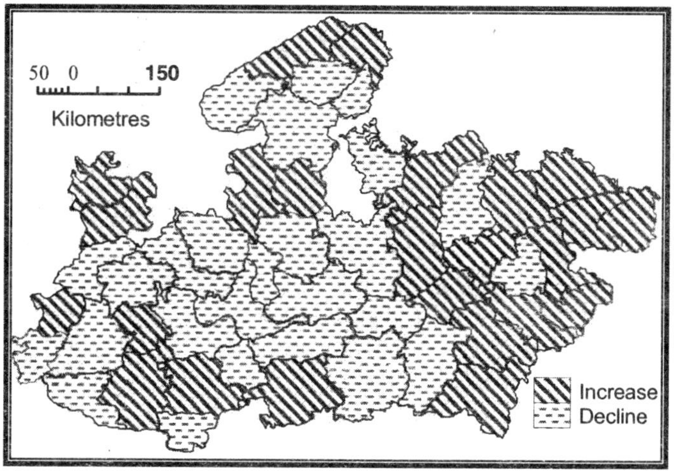

Fig. 7.6: Madhya Pradesh Trend of Rainfall 1970 to 2014

Groundwater Draft

The most serious problem is of rapid lowering of water level and depletion of groundwater in the state. It can be sensed from the fact that per hectare of net sown area there are 14.53 dug wells and 4.31 tubewells in the state in 2014-15. Groundwater draft for various uses has been calculated separately for command and non-command areas by the state Water Resource Department (2015, pp. 32-33). Total draft of groundwater for all uses in state is calculated as 19.52 BCM (18,83,352 ham) during 2014-15. It is seen that maximum groundwater drawal for all uses is 100081 ham in Ratlam district followed by Ujjain (91356 ham) and minimum is 2,679 ham in Annupur district followed by Dindori (3481 ham) in the eastern part of the state. Concentration of blocks and districts of high groundwater development is mainly in the western (Malwa region), central (Narsinghpur district), Bundelkhand region and part of Baghelkhand region of the state (Figures 7.7 and 7.8). Other parts of the state have developed low and very low proportion of their groundwater potential. Sixteen districts have developed between 60 to 90 per cent and another five districts between 90 to 100 per cent of their net groundwater potentials in the state. Indore and Ratlam districts have reached at highest level of groundwater development of 119 per cent and 126.5 per cent respectively, followed by Mandsaur (100.6%). Shajapur (98%) and Ujjain (95.3%) are very close to over-exploitation. Contrary to them, 29 districts have groundwater development index below 60 per cent. Most of them are in eastern and northern parts. Raisen, Ashoknagar, Harda, Jhabua and Alirajpur are other such districts. Annupur (7.9%), Shahdol (7.5%) and Dindori (8.9%) could develop less than 10 per cent of their potentials. The overall development of groundwater in Madhya Pradesh is 57 per cent which is in safe category of utilization. This pattern of utilization of groundwater seems to be associated, with the amount of precipitation to a great extent.

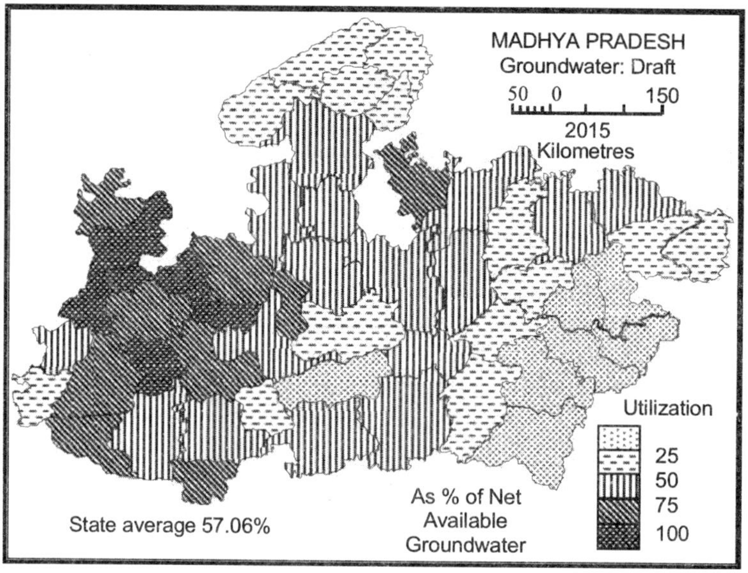

Fig. 7.7: Groundwater Draft

Over-exploitation of Groundwater

The development blocks are categorized for groundwater development based by the State Water Resource Department (2015, pp. 34-35) on two criteria– stage of groundwater development determined on the basis of the proportion of draft of groundwater in net replenishable groundwater of the block, and by long term trend of pre- and post-monsoon water levels. Four categories of development blocks/districts thus identified. In 2014-15, out of 313 blocks, 229 blocks incorporating command areas fall under safe (draft below 70% of net rechargeable resource) category of utilization. Non-command areas of 55 blocks are in semi-critical (draft between 70 to 90% of net rechargeable resource), 4 blocks are in critical (draft between 90 and 100% of net rechargeable resource), and 25 blocks are in over-exploited category (draft above 100% of net rechargeable resource). Almost all over-exploited blocks are falling in western part of Madhya Pradesh, which is known as "Malwa area" where groundwater draft increased many folds during past decades (Figure 7.8). District-wise analysis of data of groundwater availability and annual groundwater drafts indicate that three districts namely Indore, Ratlam and Mandsaur fall in "Over-Exploited" category where more than 100 per cent net potential of groundwater is already developed. Middle Narmada valley and north-eastern districts are also approaching to this state. These areas particularly the Malwa region is deprived of any medium or large irrigation projects, and therefore farmers heavily depend on tubewells for irrigation. It resulted in high density of tubewells in these districts. Density of tubewells per square kilometre of net sown in 2014-15 is more than 10 in Indore, Ratlam, Ujjain and Bhopal; and between 5 to 10 in Dewas, Dhar, Narsinghpur, Satna, Shajapur, Sehore, Vidisha, Raisen, Burhanpur, Rewa and Sheopur districts as against the average of 4.34 in the state.

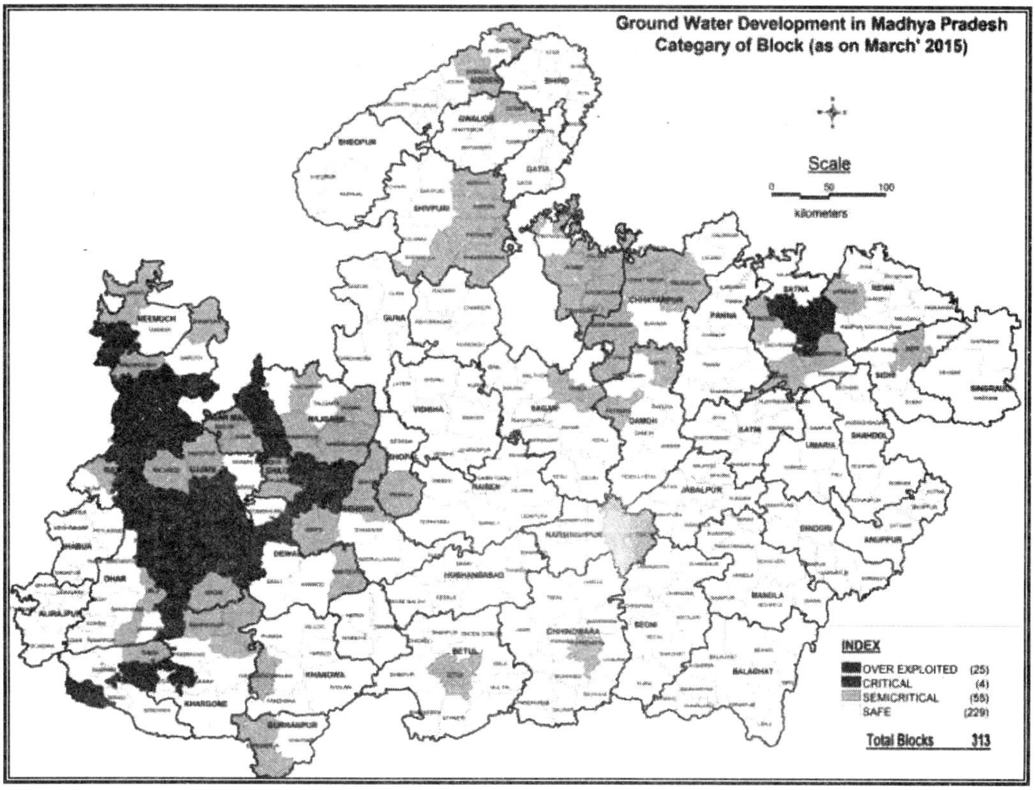

Fig. 7.8: Groundwater Development Categories of Blocks (March 2015)

Source: Based on Water Resource Department, M.P.

Water Level Fluctuation

Groundwater level refers to underground surface below which the ground is wholly saturated with water. The upper surface of the zone of saturation is the water table. Groundwater is a dynamic system and groundwater level varies significantly over time and space. Groundwater level is an important indicator for the recharge of the aquifer, groundwater abstraction and the discharge from the aquifer to surface waters and to some extent groundwater quantity. The groundwater level is subject to change due to natural and man-made causes. Decline in groundwater levels can occur as a result of natural climatic conditions (low rainfall), changes in land use, or as a result of over abstraction. Rise in groundwater levels can occur as a result of natural climatic conditions (heavy rainfall), changes in land use, or as a result of artificial recharge to groundwater by rainwater harvesting. The average depth of water level during the decade 2005-14 in pre- and post-monsoon periods (MP Groundwater Yearbook 2017-18), derived from data of 1529 observation wells of the state, is discussed below succinctly.

Trend of Groundwater Level

Groundwater is replenished from precipitation and from surface water, but the rate of abstraction (withdrawal by humans) may exceed the rate of natural recharge, leading to reduction of the resource indicated by declining trend of water level. Rising trend in groundwater levels can occur as a result of natural climatic conditions (heavy rainfall), changes in land use, construction of pond or reservoir or as a result of artificial recharge to groundwater by rainwater harvesting.

Water level of May 2015 was compared with the average water level of May during 2005-2014. About 58.8 per cent of wells showed a rise in water level. Rise up to 2 m in groundwater level was noted in 38.3 per cent and between 2 and 4 m in 12.5 per cent of monitoring wells. Rise more than 4 m in groundwater level was observed in 7.9 per cent of monitoring wells distributed in isolated manner. Fall in water level is recorded in 41.2 per cent of wells scattered throughout the state, covering mainly parts of northern districts of Morena, Bhind, Gwalior and Sheopur. Fall in groundwater level up to 2 m was observed in 29.7 per cent of monitoring wells and between 2 and 4 m in 6.6 per cent of monitoring wells. Fall more than 4 m was observed in pockets in northern half of the state, comprises 5.0 per cent of monitoring wells. Such a declining trend of water level of dug wells is indication of declining water level as a result of excessive extraction apart from poor saturation of aquifers. In fact, problem is accentuated by the tubewell with high yield of water. In 2014-15, total 6,65,658 tubewells as against 2.01 million dug wells were used for irrigation in this state. This situation taxes deep ground water resources and causing deepening of tubewells every year.

Rehabilitation and Resettlement of Affected Population

By 2015, there were 22 major, 90 minor and 4804 minor completed/under construction irrigation projects in Madhya Pradesh. Most of the major projects are located in the Narmada valley and Chambal valley. Most of these projects are fraught with serious issues related to massive displacement, forest submergence, poor rehabilitation and resettlement, tribal issues and forest rights, reservoir fishing leases, legal and procedural issues etc, all affecting land and land use in some way or the other. As an example, situation of rehabilitation and resettlement of the families displaced in the Narmada Valley Development Project, one of the world's largest multipurpose projects, is presented here. For almost three decades now agitations have been going on against Madhya Pradesh government's poor rehabilitation and resettlement. In one of the latest episodes, in Aug 2012, several affected people from Omakareshwar and Indira Sagar Dams undertook a Jal Satyagrah for 17 days together against absence of rehabilitation. The Supreme Court issued a show cause notice to MP Government. The Sardar Sarovar is one of the 30 dam-based projects of the valley. Over last 30 years, the project has been in the news due to the controversy over the enormous social, cultural and environmental costs and displacement and dislocation in large scales. It is said to be one of the most ambitious river valley projects in modern Indian history. The Sardar Sarovar dam height was raised in phases from 95 to 100 m in May 2003, 110 to 110.64 m in March, 2004, later to 119 m in March 2006 and up to 121.92 m by end of 2006, where it stood, before fresh construction started in late 2014 to take the height to 138.68 m. At its final height of 138.68

m, the reservoir itself is supposed to submerge over 37,500 hectares of land (20822 ha in M.P., 9590 ha in Maharashtra and 7112 ha in Gujarat) in 245 villages in Gujarat (19), Maharashtra (33) and Madhya Pradesh (193), including over 13,300 hectares of biodiversity rich forest land, as per official estimates. As a result of the increased height, the backwater levels have also gone up, inundating many villages and hamlets. The numbers of displaced families from these 245 villages have been conservatively estimated at over 48,000 in the three states that are yet to be settled as per guidelines of the Narmada Water Disputes Tribunal.

The Central Fact Finding Committee (CFFC 2015) after visiting the Narmada Valley presented its report regarding this issue and highlights the gravity of the situation and concerns of the affected people. As per report, many of the villages submerged/being submerged are Adivasi/tribal villages. For example, out of the 193 villages affected by SSP in Madhya Pradesh, 70 are fully tribal villages. Thousands of families, especially the tribals, fisher folk, landless poor who are under the threat of imminent inundation and thus forcible displacement, are not being given any alternative place to live, as also, alternative land and livelihood sources, in many cases. The essential components of land-based rehabilitation for even those recognized as project-affected, is not being implemented. Cultivable, irrigable lands are not located nor purchased by the Governments. Several thousands of people, whose houses and lands are in reality getting submerged during high flood levels, are not even recognized as project affected families/persons. The rehabilitation sites chosen/identified by the governments are in extremely poor conditions, having no adequate facilities for people to live there as per the NWDT norms, with poor water supply, broken roads, no electricity, and nonexistent or pathetic education and health facilities.

Inter-State Disputes

Inter-state water disputes are also one of the major issues with the management of water resources of the state. Major rivers of the state are shared by two or more states. Madhya Pradesh is located in the centre of the country and rivers flow almost in all directions and drain huge water from the state. As soon as they cross the boundary of Madhya Pradesh they are dammed in such a manner that the submerged area confines in this state but water goes to the states of lower reaches. The Rihand, Betwa, Ken, Chambal, Narmada and Wainganga are some of the rivers to be mentioned. Sometimes even dams are located within the state water is passed to other states. Bansagar located on the Sone in Shahdol district and Rangwan Dam in Chhatarpur district are glaring examples. Such a situation has created water conflicts between Madhya Pradesh and the bordering states. With the increasing demand for water which largely falls under the authority of the states, and with the states increasingly asserting their legal and political power, inter-state water disputes are in the rise and are getting more complex and contentious. At present, this state has water disputes with Andhra Pradesh, Chhattisgarh, Odisha and Karnaka for water of Wainganga and its tributaries; with Rajasthan for Chambal water, with Gujarat, Maharashtra and Rajasthan for Narmada water; with Gujarat and Rajasthan for Mahi water; with Maharashtra for Tapti water; and with Uttar Pradesh and Bihar for Son and Rihand water. At the behest of the central government agreements for the sharing river waters have been reached several times between Madhya Pradesh and concerned states. In spite of this, dispute triggers almost every year particularly in lean years or lean season.

Under the Inter-State Water Disputes Act, 1956, the Central Government constituted Narmada Water Disputes Tribunal (NWDT) on 6th Oct. 1969 to adjudicate upon the sharing of Narmada waters and for the Narmada River Valley Development. The Tribunal gave its Award on 7th Dec. 1979. The Award specified a quantum of utilizable waters at 75 per cent dependability to be shared by the four States of Gujarat, Madhya Pradesh, Maharashtra and Rajasthan. The Narmada Control Authority is an interstate high level administrative authority, set up by the Government of India, in 1980 for the purpose of securing compliance with the implementation of the decisions of the Narmada Water Disputes Tribunal by the basin States. Agreements between riparian states of Narmada were made in 1972, 1974, 1975 and 1978 besides Tribunal awards and judgments of Supreme Court of India. Similar agreements were made with Uttar Pradesh regarding Matatila dam, Rangawn dam, Jamni dam, Bhander canal complex in 1965; regarding projects of Bundelkhand in 1972, Rajghat project in 1973; and regarding Rajghat, Paisuni, Ken Canal, Kanhar, Urmil, Bansagar, and Bhander Canal in 1977. Similar agreements were made between Madhya Pradesh, Uttar Pradesh and Bihar regarding Bansagar in 1973 and regarding Kanhar waters in 1982. Disputes of the Tapti basin and Godavari basin were discussed and agreed upon several times between concerned states.

Pollution of Water

Pollution of water of rivers, ponds and groundwater is reaching to a point of crisis particularly near urban and industrial centres due to unplanned urbanization, rapid growth of industrialization and increasing use of fertilizers and other chemicals in farming. The quality of river water is deteriorating mainly due to human activities such as discharge of industrial and municipal sewage wastes and agricultural runoff, causing ecological damage and posing serious health hazards. Sharma *et al.* (2011) concluded that due to discharge of untreated sewage at Hoshangabad into the Narmada along with industrial effluents from Security Paper Mill, the water quality of Narmada has been severely deteriorated and the potable nature of water is being lost. The Kshipra, one of the sacred rivers of Malwa holding Simhasth Kumbha, carries untreated wastes and effluents from Ujjain, Dewas and Indore cities which have not only affected the quality of the river water but has aggravated the water crisis in the region. Its waters, which once quenched the city's thirst, are not even fit for bathing any longer, and government has to bring Narmada water to keep its glory. The River Betwa plays a significant role in the human life of the villages located on its banks. It has become polluted at some places of Mandideep due to industrial activities and the confluence of sewage, domestic wastes and industrial effluents of many big and small enterprises with various types of organic compounds and heavy metals. The water quality in the stretch of the river Betwa extending from its origin near Mandideep industrial area up to Bhojpur remains poor because of the regular inflow of domestic waste of the Bhopal city through the Kaliyasot river and industrial/domestic waters from Mandideep. The untreated sewage of the human habitations in Mandideep is directly flown into the Kaliasot river which flows on the one side of the town while untreated chemical waste directly mixes into river Betwa which flows on the other side of the town (Margade 2009, Vishwakarma *et al.* 2013).

Most of the water bodies in and around Bhopal are presently under great environmental stress due to pollution from point and non-point sources, eutrophication, fast growth of aquatic macrophytes, enrichment of nutrients, and human encroachments. Due to the joining of untreated domestic sewage, washing activities, and so forth, the water quality of these water bodies has deteriorated to a great extent thereby reducing the carrying capacity of the system. The sewage disposal system in the state capital covers only 40 per cent of the city's area. Everyday more than lakh litres of waste water and sewage are being discharged to the Upper Lake water and Patra, Halali and Betwa rivers. Water of water-bodies of the city is highly polluted. These domestic and industrial waste waters have polluted not only the surface and lake waters but groundwater also. Irrigation reservoirs constructed near Bhopal city – Hathaikheda, Kaliasot, Kerwan and Laharpur—are also in grip of sewage pollution. Sonam (2012) found that highly polluted water of Shahpura lake percolates downwards and makes well, hand pump and tubewell water polluted, water of which is used for irrigation.

In Indore, 202.2 million sewage is generated every day and most of it is disposed off in the Khan river without treating it. Textile and chemical industries of the city leave their wastes in the Khan river which ultimately poisons the Kshipra river. Sohani and Iqubal (2012) investigated the quality of surface water of Indore and concluded that surface water is full of pathogens which are percolating in the groundwater. They emphasized to save surface water from sewages. Gwalior drains its 50 million litres after treating and remaining 66 million untreated sewerage to the Swarnarekha river. Morar river receives sewage of the Morar town. Similarly, effluents of the Orient Paper Mills, Amlai, have polluted the Son river for long distance (Shrivatava *et al.* 1988) and of News Print Paper mill of Nepanagar of the Tapti. In this context the statement of Sunita Narain from the Centre of Science and Environment made at the occasion of opening of the 7th State of India's Environment Report at Bhopal, that 'we are killing our rivers and water bodies', is true.

Other Emerging Issues

Sedimentation of reservoirs is again severe problem. Sedimentation is the process by which larger sediments in water entering a reservoir are deposited at its upper end forming a delta and steadily raising level of the upper reaches of the reservoir. This process shortens the utility, capacity and life span of the dam. In case of Sardar Sarovar dam there is the possibility of premature siltation (Alvares, Billorey 1987, 69). Central Water Commission (2015, p. 98) in its report has surveyed the status of siltation in the Gandhi Sagar dam constructed on Chambal and found that the dam has lost capacity of 519.85 million cubic metres during 41 years from 1960 to 2001. Average rate of siltation thus arrives at 12.679 M cum per year. It is slightly lower in the Rana Pratap Sagar below the Gandhi Sagar. The Halali dam in Raisen district lost 16.91 per cent live storage capacity, from 226.940 M cum in 1976 to 188.583 M cum in 2003, yielding annual rate of siltation of 1.42 M cum. Because of loose soil present in most of the state problem of siltation exists in every project area.

Waterlogging is another serious problem which can easily be witnessed in command areas of Tawa, Chambal, Barna, Harsi projects. About one-third of the command area of

Tawa is suffering from waterlogging. The Indian Institute of Social Science estimates that 40 per cent of the command area of the Sardar Sarovar dam will become waterlogged. This area contains black soils which are particularly prone to waterlogging under perennial irrigation due to high water retentive capacity. Rich soils in Punjab and Haryana have been robbed of their use because of waterlogging. Similar situation is emerging in this state also. Salinity is also infiltering in the canal irrigated area. It is estimated that 4410 hectares are salt affected. Irrigation water unlike rainwater contains considerable amount of salt in solution which is deposited on soils to make them salty. Again in Tawa dam command a large tract of good quality cultivable land has been affected by salinity and became infertile.

Summing Up

The overview emerging from this survey is that the exploitation of the water resources has emanated such issues as of ownership, institutional arrangement and uncoordinated utilization ignoring comprehensive integrated approach and isolating water from other resource systems including land resource. Most of uses of water are concerned with immediate needs and problems and issues of long-term sustainable water management like declining water level, land degradation, pollution, waterlogging, salinity and uses of water for recreation and environmental purposes have not been taken care of. Systematic studies of local and regional water resource problems creeping from improper use and misuse of water, and perceptions of water users, particularly of farmers, have not been thought of. Like other resources, availability of water to different sections of society is exorbitantly uneven which deserves systematic examination not only for meeting the challenges for food security and sustainable agriculture but also for achieving domestic and sanitation targets. Only certain regions are privileged to get timely and sufficient water for agricultural purposes which manifests in wide variation in modernization of agriculture.

In fact, water resource management cannot be isolated from the management of other natural resources. It is because all resources are intricately interlinked. Therefore, water management is simply a component of resource management. Further, there is dual ownership of water resources. Most of the sources of surface water are under direct control of the government. Contrary to this, most of the useable reservoirs of groundwater are owned by landowners. Under such conditions, cohesive use of these two major sources of water cannot be planned, unless legal provisions and governmental interventions are made. There is lack of coordination between institutions managing not only these two segments of water resources but also between different components of land resource system. There is no provision of assessing water requirements of all sections of the society and all sectors of the economy. This is possible through the evaluation, development, and management of water resources in conjunction with land resources.

References

Alvares, C. and R. Billorey (1987). Damming the Narmada: The Politics behind the Destruction. *The Ecologist*, Vol. 12(2), p. 69.

Barlowe, R. (1960). *Land Resource Economics.* England Cliff, NJ: Prentice Hall.

Central Ground Water Board (2013). *Aquifer Systems of Madhya Pradesh.* Bhopal, North Central Region

Central Water Commission (2015). *Compendium of Silting of Reservoirs in India.* New Delhi: Government of India.

Christian, C.S. and G.A. Stewart (1953). *General Report on Survey of Katherine Darwin Region.* CSIRO: Melbourne, Land Resource Series I.

Christian, C.S. and G.A. Stewart (1968). *Methodology of integrated surveys,* pp. 233-280 in UNESCO, *Aerial Surveys and Integrated Studies.* Proceedings of the Toulouse Conference.

FAO (1976). *A Framework for Land Evaluation.* Soils Bulletin 32. Rome. 72 p.

Goswami, B.N., Venugopal, V., Sengupta, D., Madhusoodanan, M.S., Xavier, P.K. (2006). "Increasing trend of Extreme Rain Events over India in a Warming Environment". *Science,* 314, 5804, 1442-1445.

Graham, E.H., (1967). Landuse Principles and Needs, pp. 170-186 in C.H. Collison, ed., *America's Natural Resources.* New York: Ronald Press.

Margade V; S.A. Iqbal, S. Pani and N. Iqbal (2009). Environmental impact assessment of industrial effluents on water quality of Betwa river discharged from Mandideep industrial area in Madhya Pradesh. *Curr World Environ* 4(2):327-334.

Mishra, Vimal; Reepal Shah and Amit Garg (2016). *Climate Change in Madhya Pradesh: Indicators, Impacts and Adaptation.* pp. 1-95. W.P. No. 2016-05-05, Indian Institute of Management, Ahmedabad, India.

Rathore, L.S., S.D. Attri and A.K. Jaswal (2013). State Level Climate Change Trends In India. *Meteorological Monograph No. ESSO/IMD/EMRC/02/2013* India Meteorological Department Ministry of Earth Sciences Government of India. Ministry of Water Resources, Government of India. Plates III and IV.

Robertson, V.C. (1970). Land and water resource planning in developing countries, *Outlook on Agriculture,* Vol. 6, No. 4, pp. 148-157.

Sharma, Shraddha; Vishwakarma, Rakesh; Dixit, Savita and Jain, Praveen (2011). Evaluation of Water Quality of Narmada River with reference to Physico-chemical Parameters at Hoshangabad city, MP, India. *Research Journal of Chemical Sciences,* Vol. 1(3) June (2011), pp. 40-48.

Sharma, S.K. (2013). Major Issues and Challenges to uses of Water Resources in India: Current Assessment (Hindi). *Raashtriya Bhaugolik Patrika,* (BHU Hindi Journal), Vol. 2, December 2013; pp. 113-148.

Sharma, S.K. (2016). Water Resources, pp. 62-127 in L.S. Bhat (ed.), *Economic Geography, Vol. 1 Land, Water and Agriculture. ICSSR Research* Surveys and Explorations. New Delhi: Oxford University Press.

Sharma, S.K. and C.K. Jain (1994). Water Resource: Potentials, Utilization and Problems in Madhya Pradesh. *Journal of Water and Land Use Management,* Vol. III, (3), pp. 232-253, 1994.

Sohani, Smruti and Iqubal Sanjeeda, 2012, Microbiological analysis of surface water in Indore, India, *Research Journal of Recent Sciences,* Vol. 1, pp. 323-325.

Sonal T, Kataria HC. (2012). Physico-Chemical Studies of Water Quality of Shahpura Lake, Bhopal (M.P) with Special Reference to Pollution Effects on Ground Water of its Fringe Areas. *Curr World Environ* 2012 (7):139-144. Available from: http://www.cwejournal.org/?p=2016

Srivastava, R.K., W.S. Fargo, V.S. Sai and K.C. Mathur. (1988). Water quality along the Sone river polluted by the orient Paper Mill, *Water, Air, and Soil Pollution*, May 1988, Volume 39, (1), pp. 75-80.

United Nations. (1994). *Convention on Desertification*. Information Programme on Sustainable Development, United Nations, New York. 44 p.

Vishwakarma, Santosh, Alok Varma and Geeta Saxena (2013). Assessment of water quality of Betwa River, Madhya Pradesh, India. *International Journal of Water Resources and Environmental Engineering*. Vol. 5(4), pp. 217-222, April. http://www.academicjournals.org/IJWREE

Water Resource Department, Madhya Pradesh (2015). *Dynamic Ground Water Assessment-2017*. Central Groundwater Board North Central Region and Groundwater Survey, Water Resource Department, Madhya Pradesh, Bhopal, January 2018.

Whyte, R.O. (1976). *Land and Land Appraisal*. Dr. W. Junk Publishers.

Young, A., (1975). Rural land evaluation, pp. 382-423 in *Tropical Soils and Soil Surveys*. Cambridge University Press.

Zobler, L. (1962). An Economic Historical view of Natural Resource use and Conservation, *Economic Geography*, Vol. 38, pp. 189-194.

8

Anthropogeny and Land Water Resource Dynamics in an Urban Scape

A Few Illustrations from Telangana Plateau Region

S. Padmaja and *Md. Akther Ali*

Introduction

Natural processes culminating into denudation have a varying impact on surface forms. These may range from slow processes of creeps and flows to massive processes of falls and slides accompanied by erosional and depositional forms in various morpho-climatic regions of the world. Apart from natural causes, even man-induced causes too can trigger off massive destructions as we have seen in the recent past in cases like Uttarakhand and Jhelum floods in Himalayan region. However, the same role of humans gets more significant (assumes significance) in an urban milieu due to increased demand for land for various human needs. This results in subsumed natural cover with human dominated altered cover thereby enhancing the role of anthropogeny in an urban region. The dynamism of land form, land use, and land cover transformation is that much more intense in an urban environment than in its counterpart, simply due to enormous pressure on natural resources. Many cities all over the world have undergone a transformation and are still doing so, from uninuclear base to multinucleii base. The same is being witnessed in and around Hyderabad and Warangal cities representative of Telangana plateau in peninsular India.

Hyderabad is a known historical city, growing at a great pace in terms of people and area. It is the fifth ranking city in terms of population and second largest in terms of area. It is also one of the metropolitan cities of India. Hyderabad and Warangal are located on a granitic plateau with many hill remnants surrounding them as inselbergs and further disintegrated into tors. Innumerous number of small hillocks comprising of tors can be witnessed in Hyderabad and its surroundings. River Musi passes through the heart of the city dividing it into two halves. There are also a number of other waterbodies like Hussainsagar, Osmansagar, Himayatsagar, Shamirpet lake etc. lying with in the region. Climatically the area has a semiarid type of climate with erratic rainfall regime, which is shown in the Fig. 8.1 for the time periods of 2011 to 2014.

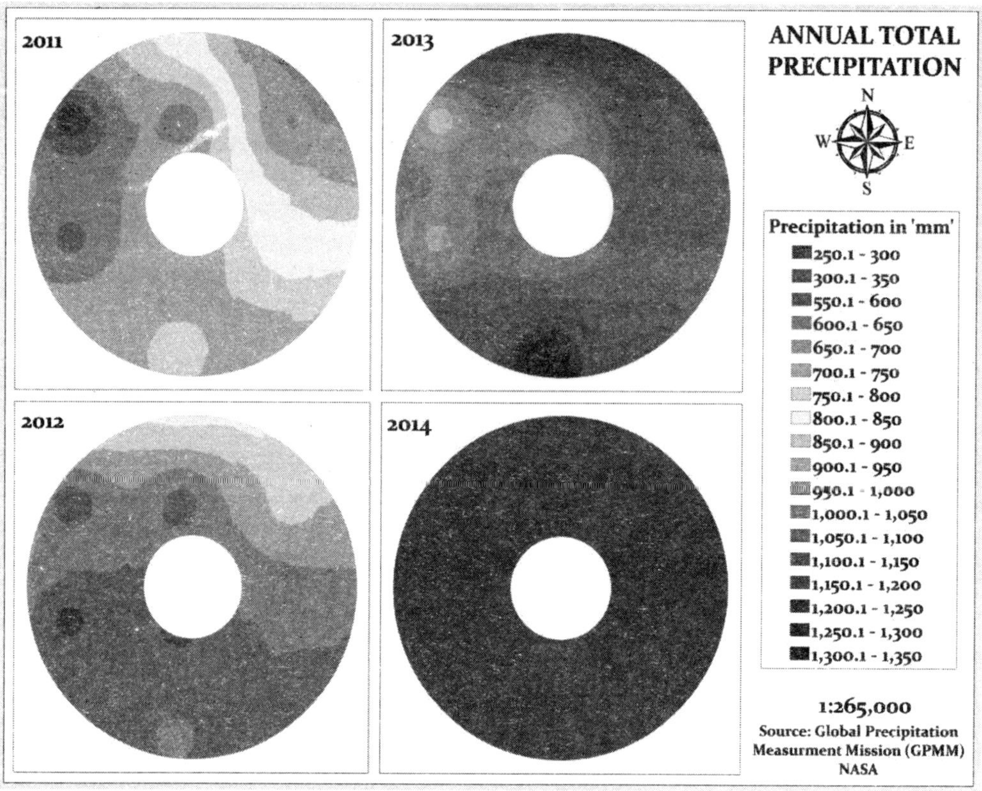

Fig. 8.1: Climate of the Study Area

Historical Past

Chronologically, the city of Hyderabad has transformed itself from a garden city status to Cyberabad a technology-based IT corridor. Quli Qutabshah founded Hyderabad and made into a trading center with their center of activity around 'Golkonda'. Subsequently Moghuls took over the dynasty in 1687, which was originally under Kaktiyas. Under their rule governor assumed importance in governing Hyderabad and first was Nizam-ul-mulk ferozjungasifjah. There were seven Nizams who ruled Hyderabad, till it was merged with Indian union after independence. In fact Nizams had a pact with Britishers and not fought them. They allowed them to have their cantonment in Secunderabad, north of Musi, which till today is known for its military prevalence (existence). In 1956 Andhra Pradesh was formed amalgamating all Telugu speaking provinces and Hyderabad was made capital.

The flourishing trading activities enhanced its status and new rail and road network allowed it to trade with other adjoining states. With this growth Hyderabad and Secunderabad became inseparable whole and were known as twin cities and still remain so. The axis of growth patterns were visible in East-West as well as North-South in response to the new avenues of trading centers joined by good road and rail network. Historical Golconda to Charminar to Koti, Abids, Ameerpet junctions grew along with Secunderabad cantonment accompanied by posh localities like Jubilee and Banjara Hills, maintaining their own, cultural identity. Through urban growth, Hyderabad grew as a cosmopolitan center accommodating all religious sects in a very amicable way. However late 1970's saw a shift in occupational structure, as service sector started overtaking due to increased IT boom and outsourcing activities provided a huge job market. A number of IT firms got established in nearby pockets around Madhapur, which came to be termed as 'Hitech' city or more appropriately 'Cyberabad' – a world of cyber networks. This further boosted up the sprawling growth of the city from an initial threshold level of 5 kms to the present 50 kms threshold encompassing all the growth centers interspersed with rural centers.

This region is a typical rurban zone with a maximum of land use activities. The penetration of urban space into rural hinterland of Hyderabad is more pronounced in the East, North East and North West leading to a dynamic transformation of these areas. The transport arteries dissect the zone upto a distance of 15 kms from the city. This zone also contains major industrial units. The shadow zone of urban influence extending from 15 to 30 kms is dotted with some of the larger and economically dynamic settlements like Ghatkesar, Ibrahimpatnam, Shamshabad, Medchal etc. Beyond this zone, the transportation arteries are farther apart and the interstitial villages increase in number. Urban centers become far and in between and the zone is predominantly agrarian.

Brief Review

A number of studies have appeared with regard to urban growth, urban sprawl and land transformation. Land use/land cover changes clearly establish a rapid transformation, in and around cities as based on 3rd level of land use classification.

Fractional impervious surfaces have been detected in urban area of Barasat town impervious surfaces like sidewalks, drive ways, roof tops and parking lots negatively impact environmental quality by decreasing water quality, groundwater reserves, habitat fragmentation, flooding and urban heat islands which can be detected through impervious surface area (ISA). As dark and bright ISA creates confusion to distinguish land covers as wetlands/water soils/cropped fields. Modified built up index (MBI) is used by authors to detect urban builtup impervious surface (UBIS) (D. Bhattacharjee and Sukla Hazra, 2014).

Urban green spaces have been identified in Chennai through NDVI analysis. Ground biomass and leaf area index have been worked out. Empirical studies assume that green spaces fulfill different functions at different levels and every level in the hierarchy of UGS system and are complimentary to each other (Narmada and Bhaskaran, 2014).

Nature and characteristics of urban sprawling in Bhubaneswar city applying Shannon's entropy model is worked out by ND Chatterji and S. Chatterji (2014). This research work is a combination of both landscape level and micro level study. Satellite images were analysed for land use categories. Patches of different land use were drawn on open source images towards to analyse the distribution and fragmented nature of landscape. Entropy model is applied to detect the randomness of sprawling. Micro level study through questionnaire has been conducted in 14 sprawling microzones. Urban sprawl has been non continuous. Most of the vacant lands are altered to residential area. Besides, the cost of land too showed an increase in both core and surrounding areas.

Spatial variation of surface energy fluxes with respect to land use changes caused by land transformation across Gulbarga is studied to improve the understanding of the impacts of land dynamics on regional climate. Carbon sources and sinks as a result of land use changes are significant for climate change. The impacts are in the representative land transformation zones, where reclamation, overgrazing, afforestation and urbanization dominates the land use land cover changes in each zone respectively. High temporal frequency information about urban landuse, how urban areas develop, whether expansive or compact, with multifamily residential complexes or single family homes, automobile dependent or enabling multiple forms of transportation with mixed use or single use zoning. Effects on transportation choices and travel behaviour and determinants of infrastructure needs and energy consumption were worked out. It is believed that urban expansion will lead to an average increase of sensible heat flux in the future. While afforestation, overgrazing and reclamation would result in decrease of sensible heat flux (Deepak Kumar, 2014).

Landscape characterization of Ranchi urban agglomeration is worked out. The study is focused on the built up area expansion and subsequent loss of open spaces like water bodies, vegetal and other landscape elements. Ecological sensitive habitats that may be affected by urban sprawl is found out (Debjani Roy, 2014).

Forest fragmentation is another aspect worked out to study impact on elephant dominated landscape. Landsat ETM for 1992 and + for 2003 and landsat 8 for 2013 were taken for change analysis. The three decadal images were then compared with 2,50,000 toposheets. The area which is high in human-elephant conflict is more fragmented than the non-fragmented landscape area (Tanilekkiya, M. *et al.* 2014).

The set objectives for the present study are (i) To identify the land transformation processes in two city regions, (ii) To identify the drivers responsible for such land transformation processes, (iii) To realise the importance of ecotope analysis for micro-regional studies.

Materials and Methods

Toposheets (SOI) on 1: 50,000 scale of 56 k series for regional and preliminary mapping and 1: 10,000/1: 25,000 Toposheets for further visual interpretation and mapping have been used. The toposheet features are corresponded with IRS LISS III and IV P6 images to acquire digital data of the same or additional or transformed features. Further, subsets are also formed based on IKONOS images of selected spots. Erdas (9.1) and Arc GIS (10.1) softwares are extensively used for final mapping and analysis. Land use land cover studies have assumed unparalleled attention in view of their relevance to land management policies. Apart from this, the land use land cover studies have become that much more popular due to spatial technological tools, like RS, GIS, GPS, which have made them more relevant due to their spatio-temporal availability at various resolutions.

(a) Bhima Basin for mining activity, (b) Tor scape changes, (c) water body shrinkage in rurban zone of Hyderabad, (d) Land transformation due to urban sprawling for Hyderabad and Warangal cities have been identified to establish the role of anthropogeny in hydromorpho resource base in urban scapes. Anthropogeny is referred as man's activities on a landscape to fulfill his demands which is occurring at a rapid pace in and around city, mining and industrial centers. In this context the above mentioned regions are selected as case studies.

Case-1: Anthropogeny and A Mining Site (Bhima Basin)

The occurrence of cement grade limestone in the Bhima group of Tandur area of Rangareddy district attracted the attention of a number of entrepreneurs and facilitated the establishment of large capacity cement plants in this area. The limestone deposits are spread over large area under agriculture lands in and around Karankote Ogipur and Malkapur villages. In addition to this numerous quarries have cropped up to tap the resource for the production of slabs required for building purposes. In the process, there is a concentrated industrial activity around these villages and human settlements were disturbed. More than 800 hectares of land was affected which forms the core zone of environmental disturbance. Regional studies have indicated a greater buffer zone of imbalance indicating land degradation.

RS Studies: IRS ID LISS III data and toposheets and cadastral revenue village maps along with IKONOS imagery from the google earth were used in preparing composite map of the area depicting the core and buffer zones. The mining areas and quarries were also depicted in the composite map to identify the regional areas of environmental impact. The precious topsoil, to the extent of 800000 m^2 either lost or degraded due to exploitation in the core zones and an equivalent area is affected in the buffer zones.

Discussion

Mining activity inherently is causing environmental degradation. Open cast mining and quarrying, wherever not practiced in scientific manner leave devastated landscape, ugly scars and wasted landscape, which have deep impression in the minds of most people that mining of minerals degrades environment especially through unscientific planning and development of limestone slab quarries. Most of the agriculture lands have been totally affected and the rejects are spilled into local drainage causing enhanced sheet wash and reducing the water carrying capacity of the streams leading to floods. Surface water is contaminated with high total dissolved solids values. The suspended material in and around areas can be correlated with the fines generated during mining activity. The chemical composition of limestone which is intercalated with argillaceous-shally. Clayey material could be the causative factor. There is a general statement by mining companies that the mined out areas would be used as water bearing tanks for future years and serve as groundwater accumulators. Geo-environmental action plan is to be formulated for all the small slab quarries to avoid socio-economic issues in the mining region. Low waste technology is the need of the hour. Transportation corridors of pollution constitute larger buffer zone in the region (Pavanaguru, R. *et al.*, 2008)

Case 2: Anthropogeny and Tor Scape: Tors are typical of tropically weathered land forms. They are the result of subsurface spheroidal weathering due to chemical decomposition along joints accompanied by surficial removal of the weathered material, exposing as Tors. These Tors take various shapes and form major topographical expressions in and around Hyderabad city. These are in a way remnants of old Precambrian massifs disintegrated into smaller forms over a vast geological time. However, these forms are being slowly tarnished (blasted off) to make the room for built up area. These forms which are due to prolonged and log lasting weathering processes in a tropical environment are a true manifestations of weathered specimens which should be well preserved rather than putting them to destruction. Tors which formed a major part of landscape till as late as 70's faced a destructive phase as the city expansion started at a rapid pace. The tors were quarried to make the way for growing city landuse needs. Today, they are left as typically shaped isolated forms within the city region and as a few hillocks in the city suburbs (outskirts).

Climate Change as an Additional Impacting Factor: Climate changes are more varied in an urban environment due to contrasting canopy and topological conditions. An increase in mean maximum temperature, along with eratic rainfall patterns, severe rainfall storms accompanied by reduced rainy days present a different picture. Vehicular emissions increase the pollution levels and accordingly increase the diurnal temperatures. Similarly also play a role in reduced rainfall conditions. Reduced rainfall along with increased temperature impacts the inflows into water bodies and reservoirs. Therefore, a clear reduction in surface and groundwater resources is observed.

A recent newsitem is newspaper reported that rainfall to be intense in south and central parts of country. After observing the rainfall patterns between 1979 and 2015 it was declared that the yearly maximum rainfall increased across the country excepting north-east and J&K. It is further stated that due to human activities, there is an increase in temperature, which is harming the environment. This means that, there is an increase in the moisture bearing (holding) capacity of the air, which will in turn lead to extreme rainfall. What determined this increase was 'anthropogenic' warming caused by man-made emissions, due to deforestation, burining of fossil fuels, causing abundance of CO_2 in the air. The amount of CO_2 produced by humans far exceeded the normal limit let out by respiration or decay (DC-3rd April 2018).

Case 3: Anthropogeny and Water Body Shrinkage (Rurban zone of Hyderabad): A number of studies have been carried out on shrinking nature of water bodies in and around Hyderabad city region whereby it is observed that major water bodies like Hussainsagar, Himayatsagar and Osmansagar shrank considerably adding to water crisis in city due to less inflows from the catchment due to increase in the built up activity increasing the imperiousness nature of the soils. Area under water body decreased as tank bottom and river beds were encroached for cultivation using conserved moisture. A few tank beds have also been filled and residential colonies have sprung up leading to filthy living conditions in these areas. [Fig. 8.2(a)–Fig. 8.2(f)].

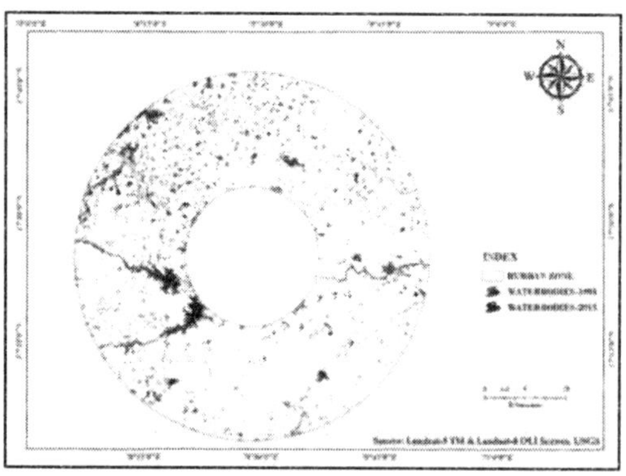

Fig. 8.2(a): Comparative View of the Waterbodies in the Rurban Zone Around Hyderabad Metropolis Between 1995-2015

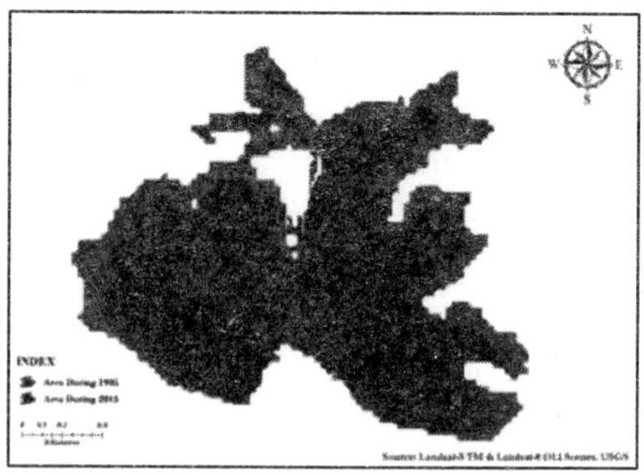

Fig. 8.2(b): Change in Area of the Idulabad Lake between 1995-2015

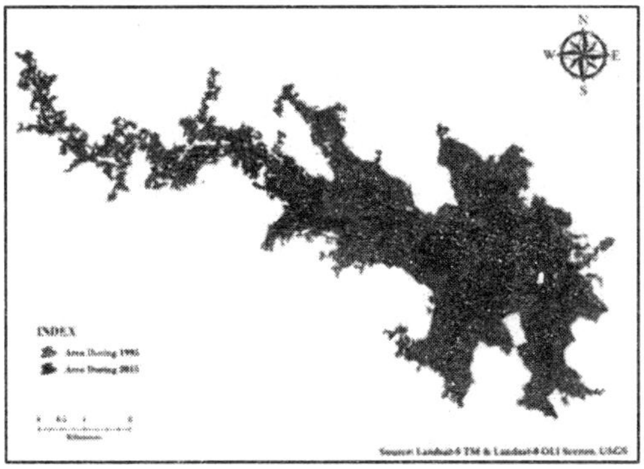

Fig. 8.2(c): Change in Area of the Osman Sagar Between 1995-2015

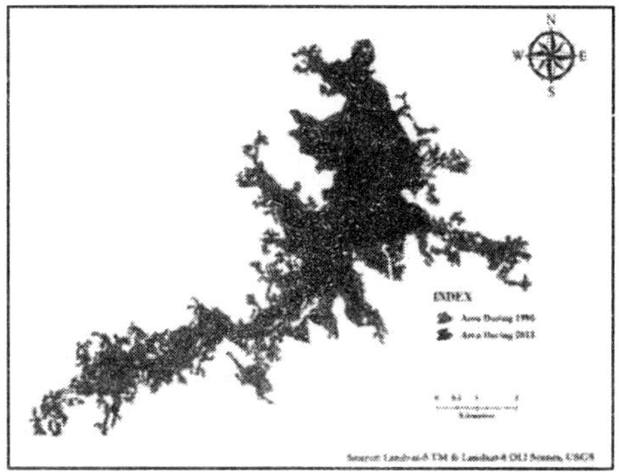

Fig. 8. 2(d): Change in Area of the Himayat Sagar Between 1995-2015

Fig. 8.2(e): Change in Area of the Shamirpe Lake Between 1995-2015

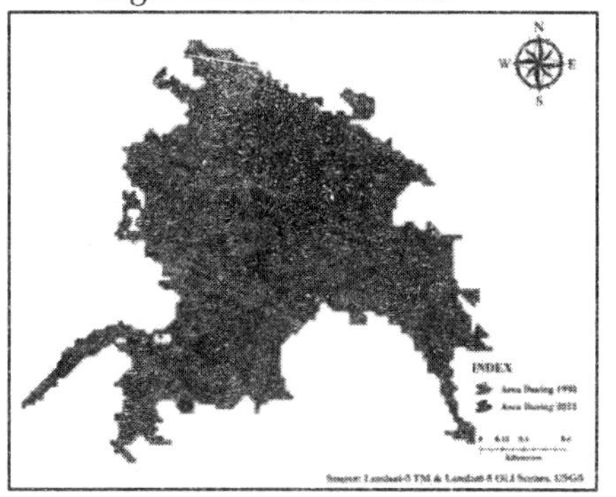

Fig. 8.2(f): Change in Area of the Ibrahimpatnam Lake between 1995-2015

The above maps clearly depict the change in the water spread between 1995-2015 in major water bodies around Hyderabad city. The map 2(a) also depicts the total change in the water bodies over an area of approximately 7000 km^2. The Table 8.1 shows the shrinkage of the major water bodies for the same period.

Table 8.1: Shrinkage in area of water bodies within Rurban zone of Hyderabad metropolis

Feature Name	*Feature Area (1995) in sq. km*	*Feature Area (2015) in sq. km*	*Change in sq. km*	*% ge of change WRT 1995*	*Remarks*
Waterbodies	318.537	134.512	184.025	57.77	Negative
Osman Sagar	25.954	10.382	15.572	60.00	Negative
Himayat Sagar	26.553	13.599	12.954	48.79	Negative
Shamirpet Lake	5.839	0.596	5.243	89.80	Negative
Ibrahimpatnam Lake	3.925	0.888	3.037	77.37	Negative
Idulabad Lake	2.423	1.797	0.626	25.82	Negative

Safety of fresh water reservoirs—Himayatsagar and Osmansagar: Analysis of hazard to fresh water reservoirs in the study area indicated that although the twin reservoirs of Himayatsagar and Osmansagar (Gandipet), which supply drinking water to the city, are not contaminated, the Manjira river and Nizamsagar located in the northwest of the city are in grave danger of contamination emanating from

Gaddapothram-Bolaram-Patancheru industrial axis as they are located within 15 kms of the fresh water source.

Apart from water bodies, drainage ways to get obliterated in a city region. A work on Musi channel within the city showed changed hydraulic geometrical characters of Musi, especially channel width due to vast encroachment reducing the carrying capacity of the river. This has a marked devastating impact on city when they culminate into floods due to sudden rainfall burst.

Case 4: Anthropogeny and Land Transformation (Rurban zones of Hyderabad and Warangal):

(A) **Hyderabad: Land Transformation (1990-2015):** Land transformation in urbanscapes is enormous. Hyderabad is no exception. Around this city region till a stretch of 40 km radius. Land transformation is as accounted in the Table 8.2.

Table 8.2: Land Use/Land Cover Change in 1990 to 2015

Sl. No.	*Land Use*	*Area Km² 1990*	*%*	*Area Km² 2015*	*%*	*Change (Area)*	*% Change*
1.	Agriculture	872.97	20.19	407.6	9.43	-465.37	-10.76
2.	Built-up	39.05	0.90	397.854	9.20	358.80	8.30
3.	Forest Cover	243.99	5.64	284.54	6.58	40.55	0.94
4.	Other land	3026.54	3171.70	73.36		145.15	3.36
5.	Waterbody	141.19	62.06	1.43		-79.13	-1.84
	Total	**4323.75**	**100**	**4323.75**	**100**	**0**	**0**

Transformation of agriculture land has been tremendous, whereby it lost almost 50 per cent of its area (872 km² – 407 km²). On the other hand builtup increased almost tenfold (39 km² – 397 km²). Forest cover increased marginally (243 km² – 284 km²), water bodies depleted to a great extent (141 km² – 62 km²). Other land inclusive of variety of land use classes, increased marginally (3026 km² – 3171 km²) [Fig. 8.3(a & b)]. This land transformation of 25 years has clearly indicated a loss of two major resources, i.e., land and water. Precious agriculture land got transformed at a fast pace accompanied by equally precious water body. These two resources, which are a means of livelihood and also a source for future food security lose their standing and areal extent, putting the stress on food availability. The less loss of green cover due to environmental awareness programmes

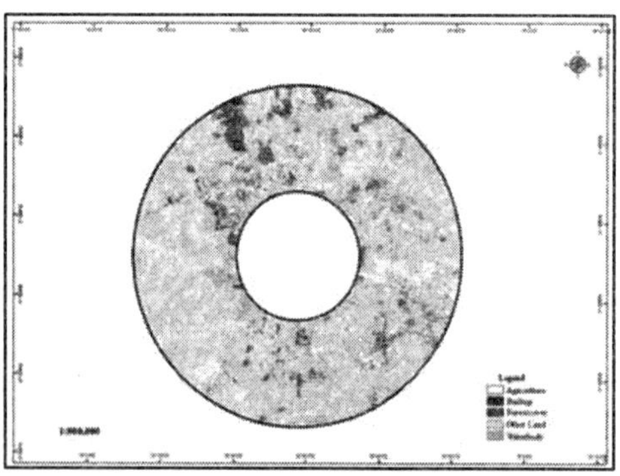

Fig. 8. 3(a): Land use/Land Cover–1990: Urban-Rural Belt (15-40 KMS) of Hyderabad Metropolis

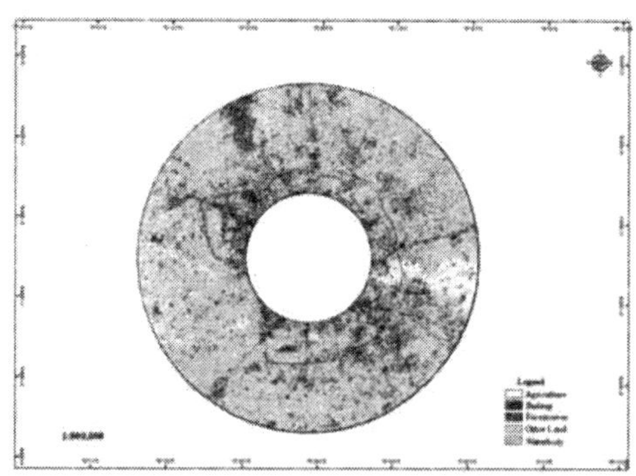

Fig. 8. 3(b): Land use/Land Cover–2015: Urban-Rural Belt (15-40 KMS) of Hyderabad Metropolis

is a welcome change in the promotion of environmental conservation, through detailed studies and inventories of carbon sequestration and other important ecological processes.

(i) Agriculture land transformation: Agriculture land transformation is maximum in other land use class followed by built up forest and water bodies. There is every possibility of agriculture land transformation to other land types due to non-availability of agriculture means either in the form of natural calamities, especially due to erratic and uneven rainfall variability leading to drought and floods but also due to increase in built up activity due to urban growth and sprawl. The fertile lands are sold for urban expansive processes getting in turn monetary compensation. It is also noted that the built-up activity leaves behind lots of pits and debris piles which subsequently impact soil structure and a natural land degradation takes place. After some time the same lands may become wastelands, fallow lands, stony or rocky waste. Similarly some pits which have been created during construction process may accumulate rainwater and may turn into small pools and water bodies.

This trend has effected the socioeconomic condition in the region where, large scale migration of farming families to urban center in search of livelihood, recurrent crop failure, high value of land, availability of credit facility, decreasing water resources, education., market economy etc. have given rise to an unsustainable trend in economic development.

(ii) Water Body Transformation: Water body lying near to an agriculture plot may slowly get altered to agriculture land due to encroached agriculture activities, especially on a dry bed. It is natural that a non-perennial water body might get converted as agriculture land, or may get converted as barren land or even scrub forest depending on the proximity factors. In the surroundings of Hyderabad it is often witnessed that the conversion of water body to other land and agriculture land is more than any other category.

(B) Warangal: Land Transformation (1990-2015): Physiographically Warangal is a plain region with an average contour values around 250 m, and with a slope varying between $0^0 - 15^0$. Warangal too has a similarity to Hyderabad in the sense that, this city too has a twin city concept. This is also a historical city thriving on its glorious past, with a domination of kakathiya regime. The Table 8.3 depicts the land use land cover statistics.

Table 8.3: Change detected from land use land cover of Warangal city during 1990-2015

	Year 1990		*Year 2015*		*Statistic*			
Feature Type	*Feature Area (sq. km.)*	*% WRT the total Area*	*Feature Area (sq. km.)*	*% WRT the total Area*	*Diff. of Area in sq. km (2015-1990)*	*% Diff. of Area (2015-1990)*	*% of Change Statistics WRT 2015 LULC*	*Remarks*
Barrenlands	152.728181	21.61	123.48029	17.47	-29.25	-4.14	23.686	Slight Decrease
Built-Up	9.623051	1.36	72.8253	10.30	63.20	8.94	86.786	High Increase
Croplands	281.30309	39.80	200.006754	28.30	-81.30	-11.50	-40.647	High Decrease
Followlands	130.069786	18.40	104.200341	14.74	-25.87	-3.66	-24.827	Slight Increase
Forest	4.241727	0.0	9.297562	1.32	5.06	0.72	54.378	Slight Decrease
Grasslands	83.251887	11.78	124.955323	17.68	41.70	5.90	33.375	High Increase
Scrublands	14.251429	2.02	47.868032	6.77	33.62	4.76	70.228	High Decrease
Waterbodies	31.355168	4.44	24.184306	3.42	-7.17	-1.01	-29.651	Slight Decrease
Total	**706.82**	**100.00**	**706.82**	**100**	**-0.01**	**0.00**	**125.956**	

Land use of 1990 is dominated by croplands fallow lands, barren lands and grass lands. Out of 706.82 km^2 area, 39.8 per cent of area is crop land, followed by 21.6 per cent of barren lands, 18.4 per cent of fallow lands and 11.78 per cent of grass lands. Only 1.36 is under built-up, followed by 2 per cent under scrub land, 0.60 per cent under forest cover and 4.44 per cent water body. In 1990s Warangal was an agrarian region, with less of forest resources [Fig. 8.4(a)].

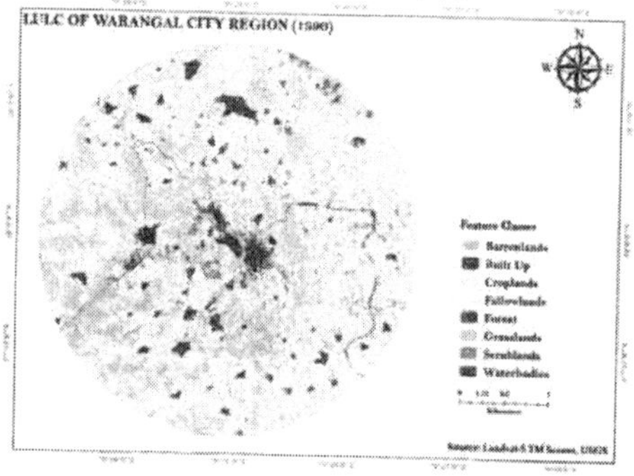

Fig. 8.4(a): LULC of Warangal City Region (1990)

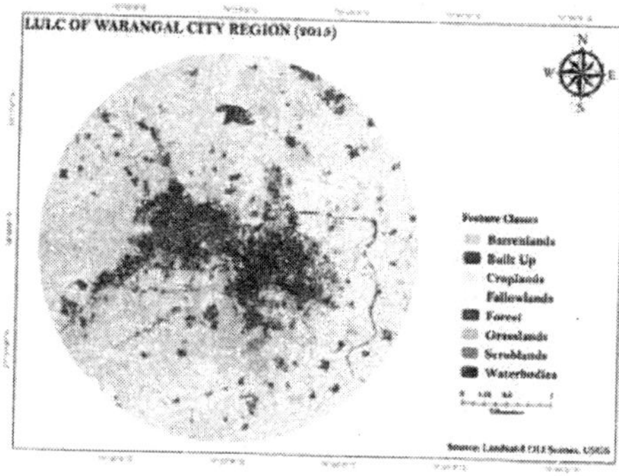

Fig. 8.4(b): LULC of Warangal City Region (1990)

Land use land cover 2015: Croplands decreased by 11 per cent, barren lands by 4.14 per cent and water body by 1 per cent, fallow lands too decreased by 3.66 per cent. A remarkable increase is witnessed in built-up area with an increase by 8.94 per cent grass lands and scrub lands increased by 5.9 per cent and 4.76 per cent respectively, With respect to class area, high increase is noticed in built up (86%), grass lands (33%) and scrub lands (70%) high decrease is witnessed in crop land (-40.65%). Slight decrease (-29%) is witnessed in water body fallow lands (-24%), barren lands (-23%). However slight increase (54%) is seen in forest cover. [Fig. 8.4(b)].

Land transformation is quite phenomenal in Warangal and after the Telangana state formation, Warangal has assumed a special significance after Hyderabad. In fact prior to state bifurcation, Hyderabad was the only major city catering to urban needs. But now with decentralization policies, satellite cities of earlier period are surfacing as prime cities of the region. This may give rise to never ending episodes of man-nature altercations evidenced more in urban scapes.

Land transformation and land degradation: Environmental degradation has occurred in the study area in several ways. Marginal agricultural lands have turned into open scrub while agricultural land have been waterlogged. Growing pressure on land results in sky-rocketing of land prices and leads to unsustainable land uses which result in degradation. The Table 8.4 depicts the same.

Evaluation of society-nature interaction for sustainable development: Due to high concentration of urban population in cities of Hyderabad and Warangal agricultural activity has suffered great losses by way of shrinking areal extent in the urban and rural belts and falling productivity in the rural area zone. For obvious reasons the percentage share of population in the agriculture in the city center of Hyderabad is a meager 0.2 per cent while the peripheral regions have over 23 per cent of adult population engaged in agricultural.

Table 8.4: Land cover land use changes and land degradation

Change (from – to)	*Type of degradation*
Agriculture –built up area	Prime agricultural land going out of production due to market forces
Agriculture – scrub	Prime agricultural land going out of production due to degradation
Agriculture – water body	Waterlogging
Scrub – agriculture	Expansion of agriculture into marginal land
Scrub – built up area	Shrinking of vegetative land
Scrub – vineyard	Expansion of agriculture into bio reserves
Scrub-water	Waterlogging
Vineyard – built up area	Encroachment of good agriculture land
Vineyard – scrub	Degradation of land
Vineyard – water	Waterlogging
Water – vineyard	Encroachment on water body
Water – scrub	Encroachment in catchment area leading to drying up of water bodies

Prime agricultural land around the city are being converted into residential, commercial or industrial use. Apart from this, large tracts are left fallow owing to various reasons like lack of adequate irrigation sources, failure of tubewells owing to fall in groundwater level, erratic rainfall, migration of land owners to urban centers, ineffective laws of tenancy, lack of input with farmers owing to poverty and shortage of labour during critical phases in cropping season, etc.

Despite massive increase in built-up area the high density of population makes available residential space scarce. In city center the available built-up area per person in 2001 was 16.3 m^2 while in urban belt it was 23.9 m^2. (Table 8.5). On similar lines there is a reduction in green space with in city area. Although builtup area is predominant only in the city center and accounts for 75 per cent of the entire region, its impact on the natural resource in the region is massive. Urbanization has altered the land use Land cover in the region with agricultural land being converted into settlements, roads, railways and commercial complexes.

Table 8.5: Percentage share of built-up area, urban green, agricultural land and other land (m^2)

Zones (distance from centre)	*Total population (million)*	*Share (m^2/person)*				
		Built-up area	*Vegetation*	*Agriculture land*	*Other land*	*Water body*
Rural Area (25-50 km)	1.13	234	1117.3	2665.9	1146.9	21.8
Rural Belt (10-25 km)	1.98	81.9	138.2	443.8	148.6	15.7
Urban Belts (5-10 km)	3.41	23.9	15.5	21.7	6.4	1.11
City Centre (0-5 km)	3.63	16.3	5.9	0.0	0.5	1.27

The above table clearly reflects the disparities in resource's availabilities in various zones from city centre. The percentage share of various land use categories as per capita availability increases from city centre to outer rural areas excepting the builtup area. Therefore, the sprawling urban activities will diminish the resource availability and there up on the total ecological services around the city centres. In this view a proper long term perspective planning is required taking into consideration the available natural resource base and the future demand in view of urban growth process.

References

Anonymous (2002): Report on estimation of ground water resources in Andhra Pradesh 2002. Ground Water Department, Government of Andhra Pradesh. pp. 356.

Anonymous (2003): Strategic overview of drinking and industrial water use, Draft Volume II Chapter 2, pp. 2-5.

Anonymous (2004): A plan for sustainable development (Hyderabad 2020) draft master plan for Hyderabad metropolitan area, HUDA. p. 322.

DE&S (2003): Trends of rice production in Andhra Pradesh (1995-56 to 2001-2002)

Dr. U.S., and M. Rajeevan (1997): Identification of anthropogenic climate change. A Review, *Vayu Mandal*, Jan-June 2-13.

Dt. U.S. (2001): Climate change impact: Regional scenario, *Mausam* 52, 1, 201-212.

Gallopin, G.C. and F. Rijsherman (2000): Three Global water scenarios. *International Journal of Water* (Ed.) Martin O Connor, Vol. I (1).

Kausalya Ramachandran and H.P. Singh (2001): Sectoral shift in land use under changed economic scenario—A case study of Hyderabad region using IRS-ID LISS –III data. *ICORG* Feb. 2-5, 2001, JNTU, Hyderabad.

Kausalya Ramachandran (2001): Assessment of contamination of natural resource using GIS and Remote sensing – A study on Hyderabad region—GIS India, 10(1) 8-13.

Lata, K.M., Shankar Rao, C.H. Krishna Prasad, B. Badrinath, B.S. and V. Ragava Swamy (2001). Measuring urban sprawl—A Case Study of Hyderabad. GIS Development, 5(12).

Padmaja, S., (2008): *EIA Studies and Planning Perspectives*, Vipla Computer Services, Hyderabad.

Pavanaguru, R. Narasing Rao, A., Vijaykumar, J., and Venugopal Reddy, K., (2004), Studies on environmental degradation in the limestone mining area of Rangareddy district. Andhra Pradesh, India. National seminar on "Recent trends in geosciences". Geology Dept. S.V. University, Tirupathi, 2nd and 3rd April 2004.

Pavannaguru, R. Lingaswamy, T. (2006). Environmental studies in the limestone mining areas of Rangareddy district, A.P. Paper presented in the National Workshop on limestone deposits-conservation and sustainable development. Hyderabad, 25th march 2006.

Seckler, D. Amarasinghe, U., Molden, D. Radhika de Silva and R. Barker (1998): World winter demand and supply, 1990 to 2025. Scenarios and Issues Research Report No. 19 by International Winter Management Institute, Colombo, Sri Lanka, 40 pp.

Sinha, S.K. (1989): Global warming and found security in developing countries. In R.K. Pachauri and S. Gupta (eds). Proceedings of the International Conference on Global Warming and Climate Change: Perspective from Developing Countries. 101-102, New Delhi, Tata McGraw Hill.

Sudhira, H.S. and T.V. Rammachandra (2003): Urban sprawl pattern recognition and modeling using GIS. www.gisdevelopment.net/application/urban/sprawl/mi03142.html

Velayutam, M, Mandal, M., Mandal, D.K., Champa and Sehgal, J (1999): *Agro-ecological sub-regions of India for planning and development*, NBSS and LUP, Nagpur, Publication.

Yeh, A.G.O. and X, Li (2001): Measurement and monitoring of urban sprawl in a rapidly growing region using entropy, photogrammetric engineering and remote sensing, 67(1).8.

9

A Perceptive Analysis of Land Transformation in Urban Rural Belt of Hyderabad Metropolis

S. Padmaja and *T. Pratigna*

Introduction

Land and its concerns are universal in nature. Land with all its natural attributes very rarely gets unchanged as time progresses. With technological advances and increasing human demands on the land, the land transforms itself at various levels. This type of land transformation is more evidenced where there is dwindling man-land ratios, which are conspicuous in urban and periurban areas. Land transformation manifests itself in many ways culminating into fragmentation and complex land breakup. Land transformation, basically is change of one type (form) of land into another form or category, though is primarily climatic driven but often is influenced by anthropogenic impact as is evidenced around either city centers or mining centers. During last 200 years, trends of demographic growth, urbanization, technological change and commercialization have driven land use changes enormously and subsequently land transformation across a variety of ecological conditions at local, regional and global levels. These changes encompass the greatest environmental concerns of human populations today, including climate change, biodiversity loss and the pollution of water, soils and air. Monitoring and mediating the negative consequences of LULCC/LT, while sustaining the production of essential resources has therefore become a major priority of researchers and policy makers around the world.

Drivers of Land Transformation

Large and small scale land transformation is possible through a set of driving forces which may include: (i) Environmental factors like—land and soil degradation, climatic variability. Encroachment onto the vegetal canopy other than croplands and the waterbodies for the rising demand of agriculture lands. (ii). Demographic factors like-population size, population growth, migration, socio-economic and cultural aspects of population. (iii) Socio-economic factors like variability in production and productivity of agriculture lands, real estate leading to land fragmentation, conversion of agriculture lands into commercial lands, marginalization of land holdings and change in occupational status. Other driving factors may be local culture like food preference, demand for specific products, infrastructural and institutional amenities, finance services, marketability, industrial development, price and policy, availability of irrigational avenues, supplies of agriculture inputs, knowledge based approaches of the advances in agriculture activities and business strategies.

Land Use/Land Cover Change/Land Transformation as detected by RS data has been the focus of many studies center in ground city regions like Vijayawada (Sundarakumar K.), Bhubaneswar (Monalisha Misra, 2006). Peshwar (Abdur Raziq *et al.*), Udaipur (Manishikha Jain), Calcutta (Mala Mukerji), South Indian cities of Banglore, Chennai and Hyderabad (Kaushalya Ramachandran and Padmaja, Raghav swami, Neelakanthan, Sumati Kulkarni, Murali Krishna (2004-2006)). Land transformation due to human domination on earth ecosystem in widely debated by *Peter M. Ritowsek et al.* (1997). *Nasreen Islam* (2000) brought out the impact of urbanization on surface water system due to land transformation. Aerial expansion and population increase characterize land transformation process. A study on Green Heart Region by *Shahab Fazal et al.* (2012) identified land transformation drivers as (a) demography, (b) economy, (c) Institution, (d) nature and environment, (e) transport and infrastructure. *Andrew and Ganghan et al.* (2008) researched upon tourism, forest conversion and land transformation in the Angkor Basin (Cambodia). Global (Tourism climate), Regional (National Policies) and local (construction and agricultural, energy and water resources to support tourism industry) factors have been interlinked.

Baldev Sahai in his presidential address has stated that natural causes transform land slowly, but what is assuming alarming proportions of the land transformation, being brought out by man, especially due to agricultural and mining activities in ruralscape followed by built-up activity in urban space. *Wolman and Fournier* have laid stress on various agricultural practices leading to Land transformation. *Bernhard lucke et al.* explained the process of land transformation through drivers of land transformation which all increase habitat loss and isolation yet impose their own unique alterations on the landscape which include perforation, dissection, fragmentation, shrinkage and alteration. *Ayyad Mohd* talks about land transformation in western Mediterranean desert in Egypt attributing the change mainly due to grazing, rained farming and irrigated farming. Economic development associated with anthropogeny leads to land transformation. *Aggrey Daniel Maina Thuo* (2010) in a special issue on agriculture raises the issue of community and social responses to land use transformations in the Nairobi rural-urban fringe of Kenya.

Brian C. Piyanowski has explained land transformation through a model involving socioeconomic driving variables comprising of population change, economies of land ownership, transportation, agricultural economics and locations of employment. Environmental driving variables include abiotic, such as distribution of soil types and elevation and biotic such as locations of endangered and threatened species or the attractiveness of certain types of vegatation pattern in the landscape for development. The conversion of agricultural land to residential uses is leading to rapid transformations in agricultural production, spatial and social structure, land ownership and land market. Selling in parcels of land have affected land holders and in most cases have made their families destitute. *Christian Korner* (2006) in a study on ecological and socio-economic consequences of land transformation in alpine regions and ancestral agricultural activities, stated that reduced farming activities have led to massive shrub encroachments and forest expansion into formerly open lands at high elevations. Forest land cover changes seem to affect evapotranspnration (ET) and runoff exerting hydrological consequences for adjacent low lands. Urbanisation process around Anyigba in Kogistate (Nigeria) is worked by *Hatermehim et al.* (2009) where by negative environmental impact is highlighted. It is also being told that rural-urban fringes are characterized by diverse land uses which often vary in relation to their functional linkages to urban and rural sectors. *Ellise et al.* (2000, 2006, 2009) have attempted micro scale landscape transformations to identity the causes and consequences of ecologically significant long term changes in parts of China. How the ecological space is consumed by urban growth is being studied in Bangalore by *Narayanan P and Hajangi A.D. Jose M.* etc (2007) who have worked on causes and consequences of agricultural land abandonment citing three sets of drivers as ecological, socioeconomic, agricultural land systems and mismanagement. *Timothy W. Foresman* and others (1997) emphasized on human impacts on Baltimore-Cheasapeake region with the help of urban-rural gradient analysis. It is noticed that physical, Environmental and socioeconomic, ecological environments have been impacted by humans. Studies on Hyderabad metropolis are few and selective. *Kausalya Ramachandran, S. Padmaja, V. Raghavaswamy, Murali Krishna* etc have worked on sprawling process and its impact on land use and land transformation, subsequently leading to resource crunch and loss.

Today with rapid urbanization and industrialization, there is increasing pressure on land, water and environment, particularly in the big metropolitan cities. Urban sprawl may be defined as the scattering of new development on isolated tracts, separated from other areas by vacant land (*othersmann* 1977). Urban sprawl has often been criticized for inefficient use of land resources and energy and large scale encroachment onto agricultural land. There are many problems associated with fragmented conversion of agricultural land into urban use. The cities are expanding in all directions resulting in large scale urban sprawl and changes in urban land use. The spatial pattern of such changes is clearly noticed on the urban fringes or city peripheral rural areas than in the city center. Inadvertently this is resulting in increase in the built-up area and associated changes in the spatial urban land use patterns causing loss of productive agriculture lands, forest cover, other forms of greenery, loss in surface water bodies, depletion of groundwater aquifers and increasing levels of air and water pollution.

Study Area

The study area spreads over 4323.75 km^2 bounded by 17^{0}3′ 43″ N – 17^0 47′ 06″ N and 78^0 6′ 0″ E – 78^0 51′ 0″E (Fig. 9.1). As per census data, the number of mandals are 38, with 611 revenue villages of 4 neighbouring districts (Mahaboob Nagar, Medak, Nalgonda, and Ranga Reddy).

Area of Interest: Although there have been several studies on urban sprawl and urban land use changes of Hyderabad city region but scant attention has been paid to impact studies due to Land Transformation. Hyderabad has attained the status of metropolis and is a fast growing city in south India with a population of around 6.1 million. The city has shown a rapid transformation since 1970's and more so since 90's due to IT boom. Leaving the city center the sprawling activity is seen maximum in the peripheries of the city, approximately 20 kms away. That's why 15-40 km stretch centering round Hussain Sagar in city center has been selected to study land transformation. Spatially and socioeconomically, the area of interest (AOI) shows an amalgamation of both rural and urban characters.

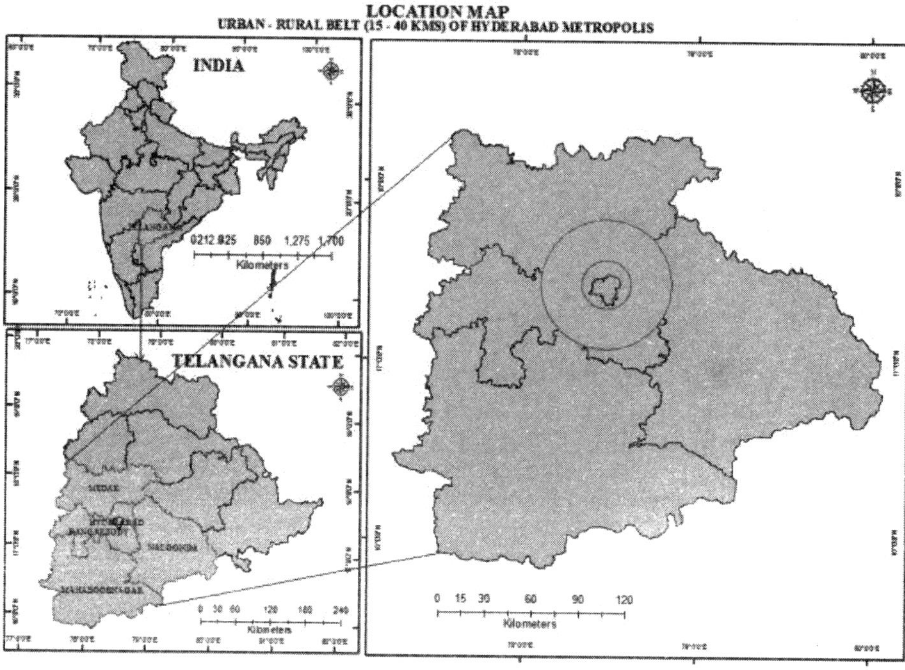

Fig. 9.1: Location Map: Urban-Rural Belt (15-40 KMS) of Hyderabad Metropolis

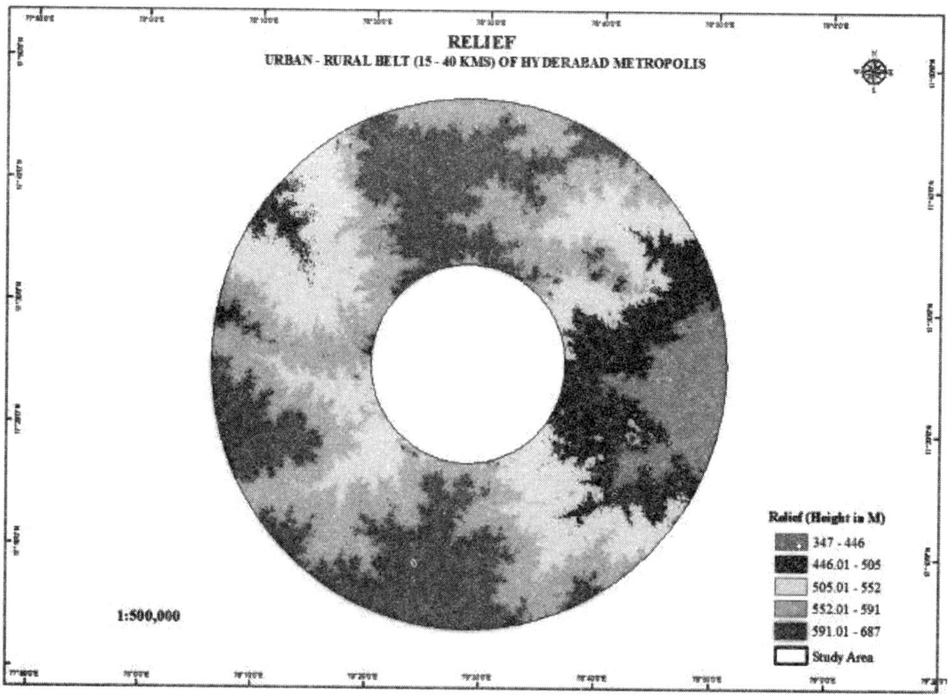

Fig. 9.2: Relief: Urban-Rural Belt (15-40 KMS) of Hyderabad Metropolis

Geographical Back Ground

Geomorphologically, study area is characterized by open grounds with preponderance of isolated hills with open scrubs and vast pediplain, studded with two major water bodies (Osman Sagar and Himayat Sagar) supplying water to the city. The average elevation is higher in the north-west with a mean elevation of 620 m a MSL, showing the topographical character with a general contour value from 360 m to 680 m trending across the region representing a dissected pediptain traversed by numerous stream lets. Fig. 9.2 reveals the general relief of the study area with North, West and South representing high relief and East representing low relief.

River Musi runs through the region widening its bank, enriched with vegetation towards the east. At present, other land use comprising of fallow lands, scrub lands, pasture lands, barren lands and other waste lands is by far the largest land use occupying approximately more than 70 per cent of the total geographical area. Agriculture land use forest cover, water bodies and settlement occupy about 10 per cent, 10 per cent, 1.5 per cent and 8.5 per cent respectively.

Climatically, the region has been experiencing erratic and irregular rainfall, accompanied with rise in temperature making the region more of a dry land region. There have been episodes of years of heavy rainfall followed by dry years [Fig. 9.3(a) to Fig. 9.3(g)].

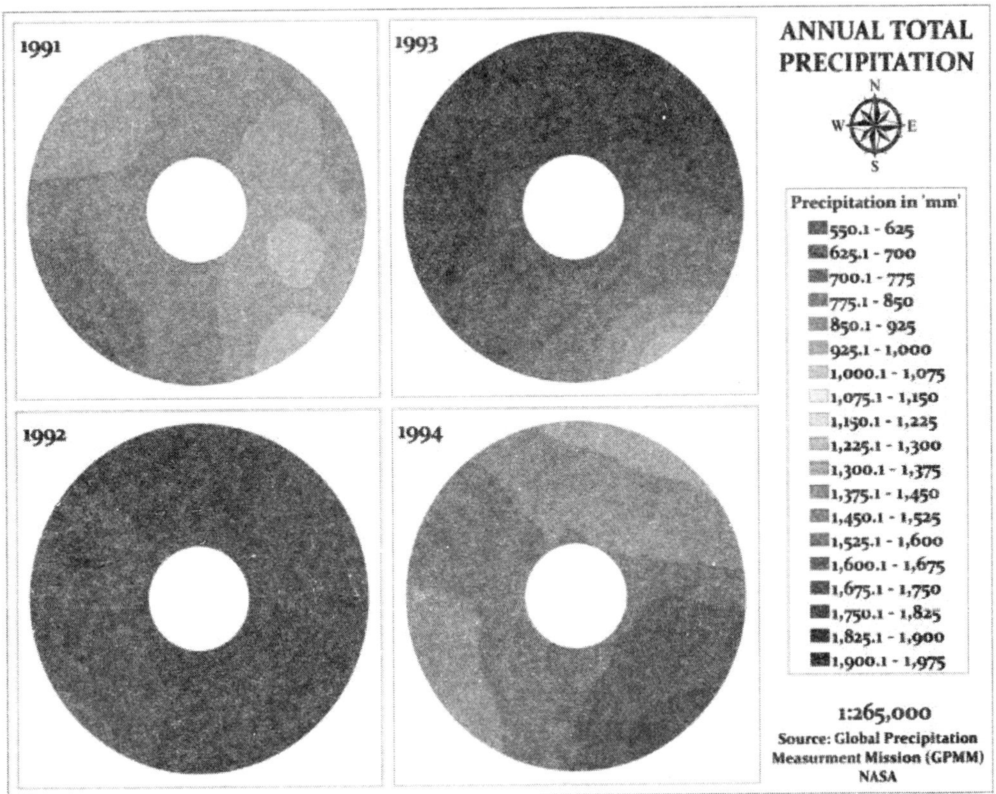

Fig. 9.3(a): Climate Conditions (1991-1994)

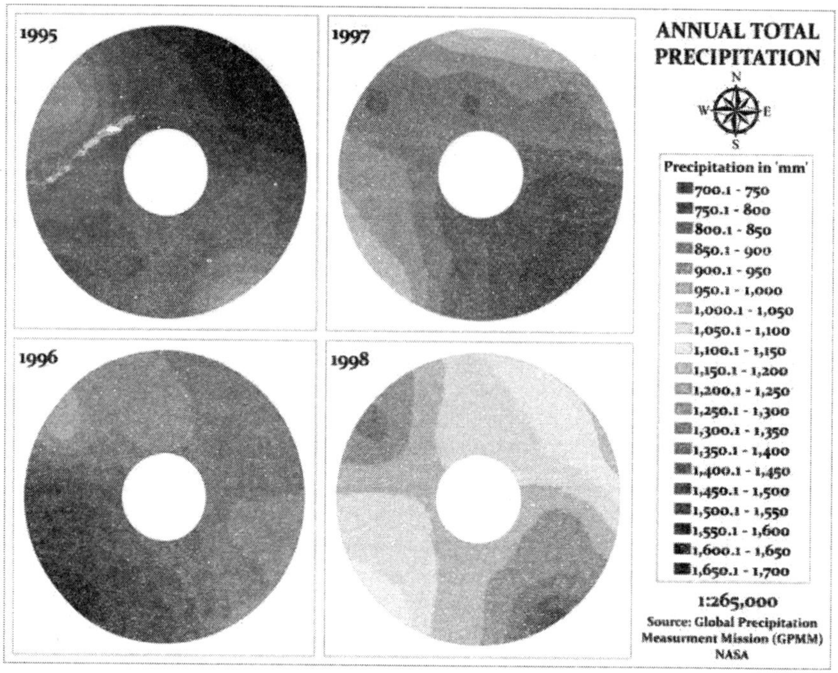

Fig. 9.3(b): Climate Conditions (1995-1998)

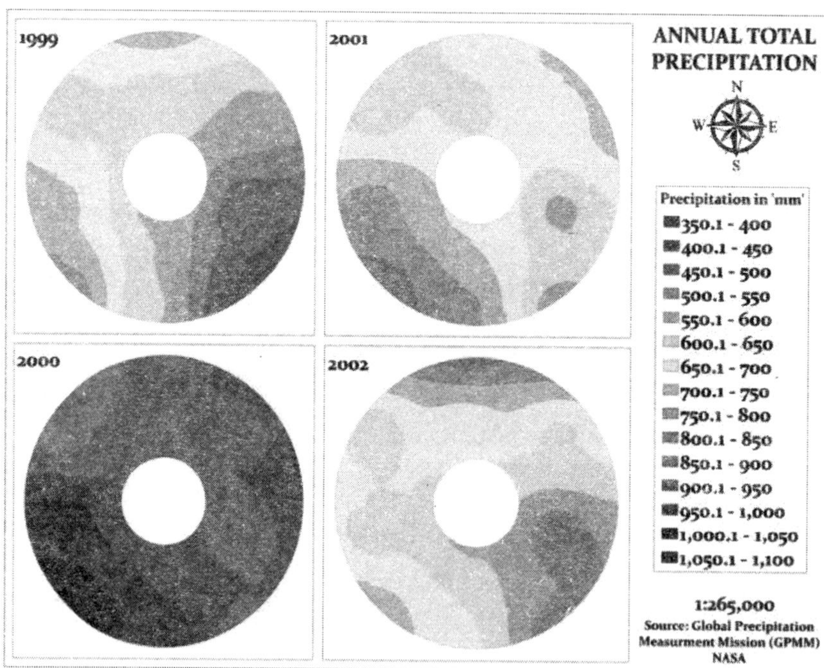

Fig. 9.3(c)

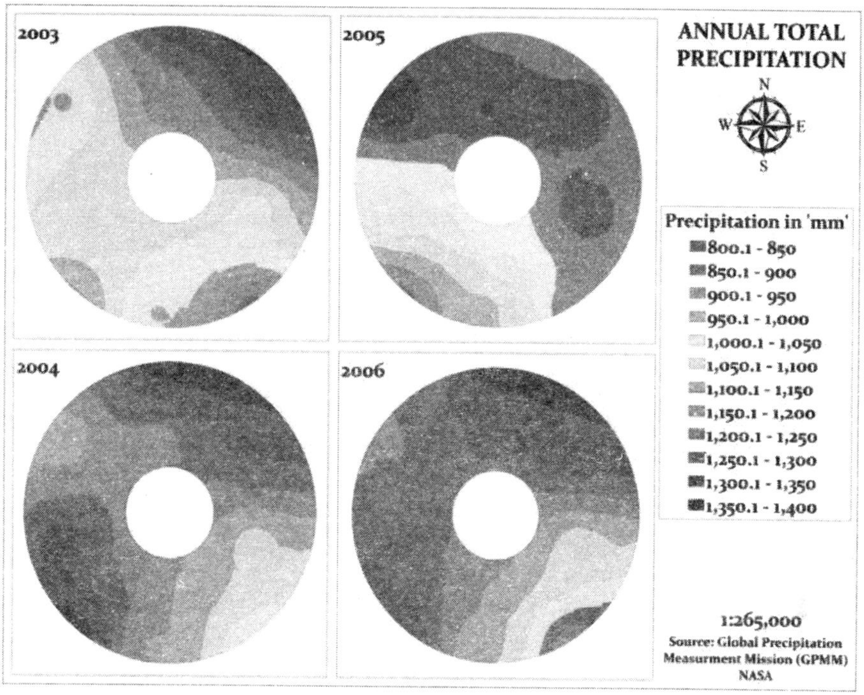

Fig. 9.3(d)

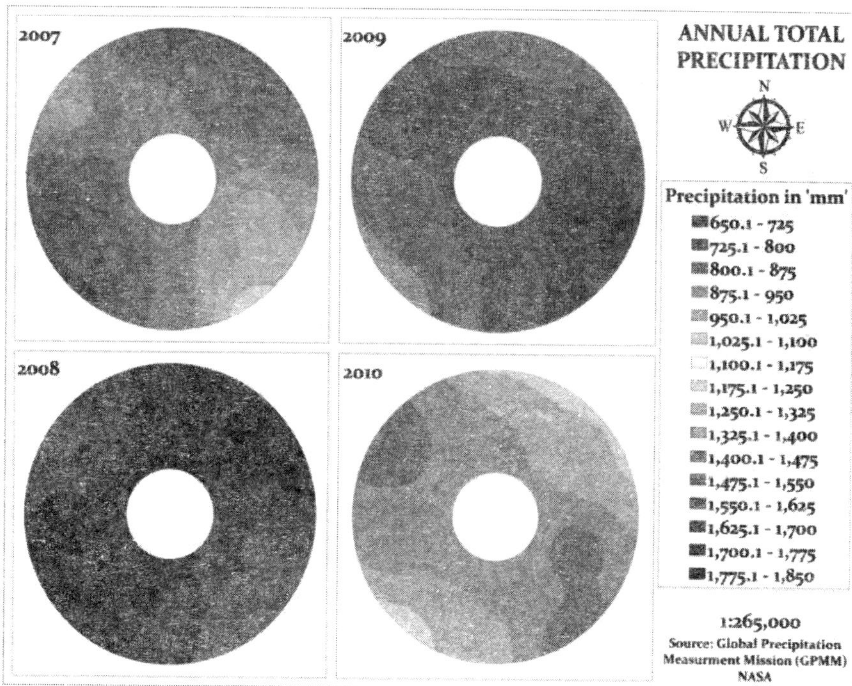

Fig. 9.3(e)

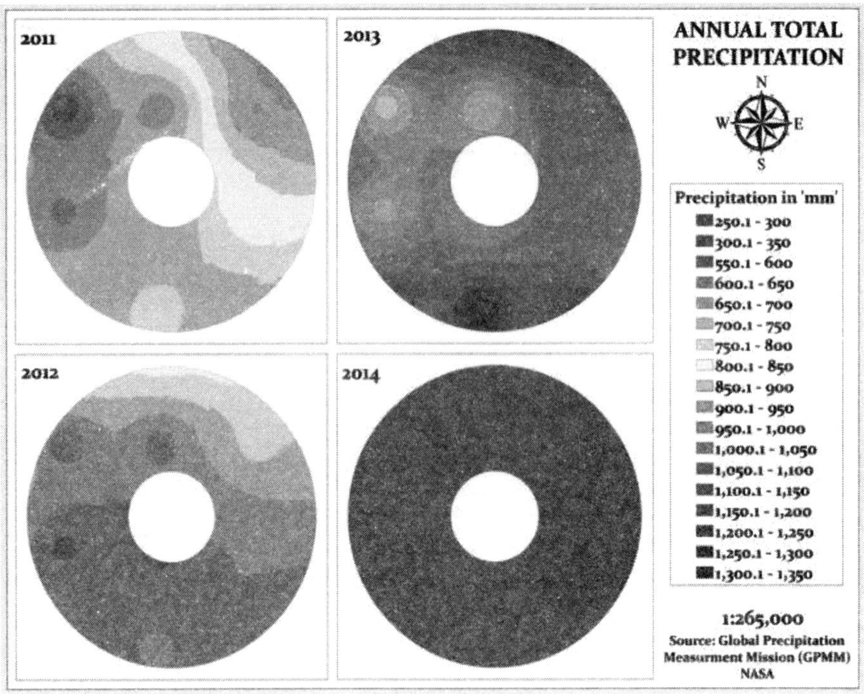

Fig. 9.3(f)

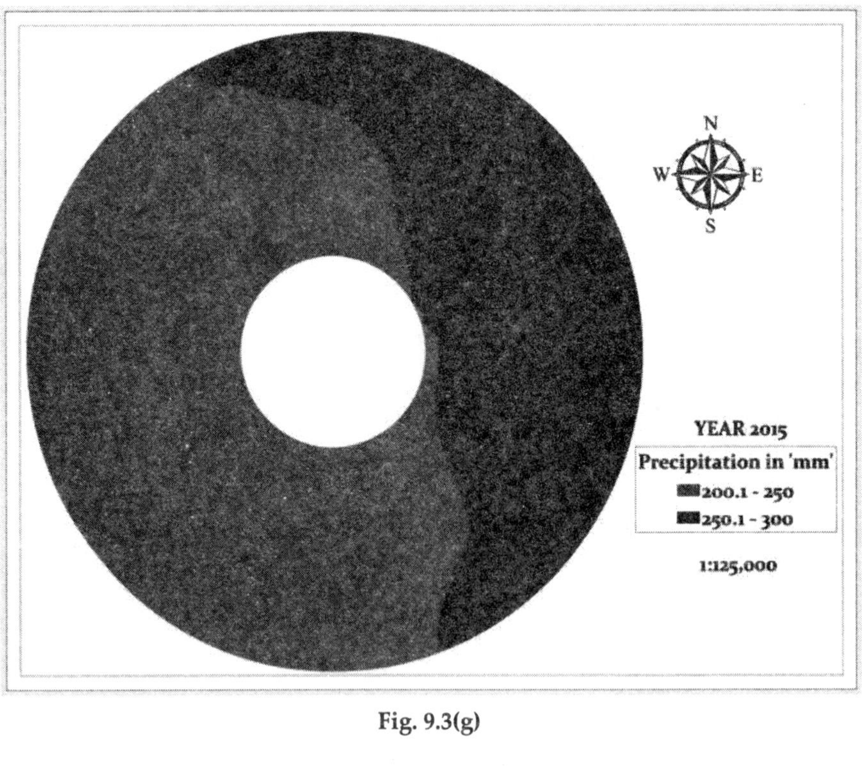

Fig. 9.3(g)

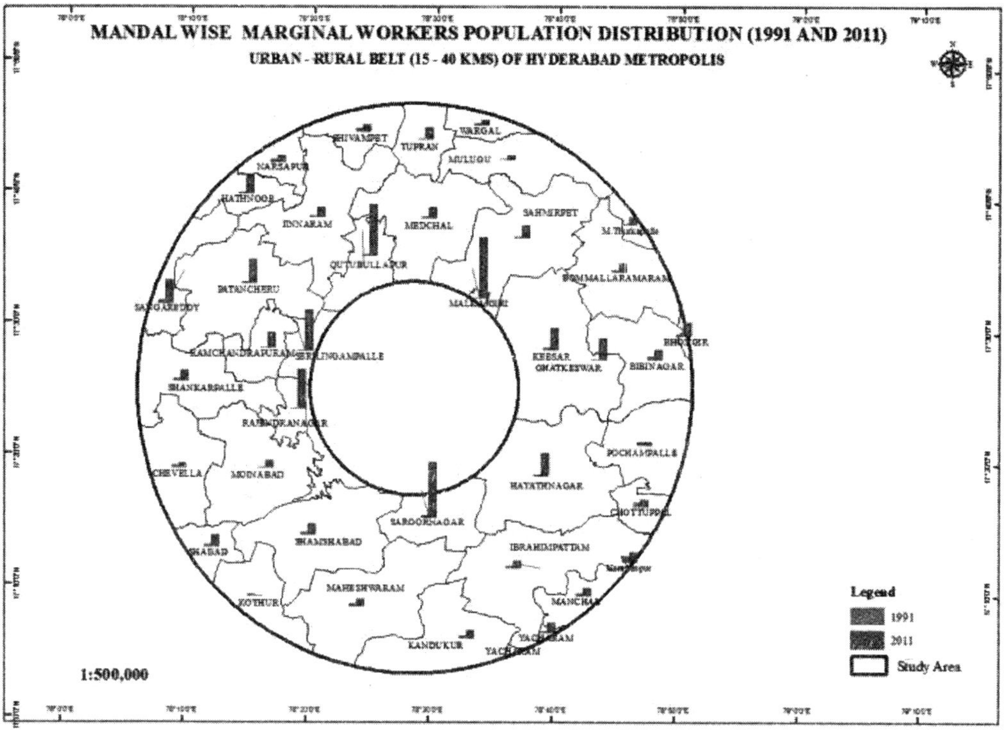

Fig. 9.4

There are three rivers partly draining the area, namely Musi, Kagna and Manjeera. Many other water bodies too like, Ramasamudracheruvu, Mella Cheruvu Bibinagarcheruvu, Gopicheruvu, Kaminenicheruvu, Masabcheruvu, Mokulcheruvu, Chintalcheruvu and Shamirpetcheruvu are found in the region. Black, red and mixed sandy, loamy and clayey soils occur in the region.

Demographically, the area, which had 4,33,291 households in 1991, increased to 11,00,176 households in 2011. Qutbullapur, Malkajgiri and Saroornagar recorded highest growth of households along with population increase and highest literacy ratio. The total population of the study area in 1991 was 22,27,634 which increased to 46,84,944 in 2011. According to 1991 census, literate population was 8,31,358 which increased to 31,04,408 in 2011. As per 1991 census, the total main workers were 9,12,714 increased to 16,10,888 in 2011. Serilingampalli, Qutbullapur, Malkajgiri recorded highest increase, whereas Kothur, Turkapalle recorded lowest increase. As per marginal workers data (Fig. 9.4), the study area had 32043 persons in 1991 increased 10 times in 2011 to 3,17,196 persons. Narayanpur, Chotuppal and Sanga Reddy recorded highest marginal workers, whereas Ramachandrapuram, Malkajgiri and Tupran recorded the lowest. Non-workers population too doubled from 1991-2011, i.e. from 12,82,063 to 27,56,860. While Serlingampally recorded the highest non Workers, Kothurmandal recorded the lowest.

Data Base and Methodology

Database: Present study is based on spatial remote sensing data as well as non-spatial data available from various sources for different periods. The primary data is collected from household survey through a questionnaire as well as from coordinates through GPS. Secondary data is collected from Survey of India toposheets of 56K series, excluding K1 and K13 (2, 3, 4, 5, 6, 7, 8, 9, 10, 11, 12, 14, 15, 16) were used for generating the base map of 1970's. Socioeconomic data is collected from Department of Economics and statistics. Climatic data is collected from Indian Meterological Department. Multi-seasonal scenes of the land sat 8 (OLI) and landsat 5 thematic map (TM) images were collected from the official United State Geological Survey (USGS) LANDSAT 5TM and LANDSAT 8OLI scenes (path 144, row 48), for the years 1990 and 2015 pertaining to the region, are used for the identification of Land Transformation. To prepare relief and slope maps the global 1°S (30 meter) DTM satellite scenes were collected from the Japanese data base of ASTER L-1 (Advanced Space Borne Thermal Emission and Reflection Radio Meter) Satellite. Climatic data is collected from IMD for a period of 50 years. Demographic and socioeconomic data is obtained from the published records, Directorate of Economics and Statistics and conducted Household Survey through Questionnaire. GIS tools like Arc GIS v10.1 and ERDAS imagine v2013 were used for Land Transformation mapping. MS Excel and IBM SPSS v22 are used to analyze the statistical data and climatic data.

Methodology: ***Preparation of base map:*** To prepare base map 1:50,000 scale toposheets bearing number 56k2 to 56k16 (excluding 56k13) are scanned and a raster five for study is created then all these toposheets are georeferenced based on the longitudinal and latitudinal coordinates. Georeferencing is carried out using ERDAS Imagine processing software. The georeferenced toposheets are further mosaiced which represents the continuity of the study area. The study area has been clipped out from this

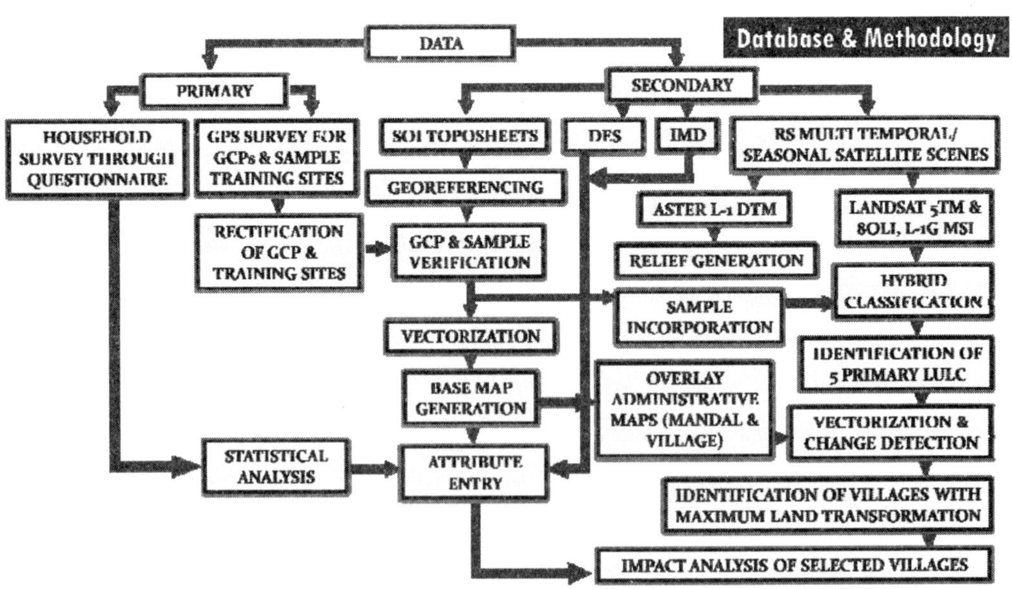

mosaiced toposheets. This base map is reprojected on WGS84 geodetic datum, universal transverse Mercator UTM map projection (UTM, ZONE 44N). The present study area contains erstwhile 4 districts of Telangana state namely Rangareddy, Medak, Nalgonda and Mahabubnagar, which have 38 revenue mandals. (2015).

Image processing software i.e. ERDASV2013 has been used for supervised classification and unsupervised classification. Each image has been classified into 5 different categories i.e Built-up area, water bodies, forest area, agriculture land and other land. For 1990 image unsupervised classification has been performed whereas on 2015 image both the classifications have been performed. 5 training sites (for supervised classification) per land use category were selected using GPS instrument (Triamble) and the values were merged to get accurate category. The spectral signatures then were given conventional colours. ArcGISV2013 software has been used for the digitisation, intergration, overlay and presentation of spatial and non–spatial data of land use change in the city region.

Land Use/Land Cover Change: A Spatio-Temporal Analysis (1990-2015): Five land use classes of agriculture, built-up, forest cover, other land and water body have been identified through digital image processing as explained in methodology, the Table 9.1 gives the statistics of land use/land cover change from 1990 to 2015.

Table 9.1: Land Use/Land Cover Change in 1990 to 2015

Sl. No.	*Land Use*	*Area Km² 1990*	*%*	*Area Km² 2015*	*%*	*Change (Area)*	*% Change*
1.	Agriculture	872.97	20.19	407.6	9.43	-465.37	-10.76
2.	Built-up	39.05	0.9	397.854	9.2	358.8	8.3
3.	Forest Cover	243.99	5.64	284.54	6.58	40.55	0.94
4.	Other land	3026.54	70	3171.7	73.36	145.15	3.36
5.	Waterbody	141.19	3.27	62.06	1.43	-79.13	-1.84
	Total	**4323.75**	**100**	**4323.75**	**100**	**0**	**0**

From the table above, it is clear that other land use class in 1990 constitutes 70 per cent of total geographical area, spreading over 3026.54 km². Agriculture land use class stands in the second position, which constitutes 20.19 per cent of total geographical area. In the third place is the forest area with an aerial extent of 243.99 km², constituting 3.27 per cent of total geographical area. Waterbodies constitute 3.27 per cent of total geographical area spreading over 141.19 km². Built-up area constituted only 0.9 per cent of total geographical area (Fig. 9.5).

When compared to the figures of 2015, there has been a marked change in land use. Agriculture area decreased by 465.37 km², accompanied by a decrease of water body by 79.13 km². Contrary to this, there is an increase of 358.80 km². contrary to this, there is an increase of 358.80 km² of built-up area, along with a marginal and slight increase of 0.94 per cent of forest area between two time periods (1990-2015). Other land too increased by 3.36 per cent (Map 9.6). Although these figures indicate a simple change of land use classes, but the actual change from which class is not depicted. For this one more table is created which indicates the actual land transformation. The Table 9.2 depicts the same.

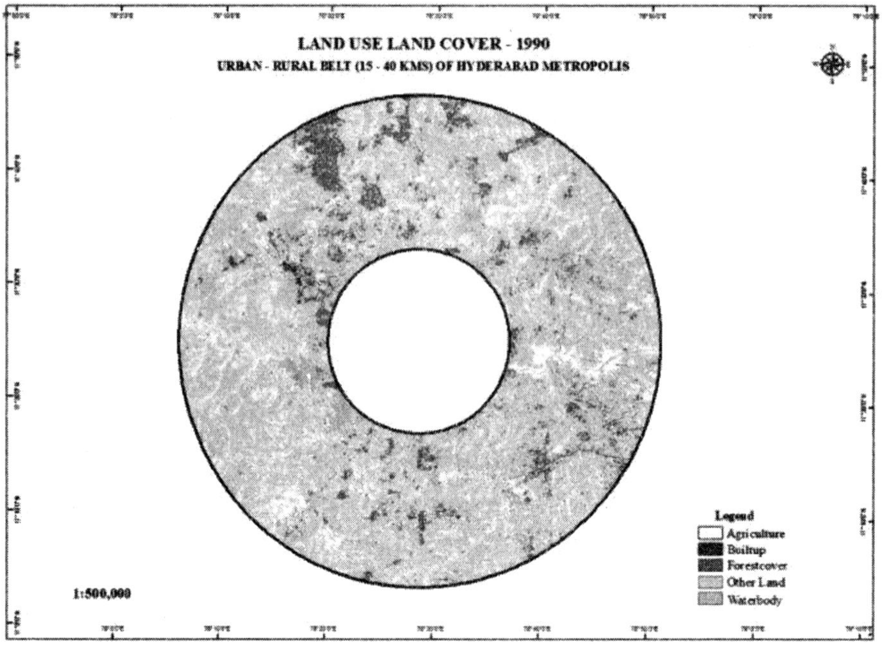

Fig. 9.5: Landuse/Land Cover–1990: Urban-Rural Belt (15-40 KMS) of Hyderabad Metropolis

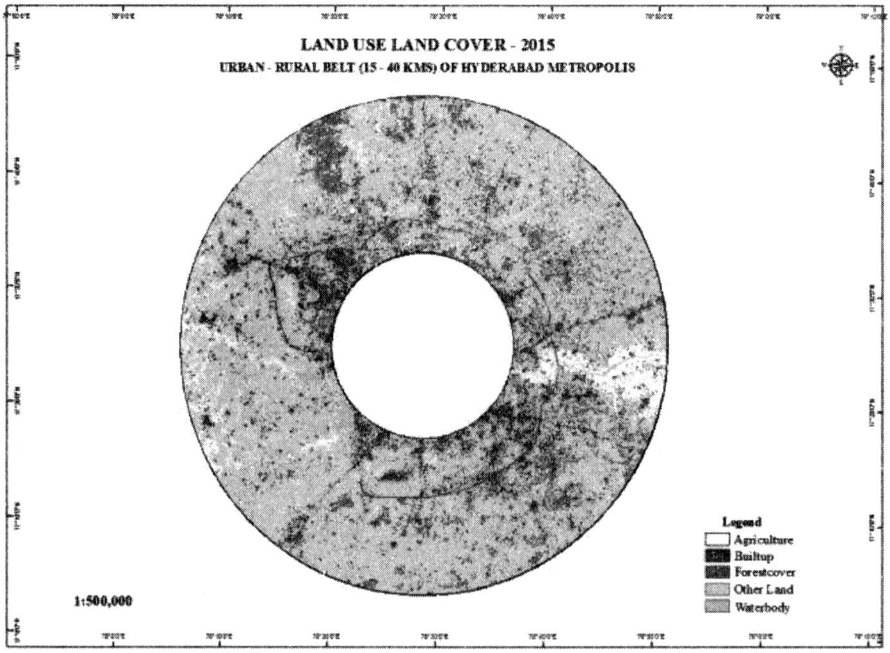

Fig. 9.6: Landuse/Land Cover-2015: Urban-Rural Belt (15-40 KMS) of Hyderabad Metropolis

Table 9.2: Land transformation in the study area

Changed Category from others	*Area*	*Changed Category from others*	*Area*
Other land to Built up	286.20	Waterbody to Agriculture	14.40
Other land to Forest cover	112.67	Waterbody to Builtup	5.31
Other land to waterbody	8.66	Waterbody to Forest cover	3.04
Other land to Agriculture	196.17	Forest cover to waterbody	0.93
Agriculture to Built up	59.28	Forest cover to Agriculture	4.93
Agriculture to Other land	582.90	Forest cover to Builtup	8.00
Agriculture to forest cover	28.63	Forest cover to Otherland	89.90
Agriculture to waterbody	10.07	**Total**	**1487.14**
Waterbody to other land	76.05		

Changed Category from same	*Area*	*Changed Category from same*	*Area*
Agricultural to Agriculture	192.09	Other land to other land	2422.85
Waterbody to waterbody	42.40	Builtup to built up	39.05
Forest to Forest	140.21	**Total**	**2797.55**

While 1487.44 km² of study area has undergone change, 2797.55 km² is without any change, retaining the original status. 192 km² of agriculture land, 42.396 km² of waterbody, 140.20 km² of forest area, 2422.85 km² of other land and lastly 39 km² of built-up area have not changed between 1990-2015. Maximum change can be evidenced in Agriculture to other land transformantion by 585 km² between 1990-2015. Similarly 286 km² of other land got converted to builtup. At the same time 112 km² of other land got converted to forest cover. At the same time 196 km² of other land got converted to agriculture land in a significant way. 59 km² of agriculture area got converted as built up land, whereas 76 km² of water body changed to other land (Fig. 9.7).

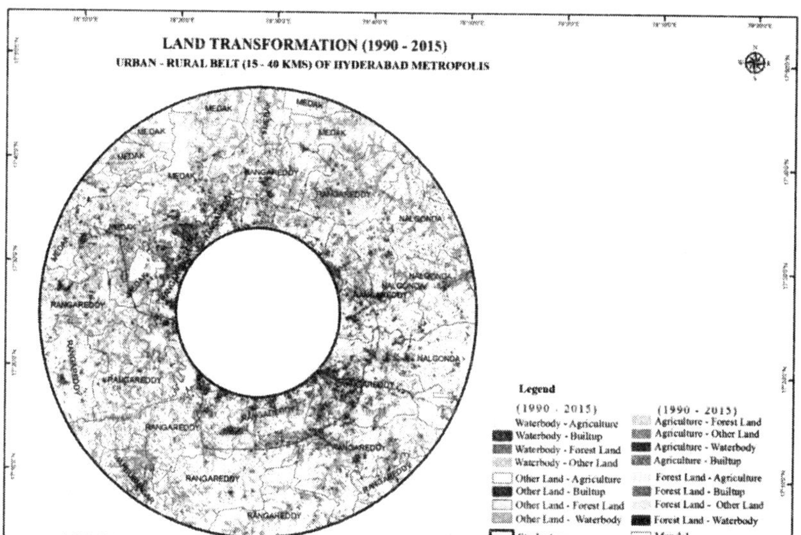

Fig. 9.7: Land Transformation (1990-2015): Urban-Rural Belt (15-40 KMS) of Hyderabad Metropolis

Land Transformation (1990-2015)

Land use/land change changes are found to become more complex and intricate, where more human interaction with the land are more change in increasing built-up area strikes us as the most obvious changed land use, followed closely by agriculture land and waterbody (Table 9.3).

Table 9.3: Land Transformation (1990-2015)

LULC Classes	*Agriculture*	*Builtup*	*Forest*	*Waterbodies*	*Otherland*	*Total Area 2015*	*Change*
Agriculture	872.97		4.93	14.404	196.162	407.604	-465.367
Builtup	59.275	39.05	8.0085	5.3067	286.195	397.846	358.796
Forest	28.626		243.99	3.0383	112.672	284.54	40.55
Waterbodies	10.0714		0.9306	141.191	8.664	62.062	-79.128
Otherland	582.8978		89.908	76.045	3026.54	317.697	145.15

Note:

1. Figure diagonally in bold area under particular land use/cover class in 1990
2. Figures in the rows of the land use/cover class are addition to that class during 1990-2015.
3. Figure in the columns of the land use/cover class are transformation (subtraction) from that class during 1990-2015.

Transformation of built-up Area

In any type of land use studies, it is evidenced that the built up area especially around city regions has been showing significant growth. The increase in builtup area in terms of urban settlement area. Commercial land use, transportational land use, recreational land use, entertainment land use and other minor and equally important land use categories can be identified, which are service oriented segments. These activities need space and therefore grow or expand on near by and convenient places which are usually agriculture lands, dry water beds, fallow lands or forest cover.

The built-up in the present study area increased by 358.8 km^2 capturing land from other land use class (286.2 km^2). Agriculture (59.28 km^2), forest area (8.0 km^2) and waterbody (5.31 km^2) spatially, Saroornagar, Shamshabad, Ibrahimpatnam mandals in the southern side, Gatkesar and Bibinagar mandals in eastern side, Rajendernagar and Qutbullapur mandals in northwestern side and Medchal in northern side.

Transformation of Agricultural Land

Agriculture land often tend to be lost either due to natural calamities [like decrease in rainfall and increase in temperature or due to human intervention. Mismanagement of agriculture land or lack of scientific farming too curb agriculture activities. Added to this, there is a heavy demand for real estate business and urban expansion based ventures, where the farmers are lured into selling of their land and move out. Accessibility and nearness to waterbodies too come as advantages. Sometimes agriculture lands turn into fallow lands or shrub lands. In a rurban zone there is a heavy demand for agriculture land.

Area under agriculture in the study area has decreased from 872.97 km^2 to 407.602 km^2. It mainly lost area to other land (582.89 km^2), Builtup (59.28 km^2), forest (28.27 km^2) and waterbody (10.07 km^2). Area under agriculture has also gained from other land (196.16 km^2), waterbody (14.40 km^2) and forest area (4.93 km^2). Overall transformation of agriculture can be seen maximum in North-eastern, North-western and South-western part of the study area, especially in Shamirpet, Patancheru, Kothur Shankarpalle and Moinabadmandals.

Transformation of Other Land

Transformation of other lands can be easily accountable, as these can easily be changed due to their non-ownership status. A certain amount of managerial skills need to be employed before they are changed or put to some utility. In the study area, other land use increased slightly from 3026.54 km^2 in 1990 to 3171.70 km^2 in 2015. It has increased by 145.15 km^2 capturing land mainly from agriculture (582.89 km^2), forest area (89.90 km^2), and waterbodies (76.045 km^2). At the same time other land use class also cost area to builtup area (286.2 km^2), agriculture (196.16 km^2), forest (112.67 km^2) and water body (8.66 km^2). Thus, other land can easily be converted to other land use classes.

Conversion of other land into agriculture can be seen maximum in eastern, western and northwestern direction of the study area in Pochampalle, Chevella, Shankarpalle, Rangareddy, Patancheru and Jinnaram mandals. Conversion of other land into builtup land can be seen maximum in Hayathnagar, Ibrahimpatnam, Saroornagar, Shamshabad, Rajendra Nagar, Qutbullapur, Medchal and Ghatkesar mandals. Conversion of other land use class into forest land is being seen maximum in southern, northern and eastern side of the study area. Especially in Ibrahimpatnam, Manchal, Yacharam, Kandukur, Maheswaram mandals of southern side and Shamirpet of northern side and parts of Hayatnagar mandal of eastern side. Conversion of other land use class into waterbody can be seen in very few places like Kothur, Sangareddy, Rajendranagar, Pochampalle, Bibinagar and Patancheru mandals.

Transformation of Forest Cover

Forest cover increased by 40.55 km^2 capturing land mainly from other land use class (112.67 km^2), agriculture (28.63 km^2) and water body (3.03 km^2), lost to built up (8 km^2), agriculture (4.93 km^2), other land (89 km^2) and water body (0.93 km^2) conversion of forest area into agriculture area can be seen maximum in Sanroornagar, Qutbullapure and Shamshabadmandals. Conversion of forest area into other land use can be seen maximum in Saroornagar, Shamshabad, Maheswaram, Medchal, Mulugu, Rajendernagar mandals. Conversion of forest area into built-up can be observed maximum in Saroornagarand Hayatnagarin south, Rajendranagar in west, Gatkesarin east, Qutbullapurand Medchalin north. Conversion of forest area into water body can be observed in Bibinagar, Chotuppal, Hayatnagar, Rajendranagar and Pochampalle Manadals.

Transformation of Waterbody

Waterbody decreased by 79.12 km^2 and lost area to other land (76.045), agriculture (14.04

km^2), builtup area (5.35 km^2) and forest (3.03 km^2). It has also gained area from agriculture (10.07 km^2), other land use (8.66 km^2) and forest area (0.93 km^2). Transformation of water body into agriculture area can be seen near Himayatsagar, Osman Sagar, Ibrahimpatnam and Shamirpet lakes in Chevella, Shankerpet, Shabad, Pochampalle and Malkajgiri mandals. Transformation of waterbody into other land use class can be seen maximum in Maheswaram, Ibrahimpatnam, Bibinagar, Shamirpet, Medchal and Rajendranagarmandals. Transformation water body into built up is observed in Saroornagar, Bhongir, Bibinagar, Maheswarammandal. Conversion of waterbody into forest area can be seen maximum in Chotuppal Malkajgiri, Shivampet and Narsapurmandals.

Impact of Land Transformation

A perception to obtain individual responses and perspectives a field survey was conducted in the study area, from which the following views are derived:

1) Land transformation did have an impact on agriculture land and farming sector got impacted a little.
2) the impact of land transformation on migration is negligent due to dispersed nature of urban amenities and well connected road network.
3) due to increase in builtup activity, there were apprehensions regarding deteriorating environmental conditions.
4) over all, land transformation seem to have less impact on available agriculture activities.
5) Non functionality of agriculture lands is attributed to increase in demand for real estate and commercial lands which is an obvious fact. For farmers, this has been a lurative business to sell their lands.
6) spatial extent of green cover has been impacted by increase in built up activity.
7) Agriculture productivity has declined due to erratic rainfall and mismanagement of the soil.
8) Under land transformation process, a few are not mindful of losing the possession of agriculture land. Whereas few are still attached to land.
9) As the agriculture based dependency is more in the region, conversion of agriculture lands will have an impact on livelihood options of the residents. These may be put to stress if the agriculture based activity gets minimal. Even daily wage labourers are more.
10) In spite of land transformation, good and balanced economy exists in the region. Land transformation seem to have boosted the economic development of the region.
11) Education and job stand out as the major preference for their future economic growth. Improving the state of present agriculture is also one of the preferences. This is keeping in view the demand for skilled workers and also the general awareness regarding the importance and relevance of education in various service sectors for which Hyderabad is famous for.
12) There has been a general betterment and improvement in standard of living of the respondents in spite of land transformation.

13) land transformation has bettered/or improved infrastructural facilities in the region.
14) 80 per cent of the respondents were of the view that the present village status was better than a possible conversion to a city or town in near future. These are willing to continue with their traditional agriculture activities along with the peace and tranquility of the village life. They are scared of pollution levels in the city. They fear land transformation will impact their environs.
15) Future land transformation is accepted provided the village environs are safe and healthy. The respondents are of the view that they should be the beneficiaries either way, i.e. they want to take advantage of urban growth process and at the same time also want the serenity of village surrounding.
16) Respondents wanted a more regulated land transformation under well laid out government rules.
17) Overall, land transformation seem to have not much of a drastic impact on socio economic sector of the study area, but it is observed that the residents are becoming aware of environmental loss especially in terms of air pollution due to land transformation.
18) In a way, land transformation seem to have benefitted the residents of urban-rural belt of Hyderabad.

It is observed from the respondents that they wanted to continue with their present occupation, i.e., farming, though agriculture production declined due to natural and manmade causes. They preferred village life with urban facilities or in other words Provision of Urban Amenities in Rural areas (Pura) seem to be the expectation of suburban dwellers of Hyderabad metropolis.

References

Abdelkader El Garouani, David j Mulla, Said El Garouani, Joseph knight (2017). Analysis of Urban Growth and Sprawl from Remote Sensing data: Case of Fez Morocco. *International Journal of Sustainable Built Environment* (2017) 6,160-169.

Abdur Raziq, Aigong Xu, Yu Li, Quanhua zhao., (2016). Monitoring of Land use/Land cover changes and Urban sprawl in Peshawar City in Khyber Pakhtunkhwa: An Application of Geo-information Techniques Using of Multi –Temporal Satellite Data. *Journal of Remote Sensing* @GIS.

Aggrey Daniel Maina Thuo., (2010). Community and social responses to land use transformations in the Nairobi. *Rural-Urban fringe*, Kenya. Speciel Issue 1/2010.

Andrea E. Gaughan, Michael W. Binford, Jane Southworth., (2008). Tourism, forest conversion and Land Transformations in the Angkor basin, Cambodia, Applied Geography xxx (1–12).

Arshad Amin, Shahab Fazal., (2012). Land Transformation Analysis Using Remote Sensing and GIS Techniques (A Case Study). *Journal of Geographic Information System*, 4, 229-236 doi:10.4236/jgis.2012.43027 Published Online June 2012.

Baldev Sahai., (1988). Remote Sensing in Land Transformation and Management, *Journal of the Indian Society of Remote Sensing*, Vol. 16, No. 4, 1988.

Bryan C. Pijanowski, Daniel G. Brown, Bradley A. Shellito, Gaurav A. Manik., (2001). Using neural networks and GIS to forecast land use changes: A Land Transforamtion Model. WWW.elsevier.com/locate/compenvurbsys.

Census of India, 1991 and 2011.

Erli Ellis, Robert Pontirus., (2010). Land use and Land Cover Change. A Resource from CAMEL Climate Adaptation Mitigation E-Learning. Published April 18, 2010, 3.06 pm. updated May 7, 2012, 4.59 pm.

Faridkhan, Vijayalakshmi, T., Santosh Kumar., (2012). Land use land cover change detection of Ghatkesar mandal, Ranga Reddy district using Remote sensing and GIS. *International Journal of Scientific and Engineering Research*, Volume 3, Issue 11, November-2012 1.

Farooq, S., Ahmad, S., (2008). Urban Sprawl Development around Aligarh City: A Study Aided by Satellite Remote Sensing and GIS. *Journal of India Society of Remote Sensing*, 36, pp. 77-88.

Fazal, S., Amin, A., (2011). Impact of Urban Land Transformation on Water Bodies in Srinagar City, India. *Journal of Environmental Protection*, Vol. 2, No. 2, pp. 142-153.

Fazal, S., (2000). Urban Expansion and Loss of Agricultural Land: A GIS Based Study of Saharanpur City, India. *Environment and Urbanization*, Vol. 12. No. 2, pp. 133-149.

Gautam, N.C., (2015). National Land Use/Land Cover Classification System, CLUMA Publication Series-13, 2015.

Handbook of Statistics (2011). Mahabubnagar District: Chief Planning Officer, Government of Andhra Pradesh.

Handbook of Statistics (2011). Medak District: Chief Planning Officer, Government of Andhra Pradesh.

Handbook of Statistics (2011). Nalgonda District: Chief Planning Officer, Government of Andhra Pradesh.

Handbook of Statistics (2009-10). Ranga Reddy District: Chief PlanningOfficer, Government of Andhra Pradesh.

Hussien Ali Oumer., (2000). Land Use and Land Cover Change, drivers and its impact: A comparative study from Kuharmichael and Lenchedima of Blue Nile and Awash Basins of Ethiopia, A thesis presented to the Faculty of the Graduate School of Cornell University in partial fulfillment of the requirements for the degree of master of professional studies.

Ifatimehin, Olarewaju Oluseyi, Musa, Salihu Danlami, (2009). An Analysis of the Changing Land Use and Its Impact on the Environment of Anyigba Town, Nigeria.

Kausalya Ramachandran, Padmaja, S., (2008). Envoronmental migration from Rain-fed regions in India forces by poor returns from Watershed Development Projects. *International Conference on Environement, Forced Migration and Social Vulnerability* (EFMSV 2008) Oct. 9-11-2008, UNU, Bonn, Germany.

Kausalya Ramachandran, Padmaja, S., (2005). Evaluating Ecological Sustainability of Cultural Landscapes in Southern India. *The Jahangirnagar Review*, Part II: *Social Science*. Vol. XXVIII, 2004, Printed in June 2005. Jahangirnagar University, Savar, Dhaka-1342, Bangladesh. pp. 1-14.

Kausalya Ramachandran, (2004). Land Use Changes and Long term Outlook for Hyderabad Urban Agglomeration. Indi-Norwegian Workshop on ' Sub-surface Assessment of contaminated sites' organised by NGRI, Hyderabad, 9-11 December, 2004.

Lata, KM., Shankar Rao, CH., Krishna Prasad, B., Badrinath, B.S. and Raghav Swamy, V., 2001. "Measuring Urban Sprawl-A Case Study of Hyderabad", *GIS Development*, 5(12).

Leonard Gaydos, William Acceredo, Cindy Bell., (2007). *Human Induced Land Transformation*. Oxford University Press.

Mala Mukherjee, (2012). Urban Growth And Spatial Transformation of Kolkata Metropolis: A Continuation of Colonial Legacy. *APRN Journal of Science and Technology*, Vol. 2, Special Issue. ICESR.

Madhavi Lata, K., Krishna Prasad, V., Badarinath, K.V.S., Raghavaswamy, V., (2009). Measuring urban sprawl: A case study of Hyderabad. National Remote Sensing Agency (NRSA), Department of Space Government of India, Hyderabad, September Geospatial World.

Manisha Garg, Arun Joshie, Sandhya Choudhary, (2013). Land Use/Land Cover Classification in Jamuna-Kotma Coal Field Region, Anuppur District MP. *IOSR Journal of Agriculture and Veterinary Science (IOSR-JAVS)*. Volume 6, Issue 2 (Nov.-Dec. 2013), pp. 38-43.

Meyer, W. B., Turner, B. L., (1992). Human Population Growth and Land Use/Cover Change. *Annual Review of Ecology and Systematics*, Vol. 23, pp. 39-61.

Monalisha Mishra, Kamal Kant Mishra, Subudhi, A.P., (2006). Urban Sprawl mappimg and Land use Chnage Analysis Using Remote Sensing and GIS monalishaiirs@gmail.comand monalisa_bhu2006@yahoo.co.in.

Murali Krishna Gumma, Irshad Mohammad, Swamikannu Nedumaran, Anthony Whitbread, Carl Johan Lagerkvist, (2017). Urban Sprawl and Adverse Impacts on Agricultural Land: A Case Study on Hyderabad, India. *Remote Sens.* 2017, 9, 1136; www.mdpi.com/journal/remotesensing.

Nasreenislamkhan, (2000). Temporal mapping and spatial analysis of land transformation due to urbanization and its impact on surface water system: A case from Dhaka metropolitan area, Bangladesh. *International Archives of Photogrammetry and Remote Sensing*. Vol. XXXIII, Part B7. Amsterdam.

Neelakantan, K., Sumati Kulkarni, Raghavaswamy, V., (2007). Economy, Population and Urban Sprawl: A Comparative Study of Urban Agglomerations of Bangalore and Hyderabad, India Using Remote Sensing and GIS Techniques, Paper to be presented at PRIPODE workshop on Urban Population, Development and Environment Dynamics in Developing Countries. Jointly organized by CICRED, PERN and CIESIN with support from the APHRC, 11-13. June, Nairobi, Kenya.

Nobi, E.P., (2009). Land Use and Land Cover Assessment along Pondicherry and its Surroundings Using Indian Remote Sensing Satellite and GIS, *American-Eurasian Journal of Scientific Research* 4 (2): 54-58, ISSN 1818-6785.

Prakasam, C., (2010). Land Use And Land Cover Change Detection Through Remote Sensing Approach: A Case Study of Kodaikanal Taluk, Tamil Nadu. *International Journal of Geomatics and Geosciences*, Volume 1, No. 2.

Priya Narayan, Ashok D. Hanjagi, (2009). Land Transformation: A threat on Banagalore's Ecology – A Challenge for Sustainable Development. *Theoretical and Empirical Researches in Urban Management*, Speciel Number 1s/April: Urban Issues In Asia.

Rither., Aggrey Daniel Maina Thuo, (2010). Community and social responses to land use transformations in the Nairobi rural-urban fringe, Kenya. The Journal of Field Actions.

Roger LeBHooke, José F. Martín-Duque, Javier Pedraza, (2012). Land Transformation by humans: A review. *GSA TODAY*, December 2012.

Roy, P.S., Giriraj, A., (2008). Land use and Land cover Analysis in Indian Context. *Journal of Applied Sciences* 8(8):1346-1353,2008. Asian Network for Scientific Information.

Saroj. M.P. Sharma, (2015). Land use/Land cover change detection using Remote Sensing @ Gis in Rohtak and Jhajjar Districts of NCR, Haryana, India. *International Journal of Science, Engineering and Technology Research* (IJSETR), Volume 4, Issue 1, January.

Shahab Fazal, (2001). 'Land reorganisation in relation to roads in an Indian City", *Pergamon, Landuse Policy*. pp. 18, 19-199.

Shahab Fazal, Arshad Amin, (2012). Impact of Urban Land Transformation on Water Bodies in

Srinagar City, India, *Journal of Environmental Protection,* 2011, 2, 142-153 doi:10.4236/jep.2011.22016 Published Online April 2011 (http://www.SciRP.org/journal/jep).

Shahab Fazal, Stan C.M. Geertman, Fred J. Toppen, (2012). Interpretation of Trends in Land Transformations—A Case of Green Heart Region (The Netherlands), *Natural Resources,* 3, 107-117. doi:10.4236/nr.2012.33015 Published Online September 2012.

Shashi Chawla, (2012). Land Use Changes in India and its Impacts on Environment, *Journal of Environment,* Vol. 01, Issue 01, pp. 14-20.

Sudhira, HS., Ramachandran, TV., (2003). "Urban Sprawl Pattern Recognition and Modeling using GIS", www.gisdevelopment.net/application/urban/sprawl/mi03142.html.

Sundarakumar *et al.,* (2012). Land Use and Land Cover Change Detection and Urban Sprawl Analysis of Vijayawada City Using Multitemporal Landsat Data, *International Journal of Engineering Science and Technology (IJEST),* vol. 4 no. 01, pp. 171-178, ISSN: 0975-5462.

Sundarakumar, K., Manoj Kumar Kaluri, Leela Gouthami Gajula, Sudheesh Vanguru, Shanmuka Sai Gutta, (2015). Study of Land Use Changes and Urban Expansion ofVijayawada City using Remote Sensing Data. *International Journal of Innovative Research in Advanced Engineering (IJIRAE),* Issue 3, Vol. 2 (March).

Vemusreenivasulu *et al.,* Pinnamaneniudayabhaskar, (2010). Change Detection in Landuse and Landcover Using Remote Sensing and GIS Techniques. *International Journal Of Engineering Science and Technology,* Vol. 2(12), 7758-7762.

Virendra Kumar, Kamlesh Bhalvai, Bhawna, Dhananjay V. Rajamani, (2015). Land Transformation Analysis Using High Resolution Satellite Data, GIS and GPS Techniques. Volume 2, Issue 1, June 2015, pp. 25-33.

William Acevedo, Keith Clarke, (1994). An Analysis of Human-Induced Land Transformations in the San Francisco Bay/Sacramento Area.https://www.researchgate.net/publication/236355886.

Wolman, M.G., Fournier, F.G.A., (1987). Introduction to Land Transformation in Agriculture. Edited by M.G. Wolman and F.G.A. Fournier John Wiley and Sons Ltd.

Yeh, A G O., and X, Li., (2001). "Measurements and Monitoring of Urban Sprawl in a rapidly Growing Region using Entropy", *Photogrammetric Engineering and Remote Sensing,* 67(1), pp. 83-90.

10
Evolution of Tilla-lunga Topography, West Tripura, India

Subhajit Sen, Saptarshi Mitra and *Sunil Kumar De*

Introduction

The term *tilla-lunga*[1] is the regional terminology, which indicates particular topography found in the synclinal valley of the Tripura Fold Belt. The *tilla*[2] and *lunga*[3] indicate the relatively high land and low land respectively. The elevation of the topography is gradually decreasing from the Baramura Hill towards west and merged with vast plain of the Bengal Basin (Sen *et al.*, 2015).

The recent works of DeCelles *et al.* (1988), Michael *et al.* (2009), Islam (2009) shows that massive weathered sediments carried out by rivers from the Himalayas are deposited in the flood plain of the Bengal Basin. Dasgupta (1978) identified these sediments as Pleistocene or Post Tipam sediments, which is also found in the western valley part of Tripura. Due to neo-tectonic activity and orogeny of the Chittagong-Tripura Fold Belt (CTFB), the drainage system of the whole Bengal Basin, its flow path and deposition pattern has undergone massive changes in different geological periods. One of the significant impacts is the planform change of recent rivers and sea level rise that changes the configuration of the continental environment (Goodbred *et al.*, 1999; Bandyopadhyay *et al.*, 2013). In the same year Uddin *et al.* (1999) documented the evidences of major paleo drainage system similar to the present-day Brahmaputra River that may be "Paleo-Brahmaputra". The early river system is the indicator of paleo-environment (Willis, 1993 and Kale *et al.*, 1996).

Morphologically, Tripura Fold Belt is characterised by ridge and valley topography (Fig. 1.1). Physiography of the district is classified into four landform units, i.e., structural ridges (on Neocene rocks), dissected higher terraces (on Pleistocene sediments), which is called *tilla*, valley flats (on Holocene terrace sediments) and flood plains (on present-day flood plain deposits) are *lunga* (Rajesh, 1981).

The distribution of *lungas,* as identified in satellite images, represents a deep association with the present drainage system. The geomorphology of *tilla-lunga* topography needs to be elaborated for understanding the evolution of the landform in

1. *Tilla-lunga* is the unique evidence of the dried-up drainage basin and denuded landform in the western synclinal valley of Tripura Hill.
2. *Lunga* is the dried up channels. The *chara* are the trace of the surface water flow lines.
3. *Tilla* comprises of massive graded sequence. The *tillas* and *lungas* represent as a rib and furrow structure.

the eastern portion of Bengal Basin (Chittagong Tripura Fold Belt). There is a knowledge gap between the orogeny of Tripura Fold Belt and present-day landforms. The objectives of present work are firstly, to identify morphology of *tilla-lunga* topography and secondly, explain the process involved in it.

Geological Setting

The eastern portion of the Bengal Basin is endowed with long history of basin fill and upliftment of the Chittagong-Tripura Fold Belt. Tripura State, one of the North-eastern states of India, falls under Eastern Geotectonic Province of the Bengal Basin. The configuration of sedimentary fill up of Bengal Basin took place in Early Eocene, Middle Miocene and Plio-Pleistocene (Alam *et al.*, 2003). The Indian Plate remains dynamic throughout the Tertiary and the Quaternary Period. It was drifted at the rate of 18-20 cm/year in the early Tertiary period, later it decreased to 10 cm/year (Curray *et al.*, 1974). The recent work of Jade *et al.* (2007) documented different rate of drifting of the Indian Plate. Now it behaves like a single rigid plate. Some published works of Smith *et al.* (1970), McKenzie *et al.* (1971), Falvey (1974), Lee *et al.* (1995), Varga (1997), Acharyya (1998) and Sen *et al.* (2014) discussed about the micro to meso scale landform evolution. In the western synclinal valley several concealed relict domal structures e.g. the Agartala Dome, Kunjaban Dome, Rokhia, Tichna and Gojalia anticlines are found. These are the evidences of paleo-geomorphic height (Jena *et al.*, 2011).

The Chittagong Tripura Fold Belt originated through the sedimentary basin fill process and upliftment of the large sedimentary rocks. The fold is broad arc shaped and convex towards west following the curvature of the Indo-Burma orogeny. In coastal Burma and Chittagong Hill tracts the fold axes trend NNW-SSW; the trend being almost N-S in the Tripura-Mizoram Hill and from there the trend changes to NNE-SSW to NE-SW in the Cachar valley. Further north, in the vicinity of the Dauki Fault there are two E-W trending anticlines, the Sylhet and Chhatak structures. Many of the anticlines fragmented to form two anticline ridges with sub-parallel axial trends. In some cases, they rejoined to form elliptical synclinal valley within them. Most of the folds are doubly plunging with low regional plunge (Dasgupta, 1984). There are four major stratigraphic units in Tripura, such as (1) Barail Group (Oligocene; ~3000 m) consisting of sandstone and shale rock, (2) Surma Group (Miocene) consisting mainly of (a) Arenaceous Lower Bhuban Formation (~9000 m), (b) Argillaceous Upper Bhuban Formation (~1100 m) and (c) Argillaceous Bokabil Formation (~1000 m), overlain by (3) Tipam Group (Pliocene; ~1300 m) consisting of feldspathic sand with fossil and minor clay, fine silt and lateritic soil occupies the synclinal valleys and overlies the (4) Post Tipam (Dupitila) over a pronounced unconformity. Later, Jena *et al.* (2011) mentioned that Dupitila Formation (thickness 1000 m) belongs to Pleistocene age and originated from fluvial deposition. Tipam Formation (thickness 1500-1700 m) is also originated from fluvial deposition and belongs to Mio-Pliocene age. The Suma formation is originated from shallow marine deposits and is found in the anticlinal portion of Tripura Fold Belt.

The present study has been carried out in the West Tripura District (Latitudes 23°42" to 24°45"N and Longitudes 91°15" to 91°35"E; 983.63 sq. km) of Tripura State in India (Fig. 10.2). Physiographically, it is a fringe zone between the flood plain of the Bengal Basin and the Open Tripura Fold Belt. Major rivers are Lohar Nala and Haora originated from the Baramura Hill flow westward and merged with the Titas River in Bangladesh. Agartala City is located along the bank of the Haora River.

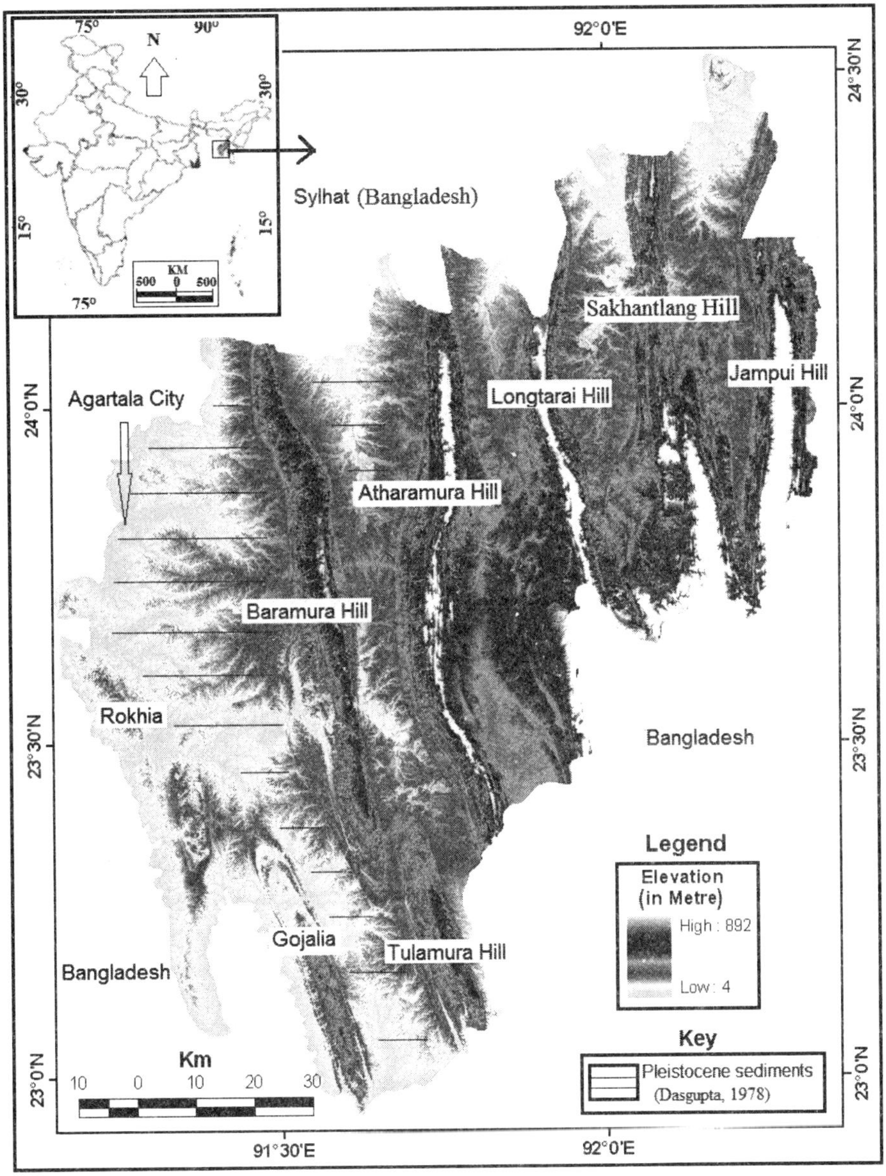

Fig. 10.1: The physiographic set-up of the Tripura Fold Belt extracted from the SRTM DEM indicates the north-south oriented major hill ranges. The Agartala Valley lies in the western portion of the Baramura Hill.

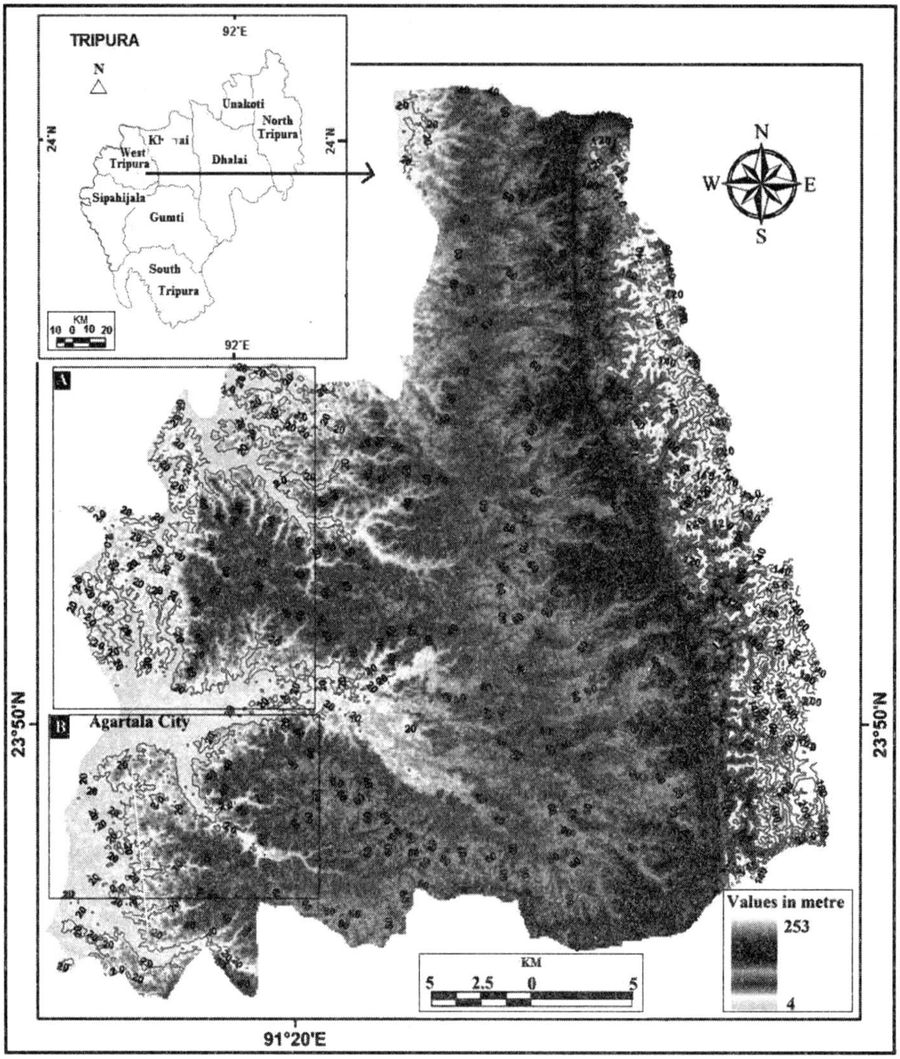

Fig. 10.2. The SRTM DEM of West Tripura District, Tripura represents location map of the study area. The focused portion of the northern (A) and southern (B) of the Haora-Kata Khal flood plain is represented in the large scale for better understanding of the spatial distribution pattern of *tilla* and *lunga* distinctly.

Materials and Method

SRTM DEM (90 m) and Survey of India toposheets (79M/1, 79M/5, 79M/6, 79M/9, 79M/10 of 1930-31; R.F 1:63,360) are used for the preparation of the geomorphology maps. GPS-receiver is used for identifying the location of the study sites.

The morphology of *tilla-lunga* topography has been studied from geomorphology map and cross-section analysis. The large scale geomorphology map provides a preliminary platform for the landscape planners (e.g. Syvitski *et al.*, 1999, 2012; Magilligan *et al.*, 2002; Griffiths *et al.*, 2008). The morphological investigation has wide range of implementation in the evolution of river system and riverine landform. As the category of *tilla-lunga* topography was not known earlier, the symbols provided by Gustavsson *et al.* (2006) has been followed for the interpretation of this peculiar topography.

The geomorphic process has been studied through litho-facies association. The new railway line has been constructed by cutting *tilla*, provide fresh outcrop of *tilla*. For this reason, the sedimentary structures and their association has been plotted with the litho-morphic sections prepared from satellite images and dumpy level survey. The thickness of the beds is measured with the help of traditional instruments. Paleo-flow direction has been measured by Clinometers and Brunton compass. The works of Miall 1973, 1977 and 1985 has been followed for the interpretation of fluvial facies.

Classification of Landforms

The *tilla* and *lunga* are genetically divided into depositional and erosional landform respectively. The *lunga* is morphologically younger than *tilla*. Landforms of the study area are divided into first and second order landform units on the basis of morphology. The first order landforms are (1) Baramura Hill, (2) *tilla,* (3) *lunga* and (4) flood plain. The Baramura Hill exists above 80 m. *Tilla* and *lunga* exist 20 m to 80 m and flood plains are below 20 m. In the lower portion of the topography (close to Indo-Bangladesh border), it is difficult to distinguish *lunga* from the flood plain, because their altitudinal variation is negligible. That is why *lungas* in this part are mostly flooded during rainy season. *Tilla* is further classified into (2.1) upper *tilla* (above 60 m), which comprises of massive graded sequence and lithified mottled clay. (2.2) middle *tilla* (40 m to 60 m) comprises of graded sequence, (2.3) lower *tilla* (below 20 m) dominantly flood deposit. (4) Flood plain comprises of massive clay (4.1) present river valley and (4.2) Chara valley.

Form and Pattern of *Tilla*

The geomorphology map represents different geomorphic units through symbols. In the northern portion of the Lohar Nala valley and the Haora River valley are separated by East-West extended structural ridge (rib structure) of upper *tilla* (Fig. 10.3). The erosion scarp has been marked in the western edge of this rib structure. The upland area is termed as Kunjabon Dome that leads to centrifugal drainage network. In the western portion levee deposits (elevation <20 m) have been identified in the intra-*lunga* areas. The shapes of *tillas* are flat top uplands and arms are plunged towards rivers. In the southern portion of Haora River, rib structure is formed in East-West direction (Fig. 10.4). The eastern portion of the River Bongeswar Gong elevation is above 100 ft (30.48 m). It is delineated by 150 ft (45.72 m) contour line. The map represents that this ridge possess linear features parallel to the intra-fluvial ridge of northern portion of flood plain of Rivers Haora and Katakhal. Western portion of the Bongeswar Gong River the topography is dissected by some dried up channel (*lunga*).

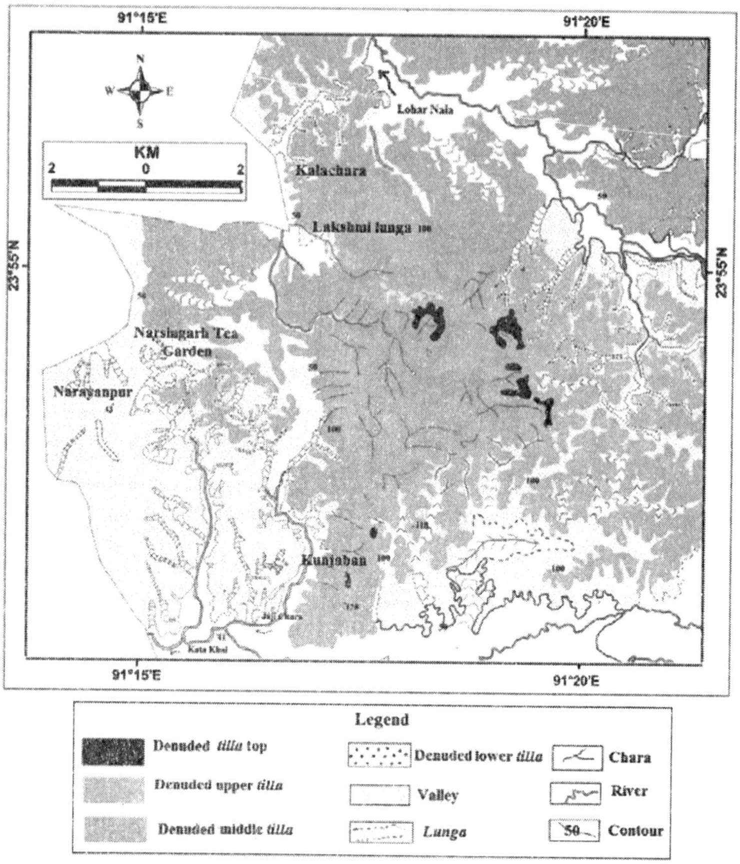

Fig. 10.3. Pattern of *tillas* and *lungas* in northern portion of the Haora-Katakhal flood plain, indicates the dissected *tilla*. *Lungas* are connected with nearby rivers.

Form and Pattern of *Lunga*

The middle portion of Rivers Lohar Nala and Katakhal, represents an intra-fluvial zone (Fig. 10.3). The tributaries are originated from this intra-fluvial zone and merging with Lohar Nala and Katakhal. In the middle portion the charas and *lungas* are forming centrifugal pattern due to the domal structure. In the southern portion of the Haora River, network of *lungas* are connected with the present drainage basin (Fig. 10.4). The *lungas* represent as a network of dried-up channel. The rivers and its tributaries (locally called *chara*) in this region are rainfed.

Analysis of the Litho-morphic Section of *Tilla-Lunga* Topography

Profile 1 (extension 23º58′1.62″N, 91º17′38.62″E to 23º44′10.65″N, 91º21′31.37″E) indicates the intra-river *tilla-lunga* topography, where rivers form the physiographic boundary. The massive graded sequence in the upper portion of *tilla* was formed through torrential

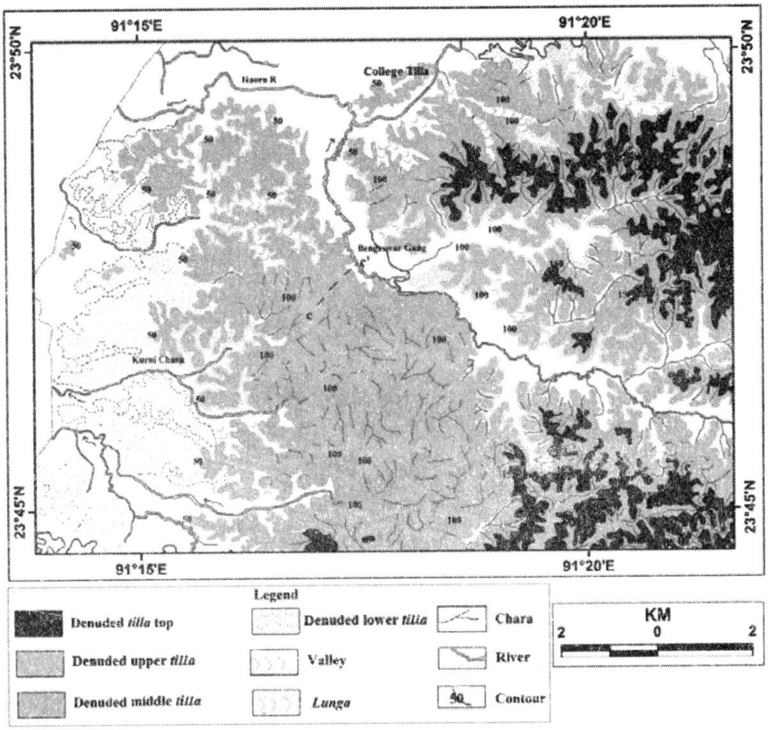

Fig. 10.4. The map represents shape of *tilla-lunga* topography in the southern portion of Haora-Kata khal flood plain. The elevation of the topography decreases towards the western portion of Tripura.

flood deposit. The litho-morphic evidence indicates that the upliftment of this area is significant along with the sedimentary basin-fill up process. Even the cross profile of the topography, as represented by profile 2 (extension 23º52′8.41″N, 91º14′24.66″E to 23º41′59.98″N, 91º32′22.67″E), indicates that the Baramura Hill comprises of the Tertiary rock structure (Surma Formation), which is extended towards west as a subsurface folded structure. The western upper portion of the folded structure is mostly covered with flood deposit and weathered sediments from hills. Profile 3 (extension 23º44′32.76″N, 91º17′40.98″E to 23º49′38.61″N, 91º21′59.96″E) indicates more detail picture of the litho-morphic variation. The elevation of *tilla* is much higher and is gradually merging with the Haora River near Ranir bazar area. The bed of *lunga* is situated much higher than the river courses. Thus, *chara* flows towards low land through *lunga*. An erosion scarp is formed in the western plunging limb of Baramura Anticline, sets the eastern limit of *tilla-lunga* topography. Profile 4 (extension 23º45′25.24″N, 91º16′57.47″E to 23º47′22.09″N, 91º17′57.02″E) indicates the cross-section of *tilla-lunga* among Rivers Sonai Nala-Bongeswar Gong-Haora. Both the limbs of the *tilla* gradually merge with the rivers plain. The lithological character of *tilla* and *lunga* plotted within the cross-section has been revealed from the artificially cut up of *tilla*. The upper portion of *tilla* is characterised by massive unconsolidated gravels, which is distributed in the whole study area.

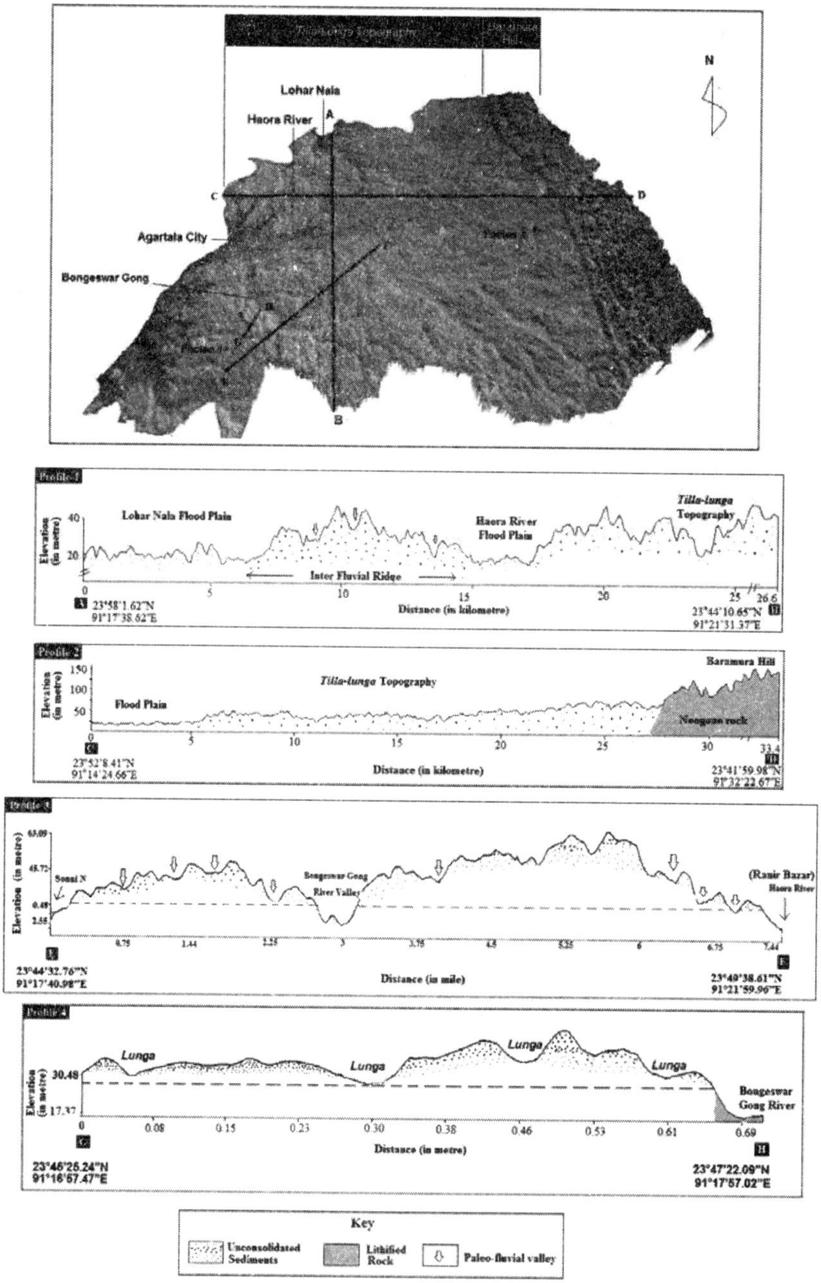

Fig. 10.5. The profile cross-section indicates the variation of the topography in different scale and lithological set-up. Periodic flood deposit in the Tertiary and Quaternary period deposited massive sediments in the synclinal valley. Cross-section indicates the variation of topography and lithologiy. The elevation of *lunga* is higher than the River Bongeswar Gong.

Analysis of Geomorphic Process from Litho-Facies Assemblage

Gani *et al.* (2004) documented four facies in north-eastern portion of Bangladesh, i.e. (i) trough cross-bedded sandstone (St), (ii) ripple cross-laminated sandstone (Sr), (iii) finely laminated mud with ripples (Fl) and (iv) massive mud with rootlets (Fm). Architectural elements include (i) channel (CH), (ii) lateral accretion (LA), (iii) sandy bedforms (SB) and (iv) overbank fines (OF). Roy *et al.* (2012) also mentioned about Dupitila facies i.e. (i) coarse-grained conglomerate facies association in relation to active current deposits of alluvial fan set up at the base of litho-succession (FAC), (ii) medium to fine-grained sandstone-siltstone-mudstone facies association in relation to strong tide (FAT) characterising the middle part of litho-succession, (iii) very fine-grained sandstone-siltstone-mudstone facies association in relation to less frequent weak tide or heterolithic facies association (FAHL) characterising upper part of litho-succession and shallow marine facies association (FASM) composing the uppermost litho-succession. In their work Sen *et al.* (2013) mentioned about four bedding patterns i.e. (i) small-scale trough cross bed, (ii) planar clay bed, (iii) graded bed and (iv) cross bed with transgression mark. For identification of the paleogeomorphic process three major fluvial sedimentological facies has been discussed in details.

Description of Facies 1

Along Agartala-Bishalghar railway track near Suryamaninagar (6 mile, south from Agartala City) fresh out-crop has been identified that represents better understanding of depositional sedimentary environment. The vertical sections indicate two distinct depositional environments. The upper portion is represented by Facies 1 and lower portion is represented by Facies 1.1 (Fig. 10.6). Facies 1 is a massive graded sequence, which is superimposed over mottled clay bed and it is represented by Facies 1.1. The total height of this facies is 16.2 m. Upper portion of the gravels of this facies are well conglomerated. In the middle and lower portion of facies, inverse gradation is identified. The massive graded bed in middle portion of *tilla* indicates high energy of sediment deposition in this region. The gradation gradually decreased down-ward. This section is exclusively identified in *tilla*. The clay bed is exposed in *lunga*. The upper portion of the graded bed is eroded and lower portion is exposed in *lunga*. The distribution of massive coarse gravel deposit with varying thickness over vast areas indicates the occurrence of massive flood. Sediments are deposited mainly in inter-fluvial areas. The sedimentary fill-up and upliftment are continuous process. The inverse gradation indicates that configuration of estuarine environment. As a result, the configuration of fluvial system in these areas has changed.

In lower portion of facies, thick molted clay deposits are identified. The paleo-current direction is NE-SW. Thin parallel lamination and cross-lamination are identified in the massive clay bed. The minor trough lamina indicates oscillation in sedimentary deposition. The schematic diagram is prepared for understanding the sequential paleo-environmental change in this area. Fluvial sedimentation has changed from alluvial river system to gravel bed river. The evidences of massive coarse grain gravel deposit indicate periodical flood occurrences. The elevation of this area increased due to upliftment of landmass. Facies architectural elements are Fsc-St-Fl-St-Fsc-Gm-Gms.

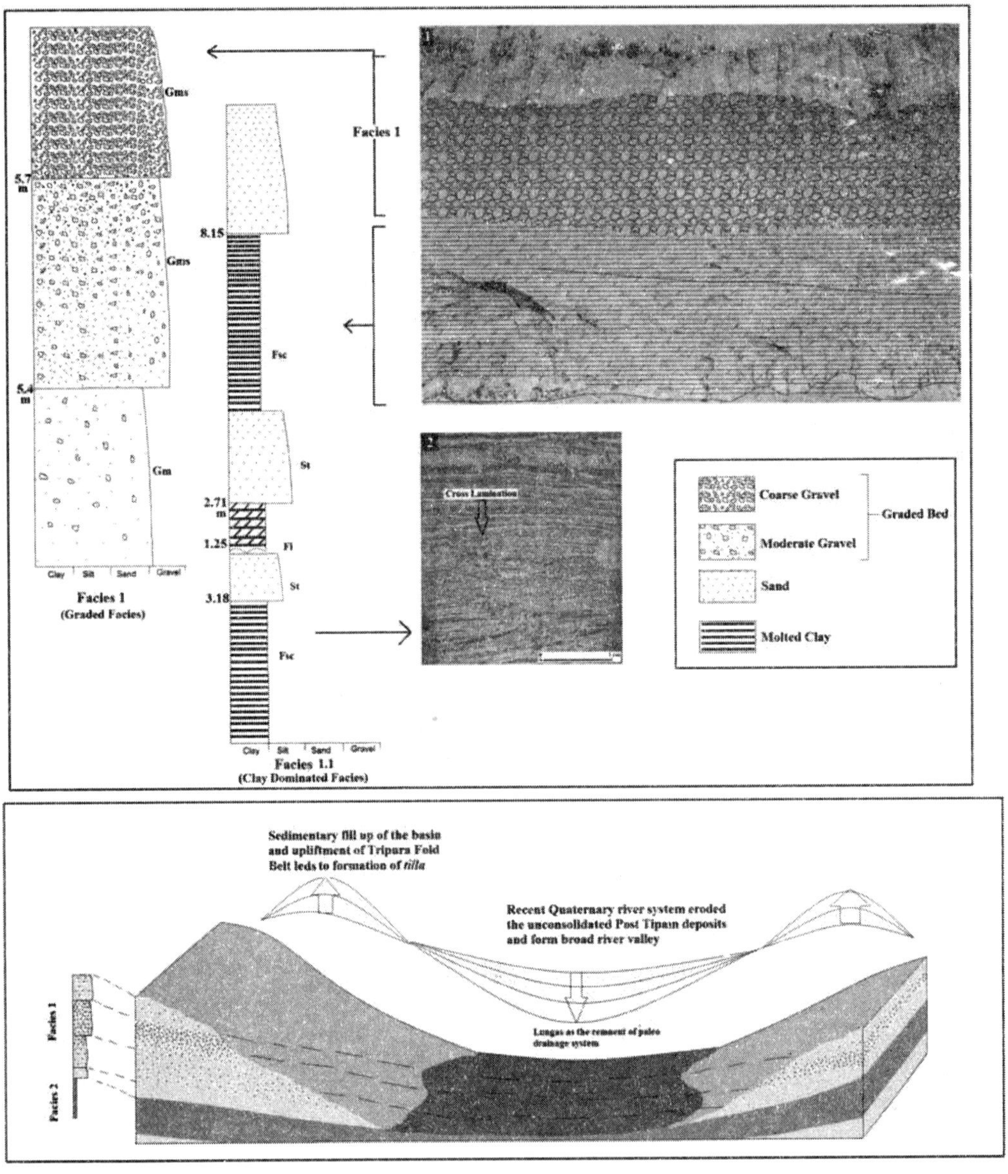

Fig. 10.6. The facies 1 and facies 1.1 are represented in the photograph 1. The facies indicate the vertical illustration of sedimentary depositions. The facies 1 overlies 1.1 indicate distinct depositional environment. Both the facies have been correlated in the diagram for understanding the complex process of sedimentary basin-fill and upliftment of the landmass.

Description of Facies 2

In the Kumbhakatha near Amtali, 7.88 km south-east from Agartala City, graded facies has been identified in the form of trough cross sloping towards *lunga* (Fig. 10.7). Facies 2 (photograph 1.1, 1.1.1) and 3 (photograph 2.1) in Fig. 10.7 are the outcrop in the lower portion of *tilla* slope. The distance between these two facies is 15.12 m. The slope of *tilla* merged with *lunga* where this gravel sequence (Gt) is completely abolished. The architectural elements are the continuation of this gravel sequence and are found in the next *tilla. Lunga* in between two *tilla* represents the existence of dried up channel. Due to gradual leaching, finer silt particles are percolated downward and deposited in lower portion of facies. The fossil wood covered with fine silt is identified in the lower portion of the facies (Fig. 10.7, photograph 2.1). In the lower portion of the facies clay cast is observed (Fig. 10.7, photograph 1.1.1). Below that clay cast small scale ripple drift is observed.

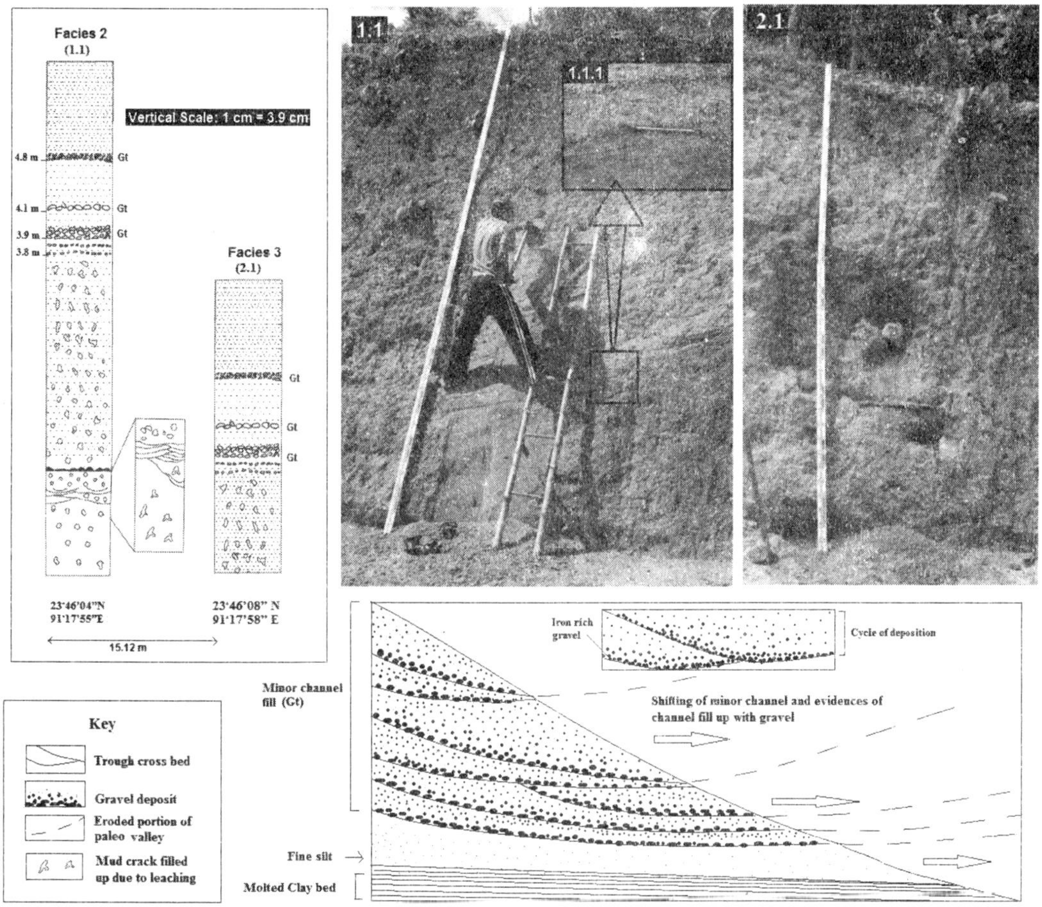

Fig. 710.. Facies 2 and 3 represent continuation of graded sequence which is forming trough cross-stratification in slope of *tilla*.

Description of Facies 4

Facies 4 is not clearly exposed excepting some small areas. Facies 4 is located in Mandai areas, 25.42 km east of Agartala City at the foothill of Baramura. The facies is important in the horizontal sequence, whereas vertical section is less important. The facies seems to be complicated as it is dominated by storm bed. The vegetation cover restricts identification of sedimentological structures. In the middle portion, facies dominated by massive sandstone. Gradually, injection of mottled clay domination in the facies is identified. The massive trough cross bed has been identified in photograph 1 (Fig. 10.8). The architectural elements are St-Sp-St. The upper portion indicates the storm incidences that lead to interruption of bed for process. The top set is eliminated due to fluvial erosion. These bed set indicates the dune migration. Photograph 2 indicates alteration of mudstone and sandstone. The enlarged photograph of 2A (Fig. 10.8) indicates alteration of transgression and regression of paleo-river. This facies is exposed in the adjacent *lunga*. The fluvial sequence indicates lateral accretion bed. These are the pene-contemporary structure where, geomorphic process and tectonic active both are playing active role. The whole sequence is exposed in *tilla* due to the upliftment of landmass.

Fig. 10.8. The photographic representation Trough cross stratification (photograph 1). The photography 2 indicates alteration of sand and clay bed. These are the pene-contemporary structure. The enlarged photograph indicates the storm bed form.

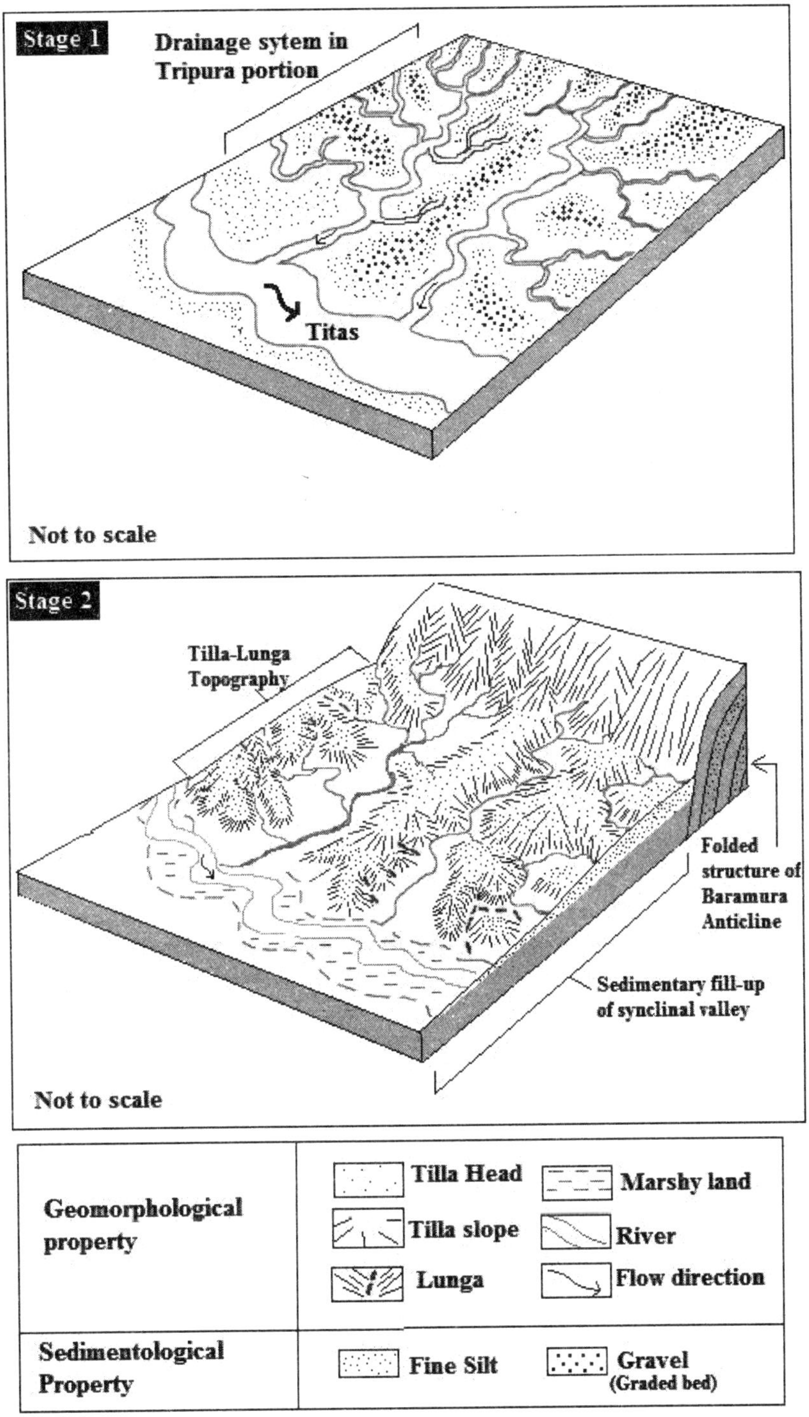

Fig. 10.9. The schematic diagram represents the stage-wise evolution of *tilla-lunga* topography.

Assemblages

The deposition process is interrupted by neo-tectonics and flood incidences. The grain size of the sedimentary structure in the sequences varies vertically from clay to gravel. Two different fluvial sedimentation environments have been found. The neo-tectonic activity in Quaternary Period plays an active role in the modification of landform. Deformation within soft sedimentary structure has also been found. Some pene-contemporary structure is developed within the fluvial sedimentary facies along with primary bed form. The graded bed is confined in the upper portion of the facies, which is absent in *lunga.* The network of paleo-flow path and massive clay in *lunga* indicates that these are the remnants of paleo-channel.

Form-Process Relation of *Tilla-Lunga* Topography

The geomorphological maps (Fig 10.3 and 10.4) of *tilla-lunga* topography indicate the Rib and furrow structure. Rib is the East-West extending axis of *tilla.* The furrows are the dried-up channels which are locally known as *lunga.* The networks of *lunga* bifurcates the arms of *tilla.* The remains of the arms are connected with the axis part of the *tilla* that portrays a unique shape of the whole topography.

From the investigation of geomorphology map, cross-section and litho-facies analysis we have developed a schematic model of evolution of *tilla-lunga* topography. Figure 10.9 represents previous (stage 1) and present (stage 2) situation of landform configuration. Stage 1 indicates continental flood plain before the Tripura orogeny phase. The thick graded sequence is uplifted due to tectonic movement. With the passage of time, surface weathering and precipitation generates rills along the slope of *tilla.* The rills are merging with each other and drainage system is developed.

Stage 2 represents the upliftment and folding of sedimentary structure. The massive inverse graded sequence in *tilla* indicates the deposition of torrential fluvial sediments and retreat of channel. Due to the upliftment of landmass base level changes and the first order tributaries of the rivers became dried up or seasonally inundated with rain water. The present-day topography might have approaching towards stability, because the vegetative cover restricts massive rill erosion. Thus the topography is neither bad land nor terrace. This young topography appears as a new discipline in morphological study.

Conclusion

From the present study the following major conclusions of the *tilla-lunga* topography in terms of form-process relationship can be drawn.

First, the first order landforms are (1) Baramura Hill, (2) *tilla,* (3) *lunga* and (4) flood plain. *Tilla* and *lunga* exist 20 m to 80 m and flood plains are below 20 m. *Tilla* is further classified into (2.1) upper *tilla* (above 60 m), which comprises of massive graded sequence and lithified molted clay. (2.2) middle *tilla* (40 m to 60 m) comprises of graded sequence and (2.3) lower *tilla* (below 20 m). The *lungas* mainly originated in middle *tilla.*

Second, the morphologically *tilla* represents the Rib structure, extended in East-West direction. It acts as water divided among the east-west flowing rivers in the west Tripura district. The middle *tilla* and lower *tilla* are sloping towards the major river courses. The major architectural elements are Fsc-St-Fl-St-Fsc-Gm-Gms (facies 1, 1.1), Gt (facies 2, 3), St-Sp-St (facies 4).

Third, *Lunga* (furrow) remains as network of dried up tributaries connected with the major river courses. Narrow channels (*charas*) are flowing through *lungas* are the tributaries of rivers. These dried up *lungas* have dissected *tilla*. Thus, *tilla* remains as denuded upland.

Four, the cross-section and facies assemblage indicates that *tillas* are characterised by massive mudstone-graded bed. In the *lunga*, the graded sequence is absent only the mottled clay bed is exposed. Thus continuity of the sedimentary sequences is interrupted by these *lungas*.

Five, the present-day geomorphology maps of *tilla* and *lunga*, cross-sections and litho-facies assemblage indicates that *tilla* is originated due to the passive impact of upliftment of the orogeny of the Tripura hills. These are the pene-contemporary structure of flood deposits and neo-tectonic activities. For this reason, upper *tillas* (east-west) are perpendicular to the anticlinal ridge of Baramura anticline (north-south). Graded sedimentary deposit in *tilla* is the indicator of sedimentary periodical floods. The deposited sediments are uplifted due to passive effect of the oronegy of the Chittagang Tripura Fold Belt. The rills and gullies emerged along the slope of this uplifted landmass later transformed into dried up network of surface flow mark that is called as *lunga*. Thus, *Lungas* are younger generation of landform than *tilla*. Finally it can be concluded that present-day *tilla-lunga* topography emerges as a unique example of earth surface processes in recent geomorphological study.

[**Acknowledgements:** Authors are especially grateful to Prof. Arabinda Ghosh, Department of Geological Science, Jadavpur University and Sri Siladitya Sengupta, Director, Geological Survey of India, Kolkata.]

References

Acharyya, S.K., (1998). Break-up of the greater Indo-Australian continent and accretion of blocks framing south and east Asia. *J. Geodyn.* 26, 149–170.

Alam, M., Alam, M.M., Curray, J.R., Chowdhury, M.L.R., Gani, M.R., (2003). An overview of the sedimentary geology of the Bengal Basin in relation to the regional tectonic framework and basin-fill history. *Sedimentary Geology* 155, 179–208.

Bandyopadhyay, S., Saha, S., Ghosh, K., De, S.K., (2013). Channel planform change and detachment of tributary: A study on the Haora and Katakhal Rivers, Tripura, India. *Geomorphology* 193, 25-35.

Curray, J.R., Moore, D.G., (1974). Sedimentary and tectonic processes in Bengal deep sea fan and geosyncline. In: Burk, C.A., Drake, CL (eds.), *The Geology of Continental Margins*. Springer, New York, 617–628.

Dasgupta, S., (1978). Pleistocene sediments of Tripura. *The Quarterly Journal Geological, Mining and Metallurgical Society of India* 51 (1), 41-46.

Dasgupta, S., (1984). Tectonic trends in Surma Basin and possible genesis of the folded belt. *Records GSI*, Vol. 113 Pt. IV, 58-61.

DeCelles, P.G., Gehrels, G.E., Quade, J., Ojha, T.P., Kapp, P.A., Upreti, B.N., (1998). Neogene foreland basin deposits, erosional unroofing and the kinematic history of the Himalayan fold-thrust belt, western Nepal. *GSA Bulletin* 110(1), 2–21.

Falvey, D.A., (1974). The development of continental margins in plate tectonic theory. *J. Aust. Pet. Explor. Assoc.* 14, 95–106.

Goodbred, S.L., Kuehl, S.A., (1999). Holocene and modern sediment budgets for the Ganges-Brahmaputra river system: Evidence for highst and dispersal to flood-plain, shelf, and deep-sea depocenters. *Geology* 27(6), 559-562.

Griffiths, J.S., Abraham, J.K., (2008). Factors affecting the use of applied geomorphology maps to communicate with different end-users. *Journal of Maps*, 201-210.

Gustavsson, M., Kolstrup, E., Seijmonsbergen, A.C., (2006). A new symbol-and-GIS based detailed geomorphological mapping system: Renewal of a scientific discipline for understanding landscape development. *Geomorphology* 77, 90–111.

Islam, M.A., (2009). Diagenesis and reservoir quality of Bhuban sandstones (Neogene), Titas Gas Field, Bengal Basin, Bangladesh. *Journal of Asian Earth Sciences* 35, 89–100.

Jade, S., Mukul, M., Bhattacharyya, A.K., Vijayan, M.S.M., Jaganathan, S., Kumar, A., Tiwari, R.P., Kalita, S., Sahu, S.C., Krishna, A.P., Gupta, S.S., Murthy, M.V.R.L., Gaur, V.K., (2007). Estimates of interseismic deformation in Northeast India from GPS measurements. *Earth and Planetary Science Letters* 263, 221-234.

Jena, A.K., Das, N.C., Saha, G.C., Samanta, A., (2011). Exploration in Synclinal Areas of Tripura fold belt, India: A Refound Opportunity. *Expanded abstract of AAPG Annual Convension and Exhibitation*, U.S.A, 2-5.

Kale, V.S., Baker, V.R., Mishra, S., (1996). Multi-channel patterns of bedrock rivers: An example from the central Narmada basin, India. *Catena* 26, 85-98.

Lee, T.T., Lawver, L.A., (1995). Cenozoic plate reconstruction of Southeast Asia. *Tectonophysics* 251, 85–138.

Magilligan, F.J., Gomez, B., Mertes, L.A.K., Smith, L.C., Smith, N.D., Finnegan, D., Garvin, J.B., (2002). Geomorphic effectiveness, sandur development, and the pattern of landscape response during jökulhlaups: Skeiöarársandur, southeastern Iceland. *Geomorphology* 44, 95-113.

Michael, H.A., Voss, C.I., (2009). Estimation of regional-scale groundwater flow properties in the Bengal Basin of India and Bangladesh. *Hydrogeology Journal* 17, 1329–1346.

Ramesh, N.R., (1981). Evidences of Stone Age culture in Pleistocene sediments of Tripura. *Indian Journal of Earth Sciences* 14 (1).

Roy, M.K., Ahmed, S.S., Bhattacharjee, T.K., Mahmud, S., Moniruzzaman, M., Haque, M.M., Saha, S., Molla, M.I., Roy, P.C., (2012). Paleoenvironment of Deposition of the DupiTila Formation, Lalmai Hills, Comilla, Bangladesh. *Journal of the Geological Society of India* 80 (3), 409-419.

Shamsudduha, M., Uddin, A., Saunders, J.A., Lee, M.-K., (2008). Quaternary stratigraphy, sediment characteristics and geochemistry of arsenic-contaminated alluvial aquifers in the Ganges–Brahmaputra floodplain in central Bangladesh. *Journal of Contaminant Hydrology* 99, 112–136.

Sen, S., Mitra, S., De, S.K., (2013). Micro geomorphic pattern of upper DupiTila formation within the *Tilla-Lunga* landform in the western portion of Tripura fold belt, Tripura, India. *Hill Geographer* XXIX 1, 1-13. ISSN 0970-5023.

Sen, S., Mitra, S., Debbarma, C., De, S.K., (2014). Impact of faults on landslide in the Atharamura Hill (along the NH 44), Tripura. *Environ. Earth Sci.* 73 (9), 5289-5298.

Sen, S., Santra, A., Debbarma, D., Mitra, S., De, S.K., (2015). Morphology of *tilla-lunga* topography in West Tripura District, Tripura, India. *Annals NAGI* XXXV (2), 77-93.

Smith, A.G., Hallam, A., (1970). The Fit of Southern Continents. *Nature* 225, 139-144.

Syvitski, J.P.M., Overeem, I., Brakenridge, G.R., Hannon, M., (2012). Floods, floodplains, delta plains—A satellite imaging approach. *Sedimentary Geology* xxx, xxx–xxx.

Syvitski, J.P., Morehead, M.D., (1999). Estimating river-sediment discharge to the ocean: application to the Eel margin, northern California. *Marine Geology* 154, 13–28.

McKenzie, D., Sclater, J.G., (1971). The evolution of the Indian Ocean since the late cretaceous. *R. Astron. Soc. Geophys. J.* 24, 437–528.

Miall, A.D., (1973). Markov chain analysis applied to an ancient alluvial plain succession. *Sedimentology* 20 (3), 347-364.

Miall, A.D., (1977). A review of the braided-river depositional environment. *Earth-Science Reviews* 13 (1), 1-62.

Miall, A.D., (1985). Architectural-element analysis: a new method of facies analysis applied to fluvial deposits. *Earth Sci. Rev.* 22, 261-308.

Uddin, A., Lundberg, N., (1999). A paleo-Brahmaputra? Subsurface lithofacies analysis of Miocene deltaic sediments in the Himalayan–Bengal system, Bangladesh. *Sedimentary Geology* 123(3), 239-254.

Varga, R.J., (1997). Burma. In: Moores E.M., Fairbridge RW (eds.) *Encyclopedia of European and Asian Regional Geology*. Chapman and Hall, London, 109–121.

Willis, B., (1993). Ancient river systems in the Himalayan foredeep Chinji Village area, Northern Pakistan. *Sedimentary Geology* 88, 1-76.

11
Role of Structure in Bandalamottu Lead Deposit, Guntur District, Andhra Pradesh

Asoori Latha and *R. Pavanaguru*

Introduction

Bandalamottu deposit (16°14': 79°43'), potentially the largest deposit in the Agnigundala belt, Guntur district, Andhra Pradesh, located at the northeastern part of the Cuddapah Basin. It primarily contains lead mineralization, stratabound in nature and is concordant to bedding. In detail the mineralization occur as lodes composed of veins, fracture fillings and disseminations. The mineralized upper dolomite horizon outcrops along the southern flank of Bandalamottu hill striking ENE - WSW and dipping 200 - 350 NNW. The regional dip of dolostone, associated phyllite and argillites is to the NNW, but there are undulations in the beds along the dip varying from <100 to as much as 350. The main structure is a low northerly plunging anticline at Bandalamottu. Systematic structural mapping has been carried out in the investigated area. The measurements include azimuth data (172 measurements of bedding and 117 measurements of foliation). The data were plotted as Poles using Equal Area (Schmidt) Net. The Fold elements were inferred through contoured pole diagrams and their control over mineralization was interpreted.

Geology of the Area

The Bandalamottu lead deposit hosts the zones of mineralization mainly confined to the upper dolostone and dolomitic limestone, belonging to the Cumbum Formation of Nallamalai Group which crop out along the southern flank of the Bandalamottu hill, striking ENE-WSW and dipping 200-350 WNW5 (Fig. 11.1). A detailed account of the geology and mineralization of the area was described earlier (King, 1872; Ziauddin, 1964; Dutt, 1975; Ramana Rao and Reddy, 1976; Nagaraja Rao *et al.* 1987; Kale, 1991). Archaeans, represented by biotite-schists and amphibolites, are the oldest rocks in this area. In addition, granites and dolerite dykes are also found. Rocks of the Nallamalai Group unconformably overlie the biotite schists/granitic basement. The Nallamalai Group is represented by Bairankonda and Cumbum Formation. The Bairankonda Formation is basically an arenaceous unit and represented by grey sandstones with intercalated shale/slate. These sandstones are predominantly exposed on the hills and ridges surrounding the granite domal structures and except for a few places, they are directly overlie the granitic basement. The Cumbum Formation is an argillaceous unit comprising shale, slaty shale, slate and phyllites interbedded with fine to medium and coarse sandstones, and dolomite/dolomitic limestones at various levels. In Bandalamottu area, chloritic and carbonaceous phyllites constitute the main country rock, in which occur a main bed of mineralized dolomite and dolomitic limestone intercalated by calcareous argillite, phyllite and pink arenaceous dolomite. Galena is the main mineral with subordinate sphalarite, chalcopyrite and pyrite. Apart from Pb, Ag, Zn, Cu and Cd are obtained as co-products from this area (Sivadas *et al.*,1985).

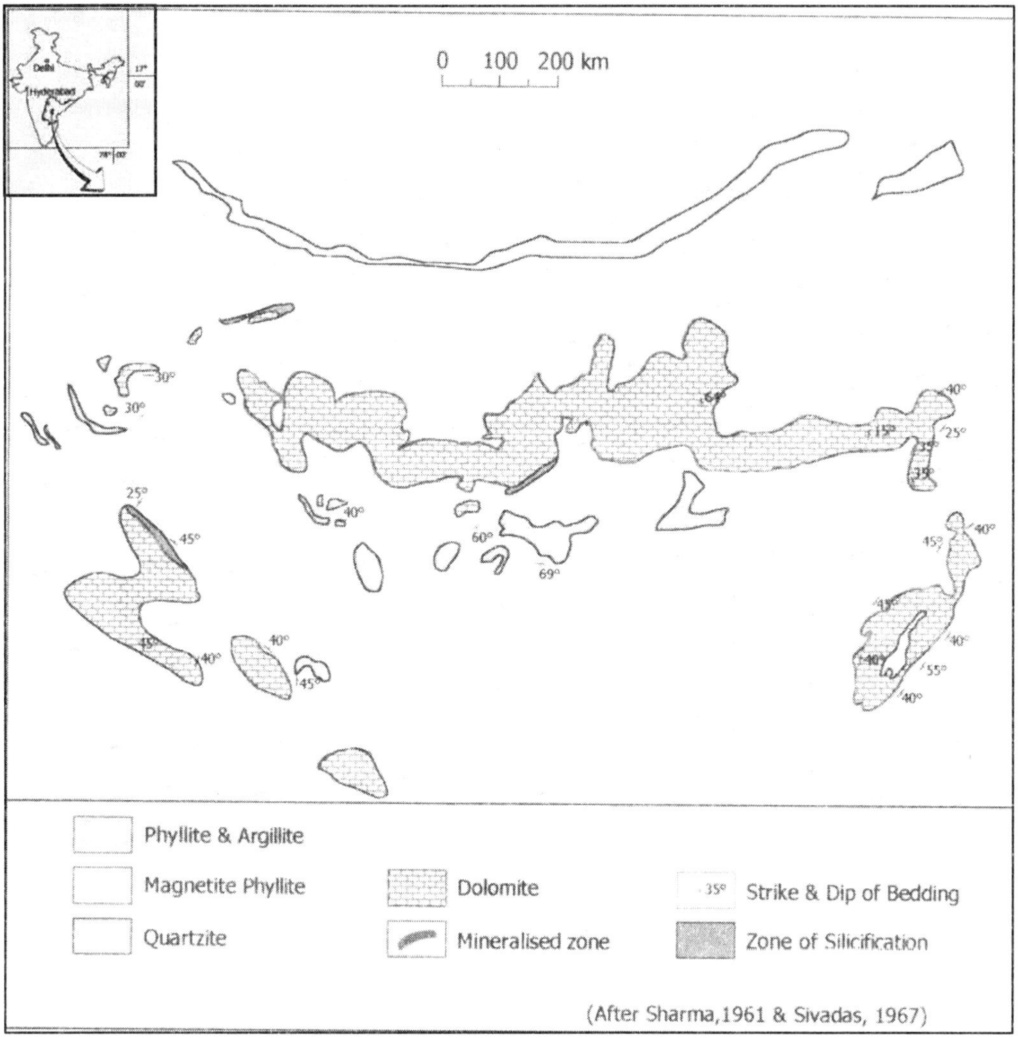

Fig. 11.1: Geological map of Bandalamottu Area

Structure of the Area

Rock sequences in the Agnigundala area exhibits two phases of deformation. NNW-SSE trending, tight and gently plunging asymmetric folds represent the first phase. The second phase is weak and represented by ENE-WSW trending gentle folds and warps. The interference of the folds has resulted in dome and basin features in the area. The main structural element in this area is the presence of Vellaturu and Ipuru domes, which are elongated, with their longer axis trending NE-SW parallel to axis of regional folding. Further, the sandstone and dolostones are also highly drag folded with axes trending dominantly N-S and E-W, probably resulting from the interaction of NE-SW trending major folds and NW–SE trending cross-folds and culminations (Sivadas *et al.*, 1985). The main structure of the Agnigundala area is a low (50-100) north plunging anticline over the Bandalamottu hill. The eastern limb of this fold constitutes part of the complimentary synclinal structure. These major folds contain conspicuous drag folds whose axes plunge toward N to N15^0 E direction. This composite structural pattern is superimposed in its southern part by well-marked N 600- S 600 E trending cross folds. Steep joints are developed parallel to the axial plane of the drag folds (N to N 150 E) and N 60^0- S 60^0 E cross folds.

In the Bandalamottu area, the main structure is a low (50-100) northerly plunging anticline showing wider outcrop width in the axial portion. The regional dip of the dolomite and the associated phyllites and argillites is to the NNW, but there are undulations in the bed along the dip, varying from less than 100 to as much as 350. Deep borehole intersections towards the NW side of Bandalamottu hill have also indicated successive pinching out of the dolomite beds down-dip at the top and their overlapping reappearance at increasing depths in the down-dip direction, probably due to either intercalation of phyllite within dolomite by simple facies variation or intense drag folding and shearing of the beds. The repetition of narrow zones of carbonaceous phyllite and magnetite phyllite in such sections of successive pinching out of dolomite seems to indicate intense drag folding and shifting of the dolomite in depth rather than facies variation. The other important structural element in the area is a main NW-SE trending steep dipping zone of silicification along the northern fringes of the lower dolomite horizon, south of the Bandalamottu hill. This appears to be a silicified shear fracture zone or fault, parallel to the NW-SE trending major regional fault extending from the NE part of Vellaturu dome. In addition to this, there are numerous E-W and NW-SE trending silicified zones within the phyllites on the Bandalamottu hill (Sivadas *et al.*, 1985).

Structural Studies

A total number of 172 bedding data and 117 foliation data of phyllite, dolostone and sandstone intercalation within the phyllite are plotted on equal-area net (Schmidt net) as poles (Fig. 11.2A and 11.3A). To bring about the more meaningful distribution of both bedding and foliation, the poles are contoured. Schmidt contouring grid data are plotted on the Schmidt net (Fig. 11.2B and 11.3B).

Contoured Pi-diagram of the poles to the bedding planes shows that the majority of beds (~65%) are striking ENE - WSW with low dips due NNW and they show swerving to WNW - ESE with dips to NNE. The beds further show a swing to NE-SW with dips due SE. The (single) maximum of the contoured Pi-diagram along with other contour trends/poles forms two girdles with two β axes, which represent two fold axes. Two β axes indicate that the beds show imprints of two phases of deformation related to two

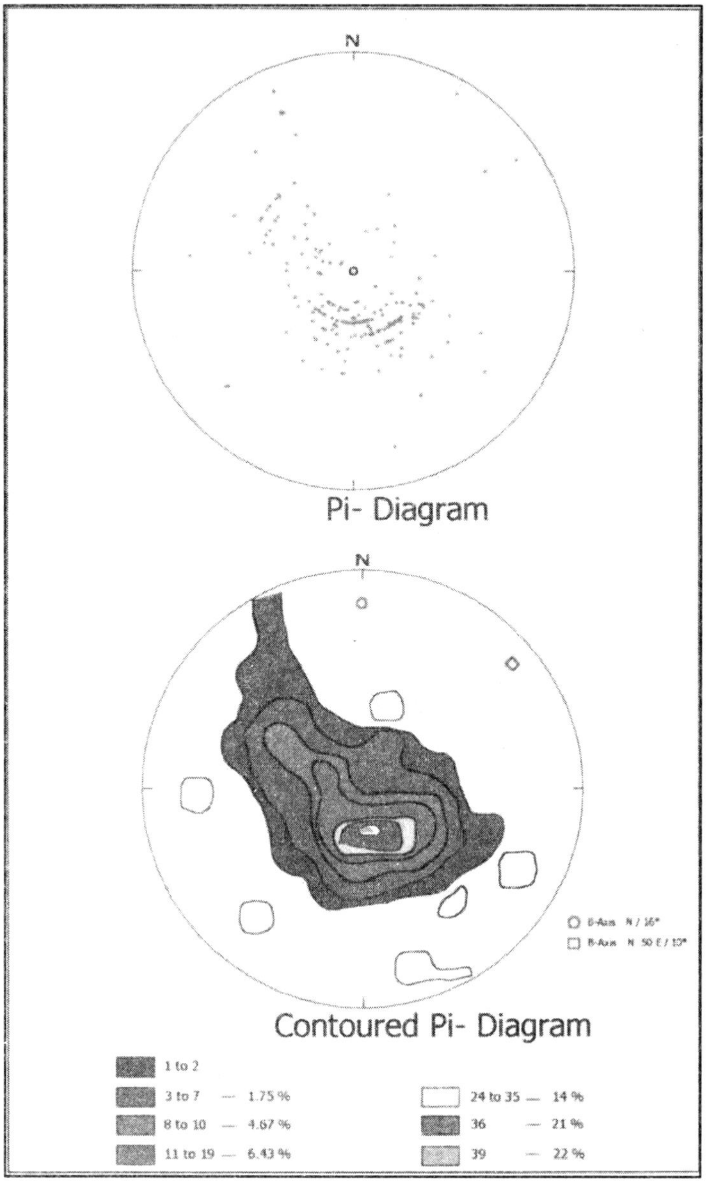

Fig. 11.2A: Pi- Diagram and 2B Contoured Pi- Diagram of Bedding

phases of folding. Early fold has N-S axis plunging in 16°due north. Later cross fold has NE-SW axis plunging 10° due N50E. Similarity of attitudes of bedding and foliation is reflected in the Pi-diagrams as similar distribution of poles to both bedding and foliation and similar pattern of contoured Pi-diagram of both bedding and foliation, i.e., bedding and foliation are parallel indicating over-turning nature of the beds. Field observations show that the attitudes of bedding defining folding on N-S axis and foliation developed on this bedding have similar attitudes, thus, supporting the above patterns obtained,

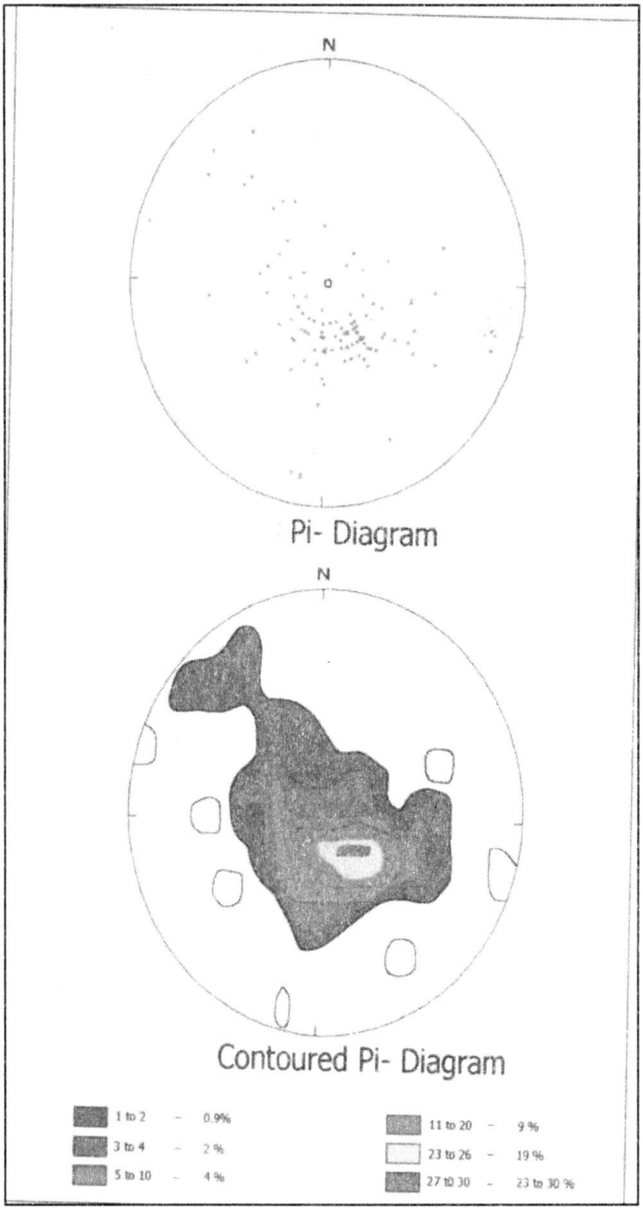

Fig. 11.3A: Pi- Diagram and 3B Contoured Pi- Diagram of Foliation

indicating overturning nature of the N-S folding. The maximum concentration of the poles, both bedding and foliation, within the SSW & SE quadrant, indicates that the beds are overturned in that direction. From the contoured Pi-diagram of poles to the bedding, it can be inferred that majority of beds form part of fold, gently plunging due North (antiform from field evidence) which is the dominant structural element of this domain of Bandalamottu block.

Decreasing order of majority

Poles Quadrant	Attitude of beds
I	
SE	ENE-WSW/NNW
I	
SW	WNW-ESE/NNE
I	
NW	NE-SW/SE

Discussion

The Bandalamottu deposit exhibits a definite structural control and shows the influence of the extensive Nallamalai folding and overthrusting. Regionally, the basin configuration itself has a definite structural control. The formation of basin by deep fundamental fractures, (Narayanaswamy, 1966; Sen and Narasimha Rao, 1967; Nagaraja Rao and Mahapatro, 1977), by en echelon type of faulting of the Moho boundary with eastward down faulting, (Kaila *et al.*, 1979), by sinking of the crustal block along the fundamental fractures, (Naqvi, 1987 and Hari Naraian, 1987) etc. all, point towards effective structural control during its formation.

Faults in the basin are mostly steep and both normal/reverse types are present. Kaila and Tewari (1982) showed that a major fault separates the deep eastern, Nallamalai basin from the shallowest basin containing older Cuddapah sediments to the west. Mantle depth increases from normal crustal thickness of 35 km on the west to nearly 50 km in the deformed Nallamalai area thus the Cuddapah basin deepens to the east. Bhattacharji and Singh (1982, 1984) showed that the downward bowing of the basement generated sufficient extensional force to cause the normal faulting. It is thus clearly evident that deep fundamental fractures have played a major role in these sedimentary units and also might have acted as channels for transporting mineralizing fluids associated along or in close proximity to these fractures.

The eastern Cuddapah basin has NNW-SSE trending faults/fractures that are subparallel to the major thrust. Minor (gravity) tension faults displace the Bairankonda sandstone at many places and they trend NW-SE, N-S and E-W. One of these faults at northern part of the Vellaturu dome traces discontinuously towards northwest through Bollapalle and extends along the southwestern part of the Bandalamottu lead deposit as inferred from a series of silicified zones in this direction. The parallelism of this fault with the trend of elongated depression between the Ipuru and Vellaturu domes and the occurrence of Bandalamottu lead deposit along this zone implies structural significance for the localization of ore mineralization.

Nallamalai Fold Belt shows increase in the intensity of folding from west to east. The major F1 folds have a curved axis, related to compression from the east. Structure has played a major role in concentrating the mineralizations in zones of intense folding, drag folding, cross folding, culminations and depressions (Narayanaswamy, 1966). The major

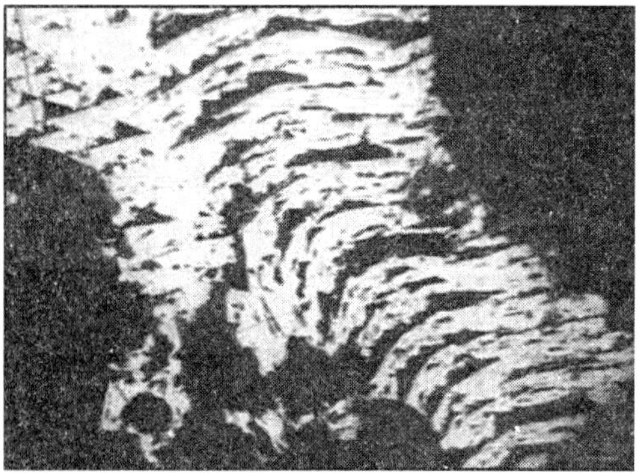

Fig. 11.4: Bending in cleavage of galena, RL, 1N

Fig. 11.5: Cofolding of host dolostone with galena, RL, 1N

Fig. 11.6: Fractured vein pyrite, RL, 1N

structural control for the ore mineralization is the N-S trending anticlinal cross-fold, which coincides with the N-S bulge of the Vellaturu dome located further south. The mineralization is further localized in E-W and WNW-ESE folds in the dolostone beds resulting from shear forces, accompanying the main N-S cross-fold flexure. Murthy (1981) opined that the Nallamalai Fold Belt lies in the collision zone between two continental cratonic masses - one stable on the western side and the other that moved from the east.

Conclusion

1. The Nallamalai Group, of Cuddapah Supergroup, of sediments belonging to compressional tectonic regime with pene-contemporaneous igneous activity represented by prominent domal structures indicates clear relationship between mineralization and tectonic/igneous activity.
2. The occurrence of deposit along a depression between the Ipuru and Vellaturu domes in parallelism with the trend of the faults in the area, and concentration of mineralizing fluids mostly in zones of intense folding, drag folding, cross-

folding, culminations and depressions along N-S trending anticlinal cross-folds, E-W and WNW- ESE folds suggest that deformation has played a major role, constituting the physical controls and structural avenues for mineralizing solution to migrate and precipitate their content.

3. Development of irregular, elongated patches of pyrite and galena, fracturing of vein pyrite grains and the bending of cleavage traces of galena (Fig. 11.4), all point towards deformation of ores by later movements.
4. The co-folding of galena along with host dolostone (Fig. 11.5), off-shooting of ore bodies by micro faults, fracturing in vein pyrite (Fig. 11.6) and displacement of twin lamellae are the structural characters that point towards syn-deformational ore mineralization.
5. The deformation resulting in the development of two generations of folds has definitely enhanced the mineral concentration related to syn- and post-deformation episodes.
6. The brecciation and folding of the dolostone constitute the physical controls and structural avenues for mineralizing solution to migrate and precipitate its contents. Thus the textural and micro-structural features of ore minerals reflect that most of the ore mineralization is syn- to post-deformation.

References

Bhattacharji, S and Singh, R.N. (1982). Mantle perturbations, thermal episodes and origin of intracratonic, Proterozoic Cuddapah Basin on south Indian shield. In S. Bhattacharji and S. Balakrishna (eds.), *Evolution of the Intracratonic Cuddapah Basin*. Inst. Ind. Peninsular Geology, Hyderabad. pp. 107-134.

Bhattacharji, S. and Singh, R.N. (1984). Thermo-mechanical structure of he southern part of the Indian shield and its relevance to Precambrian basin evolution. *Tectonophysics,* vol. 105, pp. 103-120.

Dutt, N.V.B.S. (1975). *Geology and Mineral Resources of Andhra Pradesh*. Ramesh Publications, Hyderabad, pp. 1-205.

Hari Naraian. (1987). Geophysical constrains on the evolution of Purana Basins of India with special reference to Cuddapah, Godavari and Vindyan Basins. Purana Basins of peninsular India. *Mem. Geol. Soc. India,* No. 6, pp. 5-32.

Kale, V.S. (1991). Constraints on the evolution of the Purana Basins of peninsular India. Geol. Soc. India, vol. 38, pp. 231-252.

Kaila, K.L., Roy Chowdhury, K., Reddy, P.R., Krishna, V.G., Hari Narain, Sabbotin, S.I., Sollogub, V.B., Cheunov, A.V., Kharetchko, G.E., Lazarrenko, M.A. and Ilchenko, T.V. (1979). Crustal structure along Kavali-Udipi profile in the Indian peninsular shield from deep seismic sounding. *Jour. Geol. Soc. India,* vol. 20, pp. 307-333.

Kaila, K.L. and Tiwari, H.C. (1982). Structure and tectonics of the Cuddapah in the light of DSS studies. In S. Bhattacharji and S. Balakrishna (Eds.). Evolution of the Intracratonic Cuddapah Basin, *Inst.Ind.Peninsular Geology,* Hyderabad. pp. 53-62.

King, W. (1872). On the Kadapah and Kurnool formations in the Madras Presidency. *Geol. Surv. India, Mem.* 8, pp. 269-270.

Murthy, Y.G.K. (1981). The Cuddapah Basin: A review of Basin development and basement framework relations. Fourth workshop on status problems and programmes in Cuddapah Basin. Inst. Ind. Peninsular Geology, Hyderabad. pp. 51-72.

Nagaraja Rao B.K. and Mahapatra, G.P. (1977). Geology of Palkonda and Velikonda ranges, Prg. Report F.S.1976-77. Geol. Surv. India, (Unpublished).

Nagaraja Rao, B.K., Rajurkar, S.T., Ramalinga Swamy, G. and Ravindhra Babu, B. (1987). Stratigraphy, structure and evolution of Cuddapah Basin. Purana Basins of Peninsular India. *Mem. Geol. Soc. India,* No. 6, pp. 33-86.

Naqvi, S.M and Rogers, J.W. (1987). *Precambrian Geology of India,* Oxford University Press, Oxford, pp. 70-75.

Narayanaswami, S. (1966). Tectonics of the Cuddapah Basin. *Jour. Geol. Soc. India,* Vol. 7, pp. 33-50.

Ramana Rao, N., Reddy, M.N. (1976). Nature of base metal mineralization in Bandalamottu area, Guntur district, A.P. *Indian Mineralogist,* vol. 17 pp. 39-48.

Sen, S.N. and Narasimha Rao, Ch. (1967). Igneous activity in Cuddapah Basin and adjacent areas and suggestions on the paleogeography of the basin. *Proceedings Symposium on Upper Mantle Project.* GRB & NGRI publ. No. 8, pp. 281-286.

Sivadas, K.M., Sashikumar, K.T., Subba Rao, N., Setti, D.N., Rajurkar, S.T., Sharma, R.K., Gopalakrishna, K.P., Sagar, A.K., Md. Ziauddin, Murthy, Y.G.K and Narayanaswamy, S. (1985). Lead and copper deposits of the Agnigundala Mineralized Belt, Guntur district, Andhra Pradesh. *Geol. Surv. India, Mem.* Vol. 118, pp. 1-101.

Ziauddin, Md. (1964). Origin of Agnigundala ore deposit. *J. Ind. Geo. Sci. Assn.* Vol. 4. pp. 107-111.

12

Depletion and Degradation of Land Resources in India

Baleshwar Thakur

Introduction

Land is our basic resource. Throughout history, we have drawn most of our sustenance and much of our fuel, clothing and shelter from the land. It is useful to us as a source of food, as a place to live, work and play. It has different roles. It is a productive economic factor in agriculture, forestry, grazing, fishing and mining. It is considered as a foundation for social prestige and is the basis of wealth and political power. It has many physical forms like mountains, hills, plains, lowlands and valleys. It is characterized by climate from hot to cold and from humid to dry. Similarly, land supports many kinds of vegetation. In a wider sense, land includes soil and topography along with the physical features of a given location. It is in this context that land is identified closely with natural environment. However, it is also regarded as space, situation, factor of production in economic processes, consumption good, property and capital (Barlowe, 1958).

Land is the most crucial factor in agriculture because it is the principal means of livelihood to a large number of rural people in India. The country is well endowed with cultivable land but there are regional variations in the quality of agricultural land which is determined by rainfall, access to water supply, topography and soil characteristics. With a total area of 326 mha and 1027 million people, India is a big country. However, the agricultural land-man ratio is not favourable as in many other countries like Australia, Canada, USA, Argentina, Chile and Denmark. The swelling numbers of people have put increasing pressures on land. In the last sixty years since the country gained Independence, agricultural land recorded a small increase while the population increased by 70 per cent. There is ever-shrinking availability of land due to demographic pressure and conversion of agricultural land to non-agricultural uses. At the same time, land degradation in the form of depletion of soil fertility, erosion, waterlogging, salinity and alkalinity has increased. It is estimated that about 174 mha of total geographical area suffers from degradation. The country is equally faced with problems of unequal access to land and agrarian issues like tenancy relations. The governments have not been successful in correcting the skewed pattern of land consolidation and regulation of tenancy. The study has reflected that there are weak and dysfunctional institutional arrangements for land redistribution. Scope for increase in net sown area is limited in spite of land reclamation programmes. Thus, further shrinkage in arable land can threaten India's ability to feed its growing population. The concentration of land ownership is intensifying and rural opportunities in non-agricultural work are severely limited. The purpose, therefore, is to investigate and highlight the temporal and regional changes taking place after Independence in India with respect to depletion and degradation of land resources and evaluate development challenges and management strategies. The following five questions are addressed for probing into causes and consequences of depletion and degradation of land resources: (1) To what extent Indian states vary spatially and temporally with respect to agricultural land and land-man ratio? (2) What is the spatial pattern of land degradation? Do the elements of land degradation vary across states? (3) What has been the role of reclamation and reforms in land development and management? (4) What are the challenges facing land development especially physical, social and institutional? (5) Will redistribution, reclamation, institution and information play significant role in land development and management?

Availability of Arable Land

India is well-endowed with cultivable land which has long been a key factor in the country's socio-economic development. In terms of area, India ranks seventh in the world, while in terms of population it ranks second. With a total area of 328 million hectares, India is one of the big countries. Arable land includes net sown area, current fallow, other fallow and land under tree crops. Arable land covers a total area of 167 million hectares which is 51 per cent of the total area of the country. However, the arable land-man ratio is not as favourable as in many other countries like Argentina, Australia, Canada, Chile, Denmark, Mexico and the USA. Conversely, the land-man ratio is more favourable in India than China, Egypt, Israel, Japan, the Netherlands and the United Kingdom. Land-man ratio is defined as the ratio between the habitable area and the total population of a country. The physical features in India are highly varied, complex and marked by great contrasts between high mountains and broad plains. It has a great internal variety of landform types including the Himalayan ranges in the north, the Ganga river plains and the Peninsular plateaus in the south. Across these contrasting landscapes, rainfall variations impose widely different conditions like aridity and semi-aridity in the northwest, higher rainfalls in the hilly ghats of the Peninsula and the foothills of the Himalayas and Assam and lower ranges in the eastern part of the Peninsula and Ganga river plains. They combinedly produce varied human response to the use of land resources. About 30 per cent of India's surface area is covered by hills and mountains (Fig. 12.1).

They are either too steep or too cold for cultivation. About 25 per cent of hill land is topographically usable which is scattered across the country. Plateaus constitute 28 per cent of the total surface area but only a quarter of this is fit for cultivation. Lowlands comprise 42 per cent of the total surface area and nearly 95 per cent is suitable for cultivation. Comparing the differences in proportion of surface area, this allows us to conclude that taking the country as a whole, about two-thirds is topographically usable (Dayal, 1972, 38). Moreover, soils, topography, moisture and temperature determine the limits of cultivability. As a result of this, only half of the surface area is cultivated (Brush, 1966, 11). The country is marked by significant regional variations in the quality of agricultural land depending on the availability of level land, soil quality, water availability, temperature and transportation infrastructure. The following section therefore deals with attributes of land at regional level.

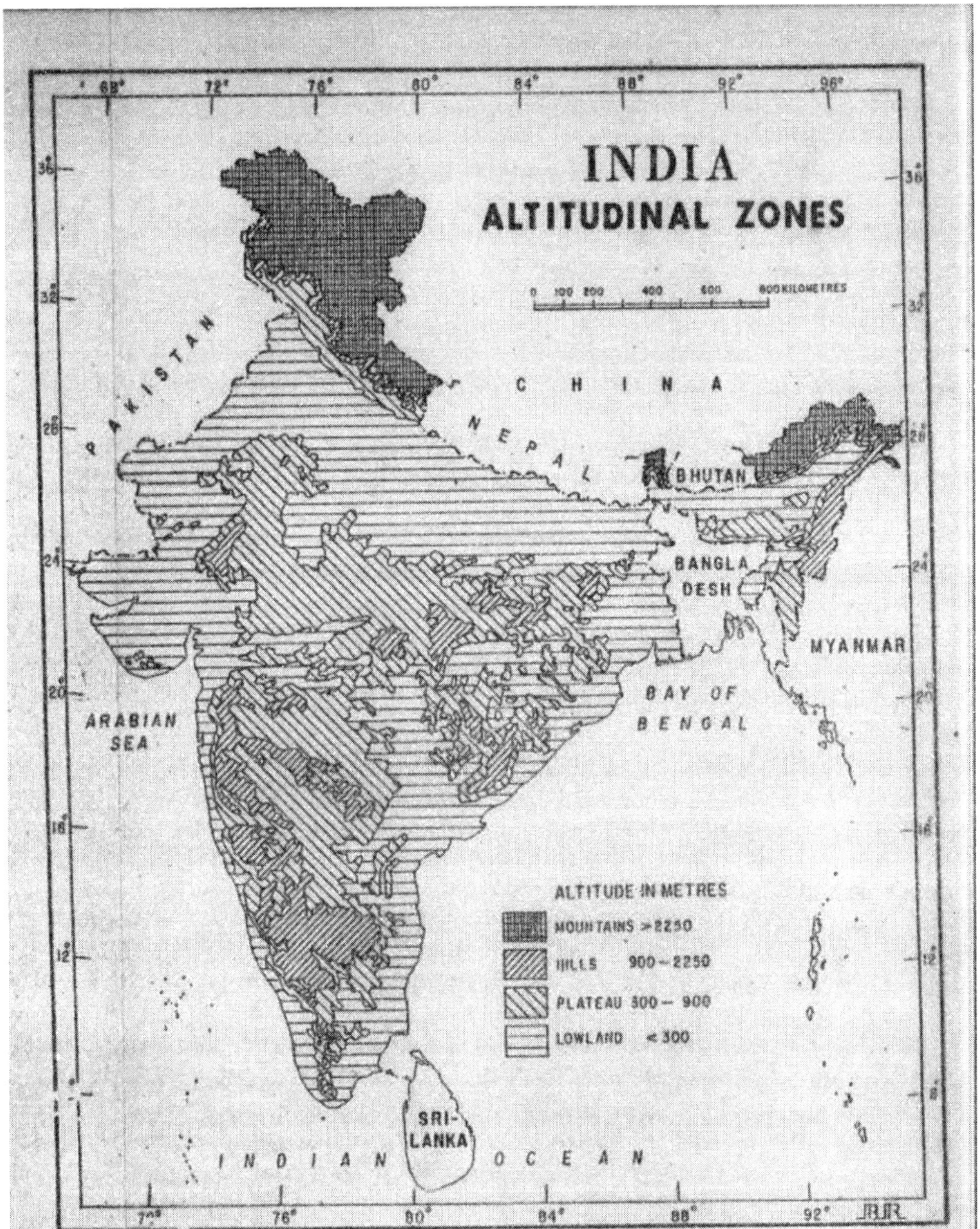

Fig. 12.1: India: Altitudinal Zones

Attributes of Land[1]

India has two contrasting land attributes: physical and social. The former relates to its age, structure, slope, elevation and physical environment and the latter consists of ownership of land by different groups of society where ethnic, religious and caste contradictions play significant role in decision-making.

Physical Attributes

The physical attributes of India vary greatly over the huge territory they encompass. The region bears resemblance to USA in geological history, topography, slope and elevation. The surface of the country can be envisioned as a plateau, bordered by fairly narrow and uniform coastal lowlands, massive Himalayan mountainous region with glaciers, forests and several high peaks and the north Indian plain (also called the Indo-Gangetic plain). From the point of view of physical attributes, India can be divided into four major units: The Himalayan mountain system, The Indo-Gangetic plain, the Peninsular plateau and the Coastal plains.

The Himalayas constitute one of the youngest fold mountain systems in the world and the youngest in India. The Himalayas are 2400 km long from Nanga Parbat in the west to Namcha Barwa in the east and 150-250 km wide from north to south. The Himalayan ranges and its offshoots cover an area of about 500,000 sq km. The massive Himalayas, rising more than 8500 meters, were formed from the collision of Indian plate with Tibetan plate, and thrust itself beneath the main mass of Asian crustal rocks. The mountain is composed of a variety of igneous, sedimentary and metamorphic rocks. Its topography is youthful, rugged and dissected. Areas of very steep slopes and escarpments are found in the northern parts. The relief is lowest in the Terai region. The landscape presents highly folded and faulted structures characterized by snow-clad peaks, large valley glaciers, deep gorges, V-shaped valleys and roaring waterfalls alongwith dense forest (Singh, 1971, 6). The three great river systems of the subcontinent, Indus, Ganga and Brahmaputra take their rise in this mountain system. The most distinctive geomorphic characteristic is the longitudinal valleys in the Indus system, transverse valleys in Brahmaputra system and transitional in the Ganga system (Noble, 1972, 18). Longitudinally, the Himalayas are classified into three parallel zones which differ in orographical, hydrographical and vegetational features: (a) Shiwalik Himalaya, flanking the Indo-Gangetic alluvial plains, is the southernmost part of the Himalaya. They are chiefly composed of Tertiary sediments and are related to the latest phase of the Alpine-Himalayan orogeny. Its average elevation is 600 metres, (b) Lesser Himalaya forms the southern boundary of the Great Himalayan range and is mainly composed of granite, quartzite, sandstone, shale and dolomite. The heights range between 2000 to 3300 metres, (c) Greater Himalaya consists of line of snowy peaks with an average height of 6100 metres. It is composed of granites, gneisses and schists and is characterized by lofty serrate ridges, cirque-indented slopes and sharp peaks produced by glacial action.

The Peninsular plateau is geologically very old and stable. It is also called a shield of old stable rocks which are rigid landmass. The plateau makes an irregular triangle in its extent. The concave base extends from the Delhi ridge to the Rajmahal hills and the apex is formed by Kanyakumari (Raza and Ahmad, 1978, 24). The plateau is bordered by the Western Ghats on the west, the Eastern Ghats on the east and the Aravalli range in the extreme northwest. It is composed of very hard crystalline igneous and metamorphic rocks. The important rocks are granite, gneiss, schist, quartzite and basalt. The lava flows are most prominent in the northwestern part of the peninsula in an area known as the Deccan plateau covering about 3 lakh sq km in Maharashtra, Gujarat, Madhya Pradesh and Karnataka. Except in the northwest, where rivers flow westward in the Arabian Sea, the predominant stream flow is eastward. While the Deccan possesses the largest share of India's minerals, agriculture is hampered by the relative absence of water. The major rivers are Godavari, Krishna, Cauvery, Mahanadi, Son, Narmada and Tapi. They have formed broad and shallow valleys with low gradients, except Narmada and Tapi. There are evidences of faults in the Damodar, Godavari and Mahanadi rift valleys and fissures in the Deccan trap region. The plateau has undergone much erosion through hundreds of millions of years and the removal of vast thickness of materials has reduced the surface to low relief. Thus, the plateau region is characterized by undulations, residual ranges, broad valleys and faulted troughs. The important residual ranges are the Western Ghats, the Eastern Ghats, the Aravalli, the Kaimur, the Satpura, the Ajanta and the outlier of Assam. The entire region south and east of Vindhya-Satpura hill is made up of several plateaus of subdued relief with elevations ranging between 300-900 metres. They are Maharashtra plateau, Karnataka plateau, Tamil Nadu upland, Telengana plateau, Baghelkhand plateau, Chotanagpur plateau, Orissa highlands, and Meghalaya or Shilling plateau. Their surface configuration consists of undulating hills with rounded, broad, low summits of peneplains and residual blocks.

Between the Himalayas and the Peninsular plateaus, a wide lowland zone crossed by three of the world's major river systems—the Indus, Ganga and Brahmaputra river, after they flow down from their Himalayan sources. They are called the Great Plains of India which are among the more recent land structures. They are crescent in shape which extends from the Indus delta to the Brahmaputra valley. The average width of these plains is between 250-400 km. The depths of the sediments deposited are the deepest, i.e. 2000 meters in the Ganga-Brahmaputra delta to 700 meters in the basin of the Indus river (Tirtha and Krishan, 2002; Geddes, 1960). Gentle gradients promote river meanders, shifting of river courses, creation of meander scars and ox-bow lakes (Tirtha and Krishan, 2002, 35). Altitude varies from 15 metres in the Bengal delta to 200 metres in the west. Annual rainfall varies from 50 cm in the west to 300 cm in the east. Because of this, the Brahmaputra lowlands are wetter than other parts of the Great Plains. The plains contain fertile sandy loam in the west to clayey loam in the east. Thus, the plains are characterized by meandering rivers, fertile alluvial soils and succession of floodplains and doabs. Geomorphologically, the Indo-Gangetic plain exhibits a surface called *bhangar* and *khadar* (Mukerji, 1963; Pascoe, 1973). Alluvial soils are of two types: *bhangar* and *khadar*. *Bhangar* soils are older and less productive while the soils formed by recent alluvium are called *khadar*. *Khadar* soils are annually inundated and are fertile from agriculture point of view. The Rajasthan desert is covered by a veneer of loose soil mixed up sometime with rocky and barren tors and wind-blown sands ridges. Sand dunes are the most important landforms in Jaisalmer, Bikaner and Barmer districts. The Ganga-Brahmaputra delta is 500 km wide and 650 km long. It is a level and low lying delta with a web of distributaries. The western part is moribund delta of abandoned channels with no annual inundation, while the eastern part (now lying in Bangladesh) is an active area of shifting meanders, wide and permanent channels, annual flooding and fertile silt. The lower part of delta is full of mangrove swamp which is called the Sundarbans.

The Peninsular plateau is flanked by coastal plains of varied width extending from

Orissa to Kachchh. The eastern coastal plain, extending from Mahanadi mouth to Cape Camorin with an average width of 60 km, is broader and economically much more important than western. The western coastal plain is narrow, except around the Gulf of Cambay and its extreme southern part in Kerala. The eastern coast is comparatively dry while the western coast receives abundant rainfall. The eastern coastal plains have been named from the north to south as Utkal, Andhra and Caromandal. The western coast has been divided from north to south into the Kachchh Peninsula, Kathiawar Peninsula, Gujarat plains, Konkan coast, Karnataka coast and Malabar coast. The succeeding section deals with social attributes of land.

Social Attributes

The regional structure of agrarian system constitutes three distinct elements, the government, land owner and cultivator (tenant) and social attributes of land relate to their interrelation. The owner-cultivator interrelation is expressed through payment of rent or sharing the crop with the owner. Basically, the interrelation among the cultivator, owner and the government refers to problems of rights and obligations of each (Thakur, 2007, 11). The examples of the inter-connectedness of the problems are the land tenure system, insecure tenancies, concentration of land in the hands of unenterprising land owners, preponderance of tiny and uneconomic holdings (Appu, 1995, 6). The succeeding section therefore deals with a review of land-society relations in India and investigates the complexities which prevent a change in the existing order to provide security to cultivators.

Tenurial systems: Through centuries, the rise of landless cultivators assumed tragic proportions in India. The dominant prevailing system of land tenure during Mughal rule was the ryotwari or peasant-proprietor system (Tirtha and Krishan, 2002, 234). The peasant was the owner of land and possessed inheritance rights to sell and transfer the land. The government collected land revenues from the peasants directly. The Mughal period exercised considerable influence in the evolution of intermediaries (jagirdars and zamindars) and revenue collection was assigned to the zamindars. The zamindar in Persian language means keeper or holders of land implying a degree of control or attachment, but not necessarily ownership (Bandyopadhyay, 1993, A-50). The British viewed land tenure system from the point of view of fixing responsibility for payment of land revenue creating intermediaries or sub-tenant of land (Thakur, 2007, 11).

There were three types of tenurial systems prevailing in the country at the time of independence, namely, the *zamindari, ryotwari* and *Mahalwari*. The three systems differed with each other with respect to mode of payment of land revenue. But, the common feature with them was that in all the three systems, the practice adopted was to get the land cultivated by the tenants. The tenants were of three types: (a) occupancy tenants, (b) tenants-at-will, and (c) sub-tenants. The occupancy tenants possessed permanent and veritable rights on land with security of tenure; conversely, tenants-at-will and sub-tenant did not enjoy such tenure and could be dispossessed from land whenever the landlord so desired (Thakur, 2007, 11).

The *zamindari* system embraced nearly three-fifths of the total area of land holdings in nine provinces of British India (Raj, 1995, 3). The system was introduced in 1793 in Bengal by Lord Cornwallis and later on became popular and prevailing in West Bengal,

Bihar, Orissa, Uttar Pradesh, Andhra Pradesh and Madhya Pradesh. The task of collecting rent from the farmers was delegated to *zamindars* or landlords, therefore, the *zamindars* acted as intermediaries between the cultivators and the government.

"However, the system had a number of drawbacks. It was based on exploitation in which the agriculturist had to pay land revenue every year whether the annual crop was successful or not. This ultimately contributed to the growth of poverty and indebtedness. Thus, there was a large-scale transfer of land from cultivating to non-cultivating owners and mass ejection of tenants from land occupied by them. The number of cultivating owners steadily diminished while that of the non-cultivating landlords grew. This process resulted into class differentiation in which the produce of land was shared without participating personally in the productive process" (Thakur, 2007, 11). The *ryotwari* system occupied one-third part of the country and was prevalent in Tamil Nadu, Maharashtra, Barar, East Punjab, Assam and Coorg. In this system there were no intermediaries between the government and the cultivator. The responsibility of paying land revenue to the government was of the cultivator (or individual ryot) himself. The ryots had full rights regarding sale, transfer and leasing of land and could not be evicted from the land as long as he paid the land revenue (Misra and Puri, 1996, 459). The *Mahalwari* system covered about five percent of the total land area. It was introduced by William Bentinck in Agra and Oudh and was later expanded to Punjab and Madhya Pradesh. In this system the payment of land revenue was completed by treating the whole village as one unit and village headman was solely responsible for collecting the land revenue. Under the *ryotwari* and *mahalwari* systems also, there were disadvantages, like the conditions of tenants-at-will and sub-tenants were poignant and distressing. On the whole, the conclusion is inescapable that the traditional tenurial system have damaged and remained detrimental to the cultivator, the agriculture and the country.

Insecure tenancies: Tenancy is a key component of agrarian structure which assumes considerable importance in India. It is a system of rights and duties associated with the occupation and use of land. In rural areas, the landlord keeps ownership rights while gives use and occupancy rights to the tenants. In return, the tenant agrees to a schedule of periodic payments in one of the three forms: (a) cash payment, (b) share of the crop, (c) or even a pledge of labour services to the owner (Goodall, 1987, 466). Thus, tenancy is a temporary possession of any place on certain conditions. Tenants are the tillers of the soil as sharecroppers. In many parts of the country, the sharecroppers pay a predetermined proportion of total production usually between 25 per cent to 50 per cent of the crop depending upon local custom and the landlord's contribution to the total inputs needed for production (Goodall, 1987, 428). More specifically, big landlords residing in urban areas, called absentee landlords, own big farmhouses in rural areas. It has been observed that tillers and sharecroppers who actually cultivate the lands of absentee landlords are not much interested in the development, modernization and management of agricultural land. They lead to low investment and low productivity. Also, most of the tenancies are oral and insecure in arrangements and are seldom entered in the land records. This practice has persisted from British times despite more than six decades of land reform implementation. At Independence, over one half of the operated area was under tenancy (Appu, 1995, 7). As a result of tenancy reform, tenants have acquired ownership rights and they have become secured. Even then, this is most burning problem of regional agrarian system. According to P.S. Appu (1995, 4), the tenants should be accorded permanent rights in the lands they cultivated subject to a limited right of resumption to be granted to the land owners.

Concentration of land: The agrarian structure of India is characterized by the preponderance of extremely small, fragmented and tiny holdings that are uneconomic and unoperational (Chadha, Sen and Sharma, 2004, 147). The causes of subdivision and fragmentation of holdings are liberal laws of inheritance, continuing pressure of population, decline of joint family system and farmer's indebtedness. Such tiny plots have resulted into low agricultural productivity and face difficulties in modernization.

The pattern of land distribution seems to be more skewed and skewness in the pattern of distribution has persisted during 1970-90 (Haque, 1995,64). In 1991, large holdings above 10 ha accounted for 1.6 per cent of the total holdings and 17 per cent of the total area, while 20 per cent holdings (2-10 ha) were medium, 19 per cent holdings (1-2 ha) were small and 59 per cent holdings less than 1 ha were marginal. The number of medium and large holdings has remained either stable or has shown declining trend. Thus, the basic character of agrarian economy with high concentration of land holdings in the hands of a few land owners, on the one hand, and growing number of marginal farmers and landless labourers, on the other, remains more or less unaltered (Haque, 1965,63). A new class of agricultural landless labourers emerged in the post-Independent period, which assumes tragic proportions in India. During British rule, its number was 28 million in 1931 and 34 million in 1948. It is estimated to be 106 million in 1991. At the country level, the incidence of landlessness among Scheduled Castes households stood at 13 per cent compared with 11 per cent non-Scheduled Castes and tribes (Chadha, Sen and Sharma, 2004,116).

Land and caste: The differential distribution of land prevalent in rural India determines specific mode of relationship among the various segments of people. These relations are very well reflected in the caste composition and land ownership, caste and control of resources, caste and land quality and caste and land security in Rampur village in Madhubani district (Thakur and Das, 2011). This complex organisation of caste and land is the major hindrance in the economic development of rural India, because of its skewed distribution in favour of the dominant caste. The farmers keep on trying to increase the number of plots under their possession as well as to possess good quality agricultural land at better locations. Consequently, they arrange, rearrange and reshuffie their agricultural plots so that the profits are maximized and the land resources are put to best use. In the process of rearrangement and redistribution of agricultural land, the dominant caste often wields their power because they are the rural elites. They occupy the best land, i.e. most fertile land, land along the canal or rail-road, while the non-dominant caste gets satisfied with the possession of the inferior quality of agricultural land. Even if the non-dominant castes possess good quality agricultural land, those lands are appropriated by dominant castes through the process of debt recovery or other fair and foul means. In fact, it is the most important process of non-dominant caste impoverishment who possess only marginal or low fertility land, which the good quality and better located agricultural lands of the village belong to dominant caste. Thus, it is evident that caste and land are intricately related in rural areas.

It is quite interesting to observe that caste-land association varies in different ecozones, because the highly priced land may not have the same location in all ecozones. For example, land located along the river is preferred in highland areas than lowland.

The three ecological regions in Bihar show distinct characteristics in terms of land-caste dynamics (Thakur and Sinha, 2008, 491-510; Sinha, 2009). The Adhwara plain is dominated by the Brahmins and other upper castes; the Gandak plain represents its imprints of past culture of Janpada, means the representation of each strata of society in the rural power structure; while in the central-south Bihar plain, the emergence of lower strata is ongoing process and in this process the naxalite movement has played an important role. The dominating castes have their land around the major places of importance in the village, so that they can hold the resources and maintain their status in the society. At the same time, the newly emerging landed castes are also trying to acquire better quality land of the village. The Ahirs, Koiri and Kurmi are the three major castes in Bihar State that are making their good hold in rural Bihar. There is poor representation of the Scheduled Castes in the state. They are still far behind in the process of getting social justice and welfare. The development of urban mindset among the upper castes or the landed castes as well as the outmigration of the labour classes are the most important factors behind the transfer of land from dominating to non-dominating castes.

Five villages in Ramgarh block of Bhabua district of Bihar display similar land-caste relations throwing deeper insight into the structure and functioning of land and caste nexus which has evolved through centuries (Thakur and Vir, 2008, 288-303). They are Bhargawan, Lasra, Baraura, Mahro and Surajpura. The village possesses multiple caste system. The bulk of the farmlands are owned and cultivated by Rajput, Brahmin, Kurmi and Yadav. The Scheduled Castes are mainly agricultural labourers. The traditional landlordism is confined to the caste of Rajput, but gradually, other middle order castes have emerged as dominant castes in many villages. The dominant caste not only possesses maximum land holding but have also command over land and other infrastructural resources available in and around the village. Moreover, the possession and distribution of agricultural land across the country must be viewed regionally because of the diversified physical environmental conditions and population pressure. The following section therefore deals with distribution of agricultural land and land-man ratio.

Land Resource Depletion

Agriculture, Agricultural Land and Man-Land Ratio

India is overwhelmingly a rural country where agriculture and agricultural land remains the mainstay of economy. India, home to more than one billion people, is characterized by agriculture representing an important primary industry. Although India is rich in agricultural land resources and has been a pervasive feature of its economic pursuits, it is comparatively inferior in accomplishments than China. India's population has surged since Independence from 361 m in 1951 to 1.02 billion in 2001. Half of India's peasants are landless and work as agricultural labourers or as sharecroppers. Despite its huge and growing population, India has achieved self-sufficiency in grain production with a four-fold increase in output since 1950. It is because some 46 per cent of India's land area is classified as cropland, compared to only 10 per cent of China.

The successes of agriculture have been mainly due to increased use of fertilizers, introduction of new high-yield varieties of wheat and rice, more labour provided by the growing rural population, better irrigation, spread of education and the development of government extension institutions to aid farmers. By its Green Revolution, India has demonstrated that high-tech agriculture, relying on hybrid seeds, intensive irrigation and fertilization, can produce dramatically higher yields in a variety of crops. In Punjab, Haryana and western Uttar Pradesh., the larger land holdings and the commercial drive of the Sikh and Jat farmers increased productivity through the large capital inputs. Here, the local infrastructure of irrigation water, 90 per cent rural electrification, 70 per cent all weather roads and good transportation provided an advantageous environment for the farmers investments. Elsewhere in India, the outcomes of the Green Revolution were less dramatic and results were negligible in areas where irrigation water was not available, farm plots were small and infrastructure was poor. Meaningful land reform has hardly begun on a massive scale. Whatever the achievements are, land reforms strengthened only the middle peasantry and their ownership of land. Because of inheritance laws, most farm plots are small, thereby, discourage mechanization and cooperative irrigation. Large areas of peninsular India have inadequate irrigation and grow coarse grains. People in the northern part of the country, especially in Middle Ganga plain, live in a poverty trap with few prospects of improvement. Thus, due to regional variations in agricultural performance, India has become a country of startling economic contrasts, which is home to hundreds of millions of desperately poor people. It suggests, therefore, there is minority of privileged and majority of poor people. An important issue is that the poor have been left out of the political process and bypassed by economic reforms. Although the new technologies have resulted in dramatically higher production rates, the increased food supplies have not eliminated hunger and malnutrition. This is basically characterized by the distinctive character of the relationship between development and agricultural land.

Agricultural land is defined as land used exclusively for agricultural production which includes the net sown area, current fallow, other fallows and land under miscellaneous tree crops (Dayal, 1984, 346). The net sown area and fallow lands account for 54 per cent of the reported area which shows the extent of the cultivable land. The difference between current and other fallows is that current fallows include lands which are lying fallow for less than one year, while other fallows have remained unploughed for one to five year (Dayal, 1972, 38). Fallows occupy 8.13 per cent of the reporting area of the country. Net sown area is of vital importance which is 46.64 per cent; however, international comparisons are strongly adverse to India. India's figure is much higher than 7 per cent in Canada, 16 per cent in Brazil, 28 per cent in former USSR and 40 per cent in USA (Tiwari, 2006, 216).

Out of the total geographical area of 328.7 m ha, land utilization statistics are available for 305 m ha only. The balance 23.7 m ha remains unsurveyed and inaccessible. The significant features of India's land utilization are: (1) high percentage of area suitable for agriculture (Fig. 12.2), (2) limited scope for further extension of cultivation, and (3) small area under pastures despite a large bovine population (Dayal, 1972, 38).

India has made remarkable progress in developing its net sown area from 118.7 m ha in 1950-51 to 142.6 m ha in 2000-2001, an increase of about 20 per cent in fifty years. This net sown area increased to 133 m ha in 1960-61, 140 m ha in 1970-71 and 1980-81, 142.4 m ha in 1990-91, and 141.40 m ha in 2000-2001. The percentage of net area to total reporting area also increased from 42 per cent in 1950-51 to 46 per cent in 2000-2001. Table 1 indicates the growth of net area sown in India from 1951 to 2006. It is apparent that

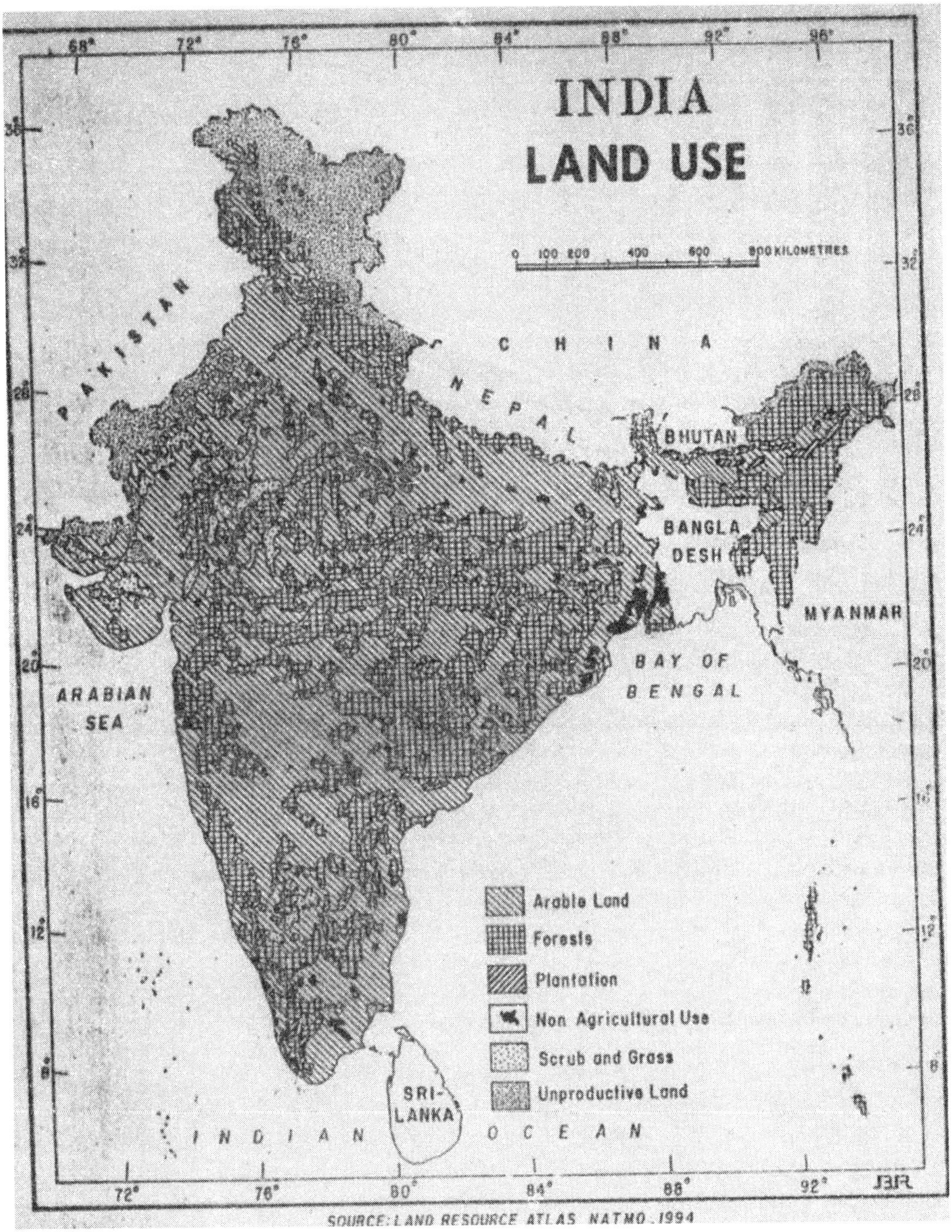

Fig. 12.2: India: Land Use, 1994.

Table 12.1: Growth of Net Area Sown in India: 1951-2006

Year	Net Area Sown	Year	Net Area Sown
1951	118.75	1979	142.98
1952	119.4	1980	138.90
1953	123.44	1981	140.00
1954	126.81	1982	141.93
1955	127.85	1983	140.22
1956	129.16	1984	142.84
1957	130.85	1985	140.89
1958	129.08	1986	140.90
1959	131.83	1987	139.58
1960	132.94	1988	134.09
1961	133.2	1989	141.89
1962	135.4	1990	142.34
1963	136.34	1991	143.00
1964	136.48	1992	141.63
1965	136.12	1993	142.72
1966	136.2	1994	142.34
1967	137.23	1995	142.96
1968	139.88	1996	142.20
1969	137.31	1997	142.93
1970	138.77	1998	141.95
1971	140.27	1999	142.76
1972	139.72	2000	141.06
1973	137.14	2001	141.40
1974	142.42	2002	141.45
1975	137.79	2003	132.69
1976	141.65	2004	140.97
1977	139.48	2005	141.14
1978	141.95	2006	141.89

Source: *Agricultural Statistics At A Glance* (2008). Directorate of Economics and Statistics, Dept. of Agriculture and Cooperation, Ministry of Agriculture, Government of India, New Delhi, pp. 264-65.

there is break in 1972-73, below which there is regular growth and after 1973 stability with fluctuations. There has been a marginal decline in fallow land accounting for 7.5 per cent in 1990-91 of the reported area as against 9.9 per cent during 1950-51. Over the last fifty years, the land under non-agricultural use has increased with the accelerated growth in economy. The process of industrialization and urbanization demands more land under roads, railways, airports, industries and human settlements. Essentially, on the limited total area must be accommodated all the cultural uses of land and one of these uses may be expanded at the expense of others. In 1950-51, the total area under non-agricultural use was 9.3 m ha which increased to 14.8 m ha in 1960-61, 16.4 m ha in 1970-71, 19.6 m ha in 1980-81 and 21.2 m ha in 1990-91. Thus, land use is a dynamic process. It changes over time due to a number of factors, including increasing population, changes in cropping system and technology. As the various sectors of the economy develop, there may be a shift in the pattern of land use. However, the bulk of the land continues to be used for raising crops. With unabated population growth, the pressure of population on arable land is bound to grow.

The increase in agricultural land and population has not been similar during the post-Independence period. Population increased from 361 m in 1951 to 1.02 billion in

2001, while agricultural land recorded a small increase from 156 m ha to 167 m ha. In other words, the growth of agricultural land could not keep up with increasing population. Agricultural land was 0.46 ha per person in 1951 which decreased to 0.42 in 1952, 0.45 in 1953, 0.41 in 1954 and 1955, 0.40 in 1956, 0.39 in 1957, 0.38 in 1958 and 1959, 0.37 in 1960, 0.39 in 1961 and 1962, 0.35 in 1963, 0.34 in 1964, 0.33 in 1965, 0.32 in 1966 and 1967, 0.31 in 1968, 0.30 in 1969 and 1970, 0.29 in 1971, 1972 and 1973, 0.28 in 1974, 0.21 in 1991 and 0.19 in 2001 (Dayal, 1984, 353 and Tiwari, 2006, 217). Thus, land-man ratio has declined remarkably. However, there are clear regional differences in land-man ratio. Regions with more than 1.5 ha of agricultural land per head of agricultural population are called high land-man ratio and with less than one ha are called low land-man ratio (Dayal, 1984, 357; Fig. 12.3). The high land-man ratios are found in the Thar desert, Chambal badlands, the Malwa plateau, the Deccan lava plateau and the Karnataka plateau. Because of infertile soils, the quality of agricultural land is poor in these regions. Here, competition among farmers grows for scarce water, hence, there is a low pressure of agricultural and total population on land resources (Fig. 3). The low land-man ratios are concentrated in well-watered and productive areas. The agricultural population is heavily concentrated in the Indo- Gangetic lowlands and the eastern fringes of the Deccan peninsula (Fig. 4). They have resulted into small holdings and intensification of land use and, therefore, labour intensive farming (Dayal, 1984, 357). Since farmers have small holdings they use the land intensively and frequently to produce enough for their requirements resulting into competition for the best use of land.

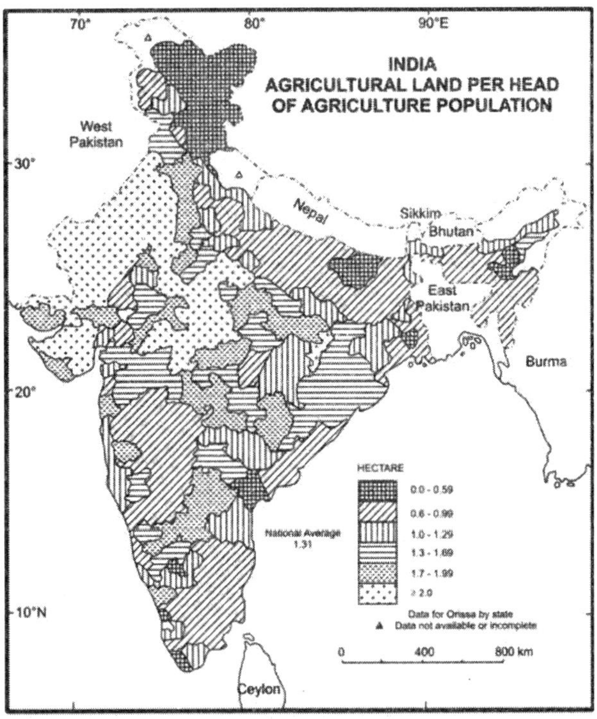

Fig. 12.3: India: Agricultural Land Per Head of Agricultural Population, 1951
Source: E. Dayal (1984), *Agricultural Land*, p. 360.

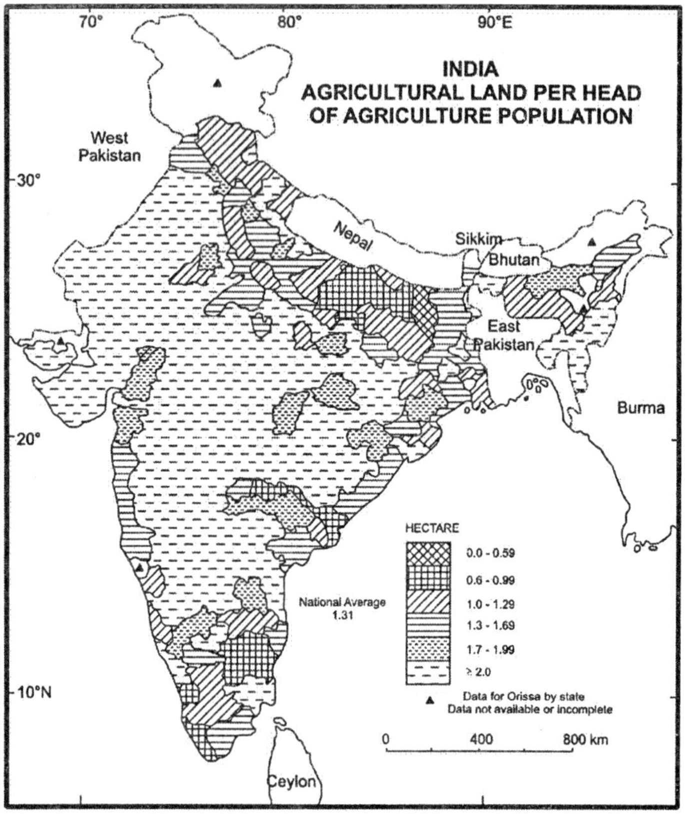

Fig. 12.4: India: Agricultural Land Per Head of Agricultural Population, 1970

Source: E. Dayal (1984), Agricultural Land, p. 358.

Table 12.2 shows state-wise distribution of per capita agricultural land in India for the year 2000-2001. Per capita land has been calculated on the basis of availability of agricultural land divided by agricultural population for twenty-nine states. Determination of such per capita land indicates the true relationship between land and man. The per capita figures obtained are arranged in descending order and are divided into high, medium and low land-man ratio. The figures above 1.00 are designated as high agricultural land-man ratio and include states like Goa (2.26), Orissa (1.76), Meghalaya (1.67), Rajasthan (1.44), Arunachal Pradesh (1.20), Gujarat (1.17), Nagaland (1.11) and Sikkim (1.04). These states are either located in the Himalayan region or cover western part of the country, Orissa is the only big state from eastern part which can be gleaned from Table 12.2. These states are either the semi-arid or the mountainous parts of the country where carrying capacity of the land is very low, and cannot sustain the large number of people. The figures between 0.50 to 0.99 are categorized as medium agricultural land-man ratio covering plateaus, mountains and fertile plains of the country. This category spreads over seventeen states like Karnataka (0.98), Kerala (0.97), Madhya Pradesh (0.94), Maharashtra (0.90), Haryana (0.87), Punjab (0.84), Uttarakhand (0.83), Andaman and Nicobar Islands (0.83), Mizoram (0.77), Chhattisgarh (0.75), Andhra

Pradesh (0.72), Assam (0.64), Jharkhand (0.61), Tamil Nadu (0.59), Jammu and Kashmir (0.56), Uttar Pradesh (0.53) and Tripura (0.52). This is the most popular and dominating group of land-man ratio which is in the process of high degree of urbanization and industrialization where there is scope for non-agricultural and service sector growth. Here, the maximum population that can be supported by agriculture and with a particular technology is comparatively higher than the first group. As a result, agricultural intensification has become important. The figures below 0.50 are categorized as low agricultural land-man ratio. This category embraces four states like West Bengal (0.44), Himachal Pradesh (0.39), Manipur (0.37) and Bihar (0.30). The low land-man ratios are concentrated in well watered and productive areas, where there is high dependence on agriculture resulting into agricultural intensification. Thus, the carrying capacity of land provides a measure of the degree of population pressure on agricultural land resources.

Table 12.2: Per Capita Agricultural Land in India: 2000-2001

State	*Agricultural Land ('000 ha)*	*Agricultural Population*	*Man-Land Ratio (ha per head)*
Goa	197	86813	2.26
Orissa	7473	4239422	1.76
Meghalaya	1058	630170	1.67
Rajasthan	22739	15696002	1.44
Arunachal Predesh	361	300391	1.2
Gujarat	12412	10600842	1.17
Nagaland	644	578285	1.11
Sikkim	155	148361	1.04
Karnataka	12905	13145274	0.98
Kerala	12323	2394004	0.97
Madhya Pradesh	17337	18439378	0.94
Maharashtra	21167	23300848	0.9
Haryana	1384	4322234	0.87
Punjab	4270	3598603	0.84
Uttarakhand	1512	1818167	0.83
Andaman and Nicobar Islands	22	26228	0.83
Mizoram	218	280661	0.77
Chhatisgarh	5590	7405197	0.75
Andhra Pradesh	15772	21722389	0.72
Assam	3224	5031814	0.64
Jharkhand	1484	6770174	0.61
Tamil Nadu	8164	13779404	0.59
Jammu and Kashmir	1050	1848233	0.56
Uttar Pradesh	19272	35777375	0.53
Tripura	310	589712	0.52
West Bengal	5749	12964101	0.44
Himachal Pradesh	804	2063601	0.39
Manipur	230	613687	0.37
Bihar	6639	21719992	0.3

Source: *Agricultural Statistics at a Glance* (2008). Directorate of Economics and Statistics, Department of Agriculture and Cooperation, Ministry of Agriculture.

Population Pressure on Agriculture Land

The number of persons engaged in agricultural activities per hactare of net sown area district-wise has been calculated from Census data and Directorate of Economics and Statistics by the National Atlas and Thematic Mapping Organization to determine population pressure on agricultural land for 2001 (Fig. 12.5). The obtained figures range from as low as less than 1 to as high as above 5 persons and are divided into five groups: (i) more than 5, (ii) between 3-5, (ii) between 2-3, (iv) between 1-2 and less than 1. These divisions are, to a large extent, a reflection of population pressure and a relationship between rural population density and the intensity of land use. It is found that there are

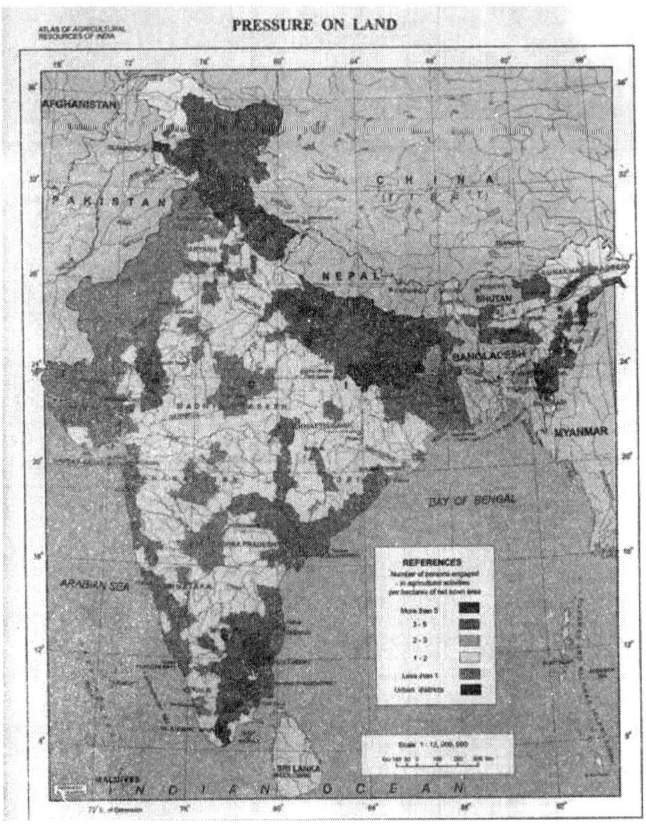

Fig. 12.5: India: Pressure on Land, 2001

Source: P. Nag (ed) (2010). *Atlas of Agricultural Resources of India*, p. 79.

large spatial variations in the effects of population pressure on agricultural land. The belt of first category includes districts of south Bihar, north Jharkhand, Himachal Pradesh, western Mizoram, southern Jammu and Kashmir, southern Rajasthan and eastern Tamil Nadu. The belt possesses fertile tracts, plenty of freshwater and a high density of agricultural workers. The districts falling in the belt are devoid of industries and urban centers. The second category with fertile soil, plenty of freshwater and multiple cropped

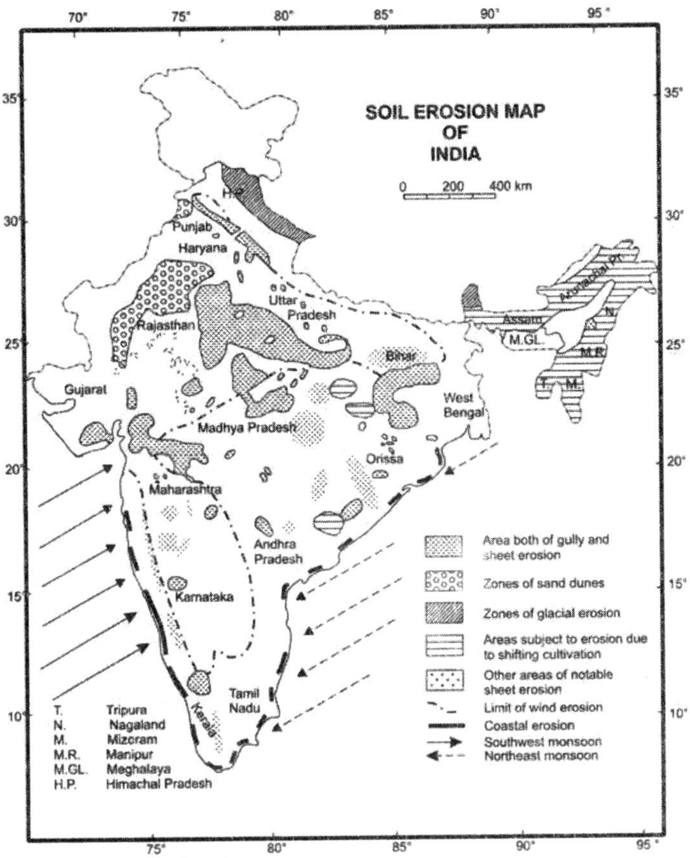

Fig. 12.6: India: Soil Erosion, 1980.

Source: F.R. Troeh, J.A. Hobbs and R.L. Donahue (1980). *Soil and Water Conservation*, p. 686.

areas sustains 3 to 5 persons per hactare of net sown area. This belt surrounds the first category with four nodes appearing around Himachal, Bihar plain, Tamil Nadu plain and Mahi plain. Some small patches are observed in the Hugli and Maldah districts of West Bengal, Puducherry, eastern Nagaland, eastern Mizoram, central Meghalaya and parts of Uttarakhand. The third category covers many of the districts of Upper Ganga plain, Lower Ganga plain, Tripura, Konkan coast, Tamil Nadu plain surrounding Puducherry and Cauvery basin, eastern Madhya Pradesh and Jammu and Kashmir having 2 to 3 agricultural workers per hectare of agricultural land. This category is characterized by moderate pressure of population on land.

Land Degradation

Land degradation denotes loss or reduction in land productivity which decreases its ability to maintain yields. It damages the physical, chemical and biological status of the land which makes the land less useful. Thus, land degradation is defined as the substantial decrease in either or both of an area's biological productivity or usefulness (Johnson and Lewis, 1995, 2). It is reversible in the short-term or irreversible in the long-term in its characteristics. The complete loss of soil though erosion is irreversible and fertility decline by loss of nutrients, which can be replaced by fertilizer inputs, is reversible (Young, 1998, 102). The overriding issue today is the perceived tension between the ecological and human- induced processes of land degradation. Land degradation is caused by soil erosion, desertification, waterlogging, salinity and alkalinity, over-grazing, deforestation, over cultivation of croplands, industrial pollution and its associated consequences. Similarly, an alternative thesis has argued for a more interconnected reasons for its development. For example, overgrazing, deforestation and over-cultivation of croplands are poverty-led degradations. By contrast, waterlogging, salinization, soil erosion and industrial pollution are linked to developmental reasons. The differences are striking, yet logical.

The causes of land degradation are complex and diverse. It is an intricate set of land system. Its rate of degradation varies from place to place and from time to time. Generally, it is paralleled by growing population densities with overcrowded, unproductive and widespread small holdings, without proper land reform and these are synonymous with developing world situations (Thakur, 2008, 2). But, population density is by no means its exclusive cause and it is not uniquely associated with densely populated land (Eden and Parry, 1996). However, there is little agreement about the sheer number of people accounting for land degradation rather mechanization of land exploitation (Thakur, 2008, 2). One theme that stands out in land degradation is the apparent advancing technologies, which cause instabilities and environment stress. Chisholm and Dumsday (1987, 27) view land degradation as the end result of any factor or combination of factors which damage the land, water or vegetation resources and restrict their use or productive capacity. Johnson and Lewis (1995, 2) focus upon land degradation as a human-induced process and identify human interference as the basic cause of land degradation. The United Nations Environment Programme (1993, 133-34) similarly discusses that it may be caused by a variety of anthropogenic pressures including deforestation, overgrazing, unsustainable agricultural practices and industrial activities. They can be broadly termed as social in character.

Processes of Land Degradation

Almost four decades of research in India have clearly established the study of land degradation as a legitimate domain for interdisciplinary enquiry. Some of this research has been pioneering work which has involved forays into unexplored domains. Recent trend in this area range from soil erosion to pollution and wasteland that emphasize the interactive effects of many endogenous and exogenous forces to uncover its underlying causes. The striking nature of this work is the absence of reliable data to support the rates, forms and trends of degradation. Leaving aside the problems of measurement, researchers generally accept that land degradation is occurring at different rates in different areas.

Accurate and reliable data on the current extent and severity of land degradation are rare and scanty. However, estimates are that out of a total area as many as 174 m ha suffer from degradation (National Commission on Agriculture, 1976; Vohra, 1982). It means on an average, at least 2 out of every 3 acres of land is in poor health. At least, half the sick land, i.e. one-third of the total, is almost completely unproductive. Another one-third is partially productive and it is only the remaining one-third which is in good health.

The chief land degradation problem is soil erosion. It is defined as the process by which soil particles are detached from their original site, transported and eventually deposited at new locations. It has become one of the major environmental problems and a serious constraint for agricultural production. It is severe, arduous and widespread in India. It results in a reduction in soil fertility and biological productivity which result in reduced cover and exposed soil surface. There are many physical and social factors which determine the extent and severity of soil erosion. The predominant physical factors are erosivity of rainfall, erodibility of soil, length and steepness of the slope and flood. The foremost social factors are deforestation, overgrazing, choice of land use and the method of cultivation. Since the terrain varies from plains to undulating to hilly and to mountainous, 38 per cent of agricultural land is situated on slopes. Steep slopes encourage quick draining of the land, especially southern slopes of the Himalayan and the Western Ghats. Our conventional agricultural practices leave the soil bare, untied and unfastened at the onset of the rains. Heavy rainfall during the monsoon, which often comes in vigorous and intense showers, moves away top soil and nutrients from the agricultural fields. The Indian rivers together transport to the seas annually about 1500 m tonnes of eroded materials (Valdiya, 1987). The Brahmaputra is one of the leading sediment transporting rivers of the world. During the monsoon the river carries an average of 2.1 m tonnes of sediment per day. The Ganga, with a drainage basin of 1.1 million sq km, carries an annual sediment load of 1.46 billion tonnes of soil. Soil erosion is categorized into four types: (a) splash, (b) sheet, (c) rill, and (d) ravines and gullies. Splash erosion is the spattering of small soil particles caused by the impact of rain drops on very wet soils. Sheet erosion involves wide flow of water moving across a sloping field. Rill erosion results from water moving through small channels in the soil surface. Gully erosion occurs where water is concentrated into a high velocity flow through a large channel that lacks vegetation. It cuts deep swathes through the soil. Gullies are the most horrifying eroded landscapes produced by different processes which have threatened vast tracts of agricultural land in India. They are the steep sided channels carrying flow only during rainstorms having few meters depth and width. They are called trenched channels or ravines depending upon environmental conditions. The ravines start as gullies that deepen owing to: (a) high intensity of rainfall, (b) loose and friable nature of the soil, (c) faulty agricultural practices, (d) deforestation and (e) overgrazing (Shafi, 2006, 411). Gullies produce badland topography famous for their rugged beauty and scenery (Singh and Singh, 1961, 153). They are produced either by overland flow, or subsurface flow and seepage or deforestation, overgrazing and ill-considered tillage. They are particularly susceptible to erosion intensity and concentration of rainfall during the monsoon and also in part due to the erodibility of deep alluvial soils. Soil erosion by ravines and gullies is widespread in India (Haigh, 1984; National Commission on Agriculture, 1976). Severe ravine zones in India are found along the margins of the Gangetic basin and in the semi-arid northwest. Gullies and ravines are concentrated within a big triangle whose nodes are Jammu, Surat and Murshidabad (Fig. 12.6). Ahmad (1970) and Sharma (1980) have identified four major areas of ravines and gullies. They are: (1) The Yamuna-Chambal ravine zone, (2) The Gujarat ravine zone, (3) The Punjab Shiwalik foothills ravine zone and (4) The Chotanagpur ravine zone. Other areas of substantial ravine erosions are in Mahi and Sabarmati, the upper Mahanadi valley, Upper Son valley, Upper Narmada and Tapi valleys located in the southern fringes of the Gangetic plain, Shiwalik and bhabar tract of the western Himalayan foothills and edges of Ganga khadar in western Uttar Pradesh. Most of these ravine lands occur in the states of Uttar Pradesh, Rajasthan, Gujarat, Madhya Pradesh, Punjab and Jharkhand. In contrast to gullies and ravines, sheet erosion is widespread over sloping deforested terrain, unterraced uplands of the Peninsular region, the Indo-Gangetic plains, coastal plains, the Western Ghats and Northeastern hills. It is a function both of slope and amount of rainfall. On an average, 1/5th inch of soil is washed away every year from India.

Several exciting developments in the understanding of causal mechanisms of soil erosion have occurred during recent decades. These include the discovery of localized soil erosion accelerated by road construction, spread of agriculture on increasingly steeper slopes and marginal lands, overgrazing and by deforestation. Landslides, which are caused by heavy and prolonged rainfall, cutting of slopes and earthquake tremors, are common in geodynamically sensitive belts in the Himalayan region. The Darjeeling Himalaya alone has recorded more than 20,000 landslides in one day in vulnerable belts (Valdiya, 1987). The Brahmaputra valley and its adjoining highlands are extremely seismic. Between 1920 and 1969, 416 shocks of magnitude 5 and above occurred in the northeast. As a corollary, over 2500 earthquakes occurred between 1970 and 1988 causing enormous quantities of silt and debris being washed down into the plains and the consequent rising of the river beds. Similarly, the earthquakes of 1897 and 1950 are among the most severe, resulting into extensive landslides on the Himalayan slopes temporarily blocking the courses of the Subansiri, Dibang and Dihang. However, involvement of geographers in this aspect of soil erosion has been relatively small.

Desertification, which is a decrease in the productivity potential of the land under arid, semi-arid and dry sub-humid climates and which may eventually lead to desert- like conditions, is prevailing in Rajasthan (Houerou, 1994, 2). Studies on desertification and land degradation appear frequently in the literature in India as geographers have effectively researched on it, especially Central Arid Zone Research Institute, Jodhpur. Desertification in Rajasthan has been studied by a team of geographers, combining remote sensing techniques, extensive field observations, national statistics and spatial modeling at CAZRI since the 1970's. Results of such studies show that although climate is one of the most important factors in the development of desertic conditions, the abuse of land and overexploitation of its vegetal and water resources are primarily responsible for the expansion of the desert (Thakur, 2007, 19). Over the last 8000 years, the Thar desert of Rajasthan has advanced 1500 km east and in the last 60 years, it has encroached upon 13,000 ha of land in Rajasthan, Gujarat and Haryana in a process known as desertification, by which arid conditions spread to areas that were previously moist (Valdiya, 1987). This is known as ecological change that converts non-desert lands into deserts. These changes include the loss of soil moisture, the loss of trees, the erosion of soil by wind and the formation of sand dunes on formerly vegetated land. These alterations can be brought on by a relatively short-term climatic variability, by long-term climatic change or by human activities that reduce moisture (Pulsipher and Pulsipher, 2006, 256).

India has many mineral resources, particularly in the older rocks of the Peninsula covered under Archaean, Dharwarian or Gondwana system of rocks. The most mineral-rich part of India is the Chotanagpur plateau of the poorer north-eastern Peninsula from where coal, iron-ore, bauxite, aluminium, mica, manganese, limestone, copper and other minerals are extracted. But during the last few decades, activities of mineral exploitation have increased substantially to fulfill the growing demands of the people. Mining is an environmentally damaging activity causing large-scale deforestation, destruction of wildlife and creates huge wastes and requires space for their dumping. It also triggers a number of environmental problems like land degradation and pollution. Land degradation is the most serious environmental impact of mining. Surface mining involves total removal of materials as in the case of sand and gravel extraction, low-grade ore, phosphates, slates, limestone quarrying and construction materials. They scar the land by leaving behind a degraded and pockmarked landscape devoid of soil cover. Land degradation results in gravel pits and huge depressions. Due to land destruction, soils are blighted and water fouled. Subsidence and collapse are common in most mining areas. Subsidence alters drainage pattern and damages surface infrastructure and buildings. Collapse drastically alters the surface configuration of the affected areas and degrades the affected areas. Water quality changes and, finally, area is abandoned unless it is restored. The degree of damage varies with the topographical setting, the type and depth of overburden, the dimension of the deposit mined and climatic conditions in which opencast mining is carried out. Land is also degraded due to unregulated mining or quarrying of coal, limestone and other minerals. The large-scale landslides near Dehradun due to indiscriminate quarrying of limestone is an important example (Govind, 1989; Thakur and Kurl, 1996). The other examples are the strip and shaft mining at Makum coal mine in upper Assam which have damaged the quality of top soil (Govind, 1989) and coal mining belt in Chotanagpur (Prasad, 1983; Prasad, 1988; Prasad, 2008). As it stands, the total land area affected by mining is of the order of 80,000 ha.

Waterlogging and soil salinization are the twin important problems in India arising from canal irrigation. They constitute the second major threat to the soil and have claimed 13 m ha and threaten many more (Bowonder, Ramana and Ravi and Srinivas, 1987). In 2001, 55 m ha of land were being irrigated of which 30 per cent was being irrigated twice a year. At state level, it ranges from 8 per cent in Tripura to 16.5 per cent in Rajasthan to as much as 76 per cent in Punjab where irrigation facilities are extensive. Canals irrigate 39 per cent of net irrigated area and is prevalent in the northern plains, eastern coastal plains and along the narrow valleys of the Deccan rivers. This includes large areas of land in Punjab, Haryana, Uttar Pradesh, Bihar, parts of southern states, Orissa, West Bengal and Assam (Fig. 12.7). It is in these areas that waterlogging and soil salinity occur. Waterlogging occurs when the level of groundwater table rises and causes a reduction of oxygen available to plant root systems (Bowonder, Ramana, Ravi and Srinivas, 1987, 331). Salinity occurs when the water table is a meter or so below the soil, water flows to the surface, evaporates and leaves salt deposits (Livernash, 1995, 90). About 3 m ha are apprehended to be influenced by salinity. The important contributory factors for waterlogging include lack of drainage in irrigated areas, overuse of water, seepage of water from canals and distributaries, siltation of drains, lack of land development before provision of irrigation, lack of lining of canals and distributaries, adoption of unsuitable cropping patterns, poor water use practices and poor conjunctive use of groundwater facilities (Bowonder, Ramana, Ravi and Srinivas, 1987, 331). Singh (1994) states that the genesis of the waterlogging problem also lies in the unfavourable topography, unfavourable sub-soil geohydrology, lithology, rainfall, temperature and evaporation. Given these factors in the Karnal district, the fertile areas have been turned into unproductive land. This suggests that extension of irrigation, the design of canal and cropping pattern should be carried out scientifically keeping in mind the degree of impregnation of the soils with salt.

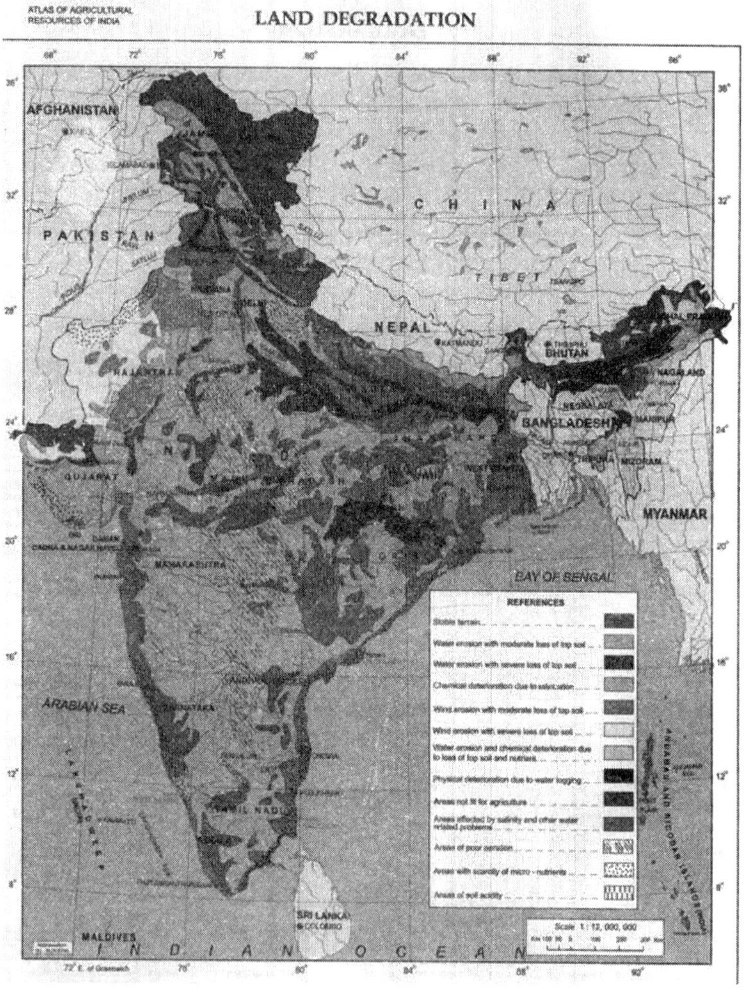

Fig. 12.7: India: Land Degradation, 2001

Source: P. Nag (ed) (2010), *Atlas of Agricultural Resources of India*, p. 41.

Patterns of Land Degradation

The sum total of physical, chemical and biological processes of land degradation is the vast expanse of wasteland. It is generally unproductive, ruinous and devastating land which is normally uncultivated. Broadly, wasteland is defined as a land which is presently lying unused to its optimum potential, thus, leading to national wastage (Government of India, 1990, 11). According to Stamp (1962), wasteland may be defined as that land which has been previously used but which has been abandoned. The economists have defined it as any land which gives less than 25 per cent of its potential may be considered as wasteland. Bhumbla and Khare (1987) have defined wasteland as the land which are ecologically unstable, have lost their topsoil and have developed toxicity in the crops and trees. The National Remote Sensing Agency has estimated 55.27 m ha (17.45%) as wasteland in the country (Fig. 12.8) (Gautam and Narayan, 1988) and has given functional and scientific definition of wasteland. However, a precise definition of wasteland has not emerged. It is a dynamic concept and the estimation of wasteland on the basis of ambiguous and conflicting definitions would be difficult as the status of land keeps changing with change in the status of technology of the region (Mohammad, 1990).

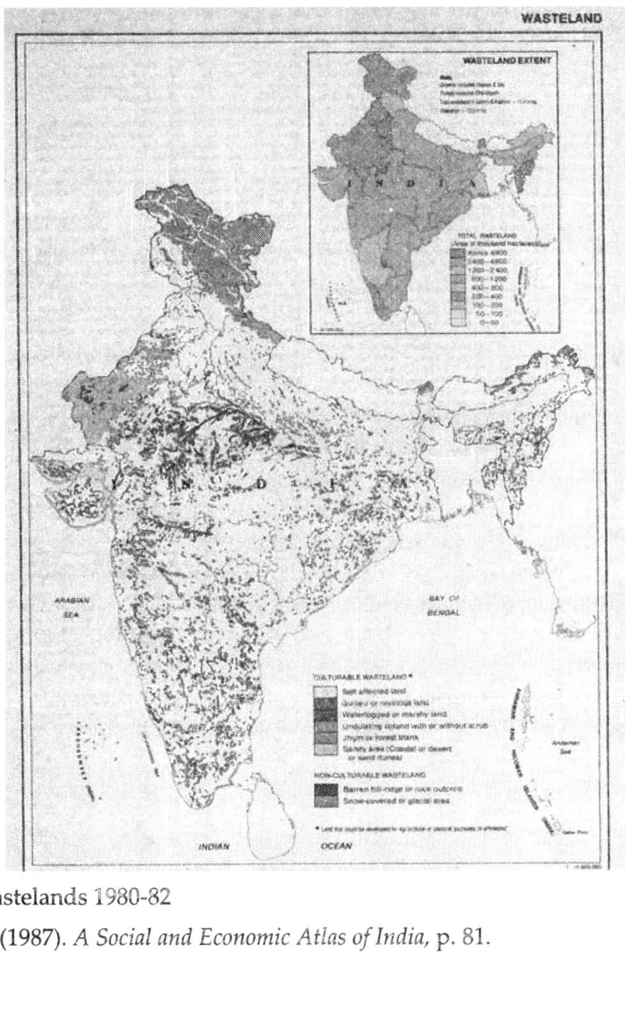

Fig. 12.8: India: Wastelands 1980-82

Source: S. Muthiah (1987). *A Social and Economic Atlas of India*, p. 81.

Wasteland is classified into two categories: culturable and non-culturable. Culturable wasteland is capable or has the potential for development of agriculture, pasture and deforestation but is not being used to its optimum level owing to constraints, such as soil erosion, salinity, alkalinity, waterlogging, deforestation, occurrence of abundant deposit of sand, unfavourable physiographic positions and human negligence. These lands are put to some productive use after proper treatment and reclamation measures. On the other hand, non-culturable land is that which is barren and cannot be put to any productive use, e.g. snow covered and glacier areas and barren outcrop hill ridges.

Assessment of Degraded Lands[2]

There are many agencies in the country which have assessed the extent of degraded lands under various categories. The estimates by these agencies are at great variance due to differences in approaches and criteria for assessment. The following are the prominent agencies:

(i) Assessment by Icar

The National Bureau of Soil Survey and Land Use Planning, Nagpur has assessed soil degradation on 1:2,50,000 scale for the country. Accordingly, about 146.82 m ha (45 per cent of total geographical area) is degraded due to different degrading agents. Table 12.3 consists of water erosion (93.6 m ha), wind erosion (9.4 m ha), waterlogging (14.3 m ha), salinity/alkalinity (5.9 m ha), acidity (16.0 m ha) and other complex problems (7.4 m ha). Water erosion is quite widespread in the country with sizeable areas in Madhya Pradesh including Chhattisgarh, Andhra Pradesh, Uttar Pradesh, including Uttarakhand, Maharashtra, Karnataka, Jammu and Kashmir, Gujarat and Odisha. The areas under waterlogging are more in Uttar Pradesh, Kerala, Bihar including Jharkhand and Andhra Pradesh. The large areas under saline/alkaline soils are in Rajasthan, Uttar Pradesh and Maharashtra. The acid soils are more prominent in Madhya Pradesh including Chhattisgarh and Arunachal Pradesh. The states suffering due to degradation are Mizoram (89%), Himachal Pradesh (75%), Kerala (67%), Nagaland (60%), Tripura (60%), Madhya Pradesh and Chhattisgarh (59%), Delhi (55%), Andhra Pradesh (54%), Arunachal Pradesh (54%), Meghalaya (54%) and Uttar Pradesh and Uttarakhand (52%).

(ii) Assessment of Wastelands by Ministry of Rural Development

The Ministry of Rural Development assigned and entrusted National Remote Sensing Agency (NRSA), the responsibility for identification of various types of wastelands and their extent in the country. For the categorization, the wasteland is described as "*degraded land which can be brought under vegetative cover with reasonable effort and which is currently underutilized and land which is deteriorating for lack of appropriate water and soil management or on account of natural causes. Wastelands can result from inherent/imposed disabilities such as by location, environment, chemical and physical properties of the soil or financial constraints*".

NRSA has used the IRS LISS-III data for generation of wastelands maps/data base. The methodology involves interpretation of enlarged satellite false colour data (1:50,000 scale) based on image characteristics such as tone colour, texture, pattern, shape, size, location and association to identify and delineate different types of wastelands. Old vector, IRS LISS-III data of 2003 and

Table 12.3: Land Degradation in India

Area ('000 ha)

Sl. No.	State	Water erosion	Wind erosion	Water logging/ flooding	Salinity/ Alkalinity	Soil acidity	Complex problems	Total degraded area	T.A.G.	Degradat Area in %
1.	Jammu and Kashmir	5460	1360	200	0	0	0	7020	22224	31.6
2.	Himachal Pradesh	2718	0	1303	0	157	0	4178	5567	75
3.	Punjab	372	282	338	288	0	0	1280	5036	25.4
4.	Haryana	315	536	146	256	0	214	1467	4421	33.2
5.	Uttar Pradesh and Uttaranchal	11392	212	2350	1370	0	0	15324	29441	52
6.	Delhi	55	0	6	10	0	11	82	148	55.4
7.	Rajasthan	3137	6650	53	1418	0	110	11368	34224	33.2
8.	Gujarat	5207	443	523	294	0	1666	8133	19602	41.5
9.	Maharashtra	11179	0	0	1056	517	303	13055	30771	42.4
10.	Madhya Pradesh and Chhattisgarh	17883	0	359	46	6796	1126	26210	44345	59.1
11.	Kerala	76	0	2098	0	138	296	2608	3886	7.1
12.	Tamil Nadu	4926	0	96	96	78	138	5334	13006	41
13.	Karnataka	5810	0	941	110	58	712	7631	19179	39.8
14.	Andhra Pradesh	11518	0	1896	517	905	156	14992	27505	54.5
15.	Goa	60	0	76	0.4	2	24	162.4	370	439
16.	Bihar and Jharkhand	3024	0	2001	229	1029	0	6283	17387	36.1
17.	West Bengal	1197	0	710	170	556	119	2752	8875	31
18.	Orissa	5028	0	681	75	263	75	6122	15571	39.3
19.	Sikkim	158	0	0	0	76	0	234	710	33
20.	A & N Islands	187	0	0	9	0	9	205	825	24.8
21.	Arunachal Pradesh	2372	0	176	0	1955	0	4503	8374	53.8
22.	Mizoram	137	0	0	0	1050	694	1881	2108	89.2
23.	Manipur	133	0	111	0	481	227	952	2233	42.6
24.	Nagaland	390	0	0	0	127	478	995	1658	60
25.	Assam	688	0	37	0	612	876	2213	7844	28.2
26.	Tripura	121	0	191	0	203	113	628	1049	59.9
27.	Meghalaya	137	0	7	0	1030	34	1208	2243	53.9
	Total	**93680**	**9483**	**14299**	**5944.4**	**16033**	**7381**	**146820**	**328602**	

Source: National Bureau of Soil Survey and Land Use Planning, 2005.

ground data information collected were used to prepare wasteland map (2005). As per study by NRSA, the total extent of wastelands in the country is 55.27 m ha compared to 63.85 m ha of the year 2000. The extent of wastelands, therefore, has decreased over the years. The extent of wastelands category and state-wise is contained in Tables 12.4, 12.5, and 12.6). The wastelands have been now classified into 28 categories compared to 13 categories in the year 2000.

The land with and without scrub is maximum, i.e. 18.4 m ha (5%) followed by degraded forest covering 10.8 m ha. The other wastelands, i.e. gullied and ravinous lands, salt affected soils, waterlogged areas, shifting cultivation areas, industrial and mining wastelands etc. also need immediate attention for increasing the productivity of these lands.

Table 12.4: State-wise Wastelands of India

(Area in Sq Km)

S. No.	State	No. of Districts	TGA	Total WL	% To TAG
1.	Andhra Pradesh	23	275068	45267.15	16.46
2.	Arunachal Pradesh	16	83743	18175.95	21.7
3.	Assam	23	78438	14034.08	17.89
4.	Bihar	37	94171	5443.68	5.78
5.	Chhattisgarh	16	135194	7584.15	5.61
6.	Goa	2	3702	531.29	14.35
7.	Gujarat	25	196024	20377.74	10.4
8.	Haryana	19	44212	3266.45	7.39
9.	Himachal Pradesh	12	55673	28336.8	50.9
10.	Jammu and Kashmir*	14	101387	70201.99	69.24
11.	Jharkhand	19	79706	11165.26	14.01
12.	Karnataka	27	191791	13536.58	7.06
13.	Kerala	14	38863	1788.8	4.6
14.	Madhya Pradesh	49	308252	57134.03	18.53
15.	Maharashtra	33	307690	49275.41	16.01
16.	Manipur	9	22327	13174.74	59.01
17.	Meghalaya	7	22429	3411.41	15.21
18.	Mizoram	8	21081	4469.88	21.2
19.	Nagaland	7	16579	3709.4	22.37
20.	Orissa	30	155707	18952.74	12.17
21.	Punjab	17	50362	1172.84	2.33
22.	Rajasthan	32	342239	101453.86	29.64
23.	Sikkim	4	7096	3808.21	53.67
24.	Tripura	4	10486	1322.97	12.62
25.	Tamil Nadu	29	130058	17303.29	13.3
26.	Uttaranchal	13	53483	16097.46	30.1
27.	Uttar Pradesh	70	240928	16984.16	7.5
28.	West Bengal	18	88752	4397.56	4.95
29.	Union Territory	20	10973	314.38	2.87
	Total	**597**	**3166414**	**552692.25**	**17.45**
	Total (Mha.)		**316.64**	**55.27**	

Source: 1:50,000 Wasteland Mps – 2003 prepared based on IRS – LISS III Data.

Unsurveyed Areas (J & K) =120849.00

Total Geographical Area (TGA) = 3287263 Sq. Km= 328.7 m ha.

Table 12.5: Summary of Degraded Lands

		Lakh ha	Million Hactare
1.	Gullied Lands (1 to 3)	19.03	1.9
2.	Land with or without Scrub (4&5)	187.95	18.8
3.	Waterlogged (6&7)	9.74	0.97
4.	Saline/Alkali (8 to 10)	12.02	1.20
5.	Shifting Cultivation (11 to 12)	18.75	1.88
6.	Degraded Forest and Agricultural Land Under Forest (13 & 14)	126.6	12.66
7.	Degraded Pastures/Plantation (15 & 16)	21.5	2.15
8.	Sands (17 & 23)	34	3.4
9.	Mining and Industrial Wastelands (24 & 25)	1.97	0.2
10.	Barren/Stony/Snow Covered (26 to 28)	121.17	12.11
	Total	**552.73**	**55.27**

Source: S.S. Singh, 2008.

Table 12.6: Category-wise Wastelands of India

S. No.	Category of Wasteland	Area (Sq. Km)	Per Cent
1.	Gullied and/or Ravinous - Shallow	10283.06	0.32
2.	Gullied and/or Ravinous -Medium	4685.43	0.15
3.	Gullied and/or Ravinous – Deep	4070.85	0.13
4.	Land with Scrub	150566.6	4.76
5.	Land without Scrub	37382.89	1.18
6.	Waterlogged and Marshy - Permanent	5341.15	0.17
7.	Waterlogged and Marshy - Seasonal	4403.82	0.14
8.	Saline/Alkaline- Strong	2569.69	0.08
9.	Saline/Alkaline-Moderate	5349.64	0.17
10.	Saline/Alkaline- Slight	4104.72	0.13
11.	Shifting Cultivation - Abandoned	12218.99	0.39
12.	Shifting Cultivation –Current	6546.87	0.21
13.	Degraded Forest - Scrub Domain	108417.76	3.42
14.	Agriculture Land inside Notified Forest	18134.05	0.57
15.	Degraded Pastures/Grazing Land	19344.3	0.61
16.	Degraded Land Under Plantation Crops	2138.24	0.07
17.	Sands -Flood Plain	1945.55	0.06
18.	Sands – Levees	32.24	0
19.	Sands – Coastal Sand	943.14	0.03
20.	Sands – Semi-Stab. To Stab. (>40 m)	2672.88	0.08
21.	Sands –Semi-Stab. Mod (15-40 m)	16380.7	0.52
22.	Sands –Semi-Stab. To Stab. Low (<15)	10262.95	0.32
23.	Sands – Closely Spaced Inter- Dunal Area	1746.74	0.06
24.	Mining Wastelands	1421.72	0.04
25.	Industrial Wastelands	555.63	0.02
26.	Barren Rocky/Stony Waste Area	57747.11	1.82
27.	Steep Sloping Area	9097.38	0.29
28.	Snow Covered and/or Glacial Area	54328.16	1.72
	Total	**552692**	**17.45**
	Total (Mha)	**55.27**	**17.45**

Source: Wasteland Atlas (2005), Ministry of Rural Development, Government of India.

Note: 1,20,849.00 Sq. Km. in Jammu & Kashmir is not mapped.

Wastelands in Agro-Climatic Zones

Table 12.7 reflects ten categories of culturable wastelands agro-climatic region collapsed from the original twenty-five categories (Table 12.2). Chadha, Sen and Sharma (2004) calculated these categories from NRSA Wasteland Atlas. As can be observed from Table 12.7, degraded forests (36.07%) and pastures (29.75%) are important in the Western Himalayan region followed by degraded plantation crop (16.68 per cent) and scrub (15.17 per cent), while shifting cultivation (41.74 per cent) is important in the Eastern Himalayan region followed by sands (18.79 per cent) and degraded forest (15.81 per cent). In like manner, waterlogging (35.05 per cent), scrub (22.47 per cent) and sands are predominantly found in both middle Gangetic and lower Gangetic plains. More specifically, a serious problem of waterlogging (20.11 per cent) and saline/alkaline soils is reported to be in the upper Gangetic plains. Similarly, the Trans-Gangetic plains display more proportions of scrub (28.09 per cent) and sands (24.18 per cent). Increasingly, a striking trend is observed with respect to degraded forests and scrubs, which are predominating in Eastern Plateaus and Hills (51.95 per cent and 43.64 per cent), Central Plateaus and Hills (29.86 per cent and 46.05 per cent), Western Plateaus and Hills (29.18 per cent and 60.42 per cent), Southern Plateaus and Hills (54.61 per cent and 40.90 per cent), East Coast Plains and Hills (35.03 per cent and 42.81 per cent), West Coast Plains and Hills (28.85 per cent and 55.70 per cent) and Gujarat Plains and Hills (13.87 per cent and 55.52 per cent). Lastly, the Western Dry region is exceptionally covered by sands (64.71 per cent) and scrub (19.74 per cent). Thus, it is clear and worth noting that agro-climatic regions are important units for the analysis and comprehension of contemporary patterns and processes of land degradation. The lasting contribution of this unit is to infuse region's individuality with a concern for the construction of spatial inequality.

Table 12.7: Agro-climatic Region-wise Composition of Wasteland, 2006 (As Proportion of Total Geographical Area)

Agro-climatic Regions	*Gullied/ Ravinous*	*Sands*	*Scrub*	*Water logged/ marshy*	*Saline/ Alka-line*	*Shifting Cultiva-tion Area*	*Degra-ded Forest*	*Degra-ded Pastur-es*	*Degra-ded Planta-tion Crop*	*Mining/ Indus-trial Waste*
Western Himalayan Region	0.83	0.71	15.17	0.2	0.01	0	36.07	29.75	16.68	0.59
Eastern Himalayan Region	0	18.79	4.32	8.26	0	41.74	15.81	11.06	0.01	0
Lower Gangetic Plains	3.16	15.94	22.47	35.05	2.42	0	13.12	6.97	0	0.87
Middle Gangetic Plains	2.34	3.22	25.46	29.63	4.84	0	31.55	2.51	0.17	0.29
Upper Gangetic Plains	11.3	2.38	13.22	20.11	41.99	0	10.38	0.35	0.2	0.07
Trans-Gangetic Plains	2.73	24.18	28.09	10.33	5.88	0	13.69	11.86	2.7	0.54
Eastern Plateau and Hills	1.89	0.14	43.64	0.09	0	0.26	51.95	0.42	0.85	0.77
Central Plateau and Hills	11.58	0.99	46.05	0.07	2.69	0	29.86	7.79	0.8	0.16
Western Plateau and Hills	7.99	0.03	60.42	0	0.03	0	29.18	2.28	0	0.07
Southern Plateau and Hills	1.2	0.34	40.9	0.02	1.64	0	54.61	0.58	0.32	0.38
East Coast Plains and Hills	2.06	3.34	42.81	5.74	7.56	0.19	35.03	2.14	0.82	0.3
West Coast Plains and Hills	0.09	0.94	55.7	4.66	1.79	0	28.85	3	4.92	0.04
Gujarat Plains and Hills	2.58	0.48	55.52	6.77	19.46	0	13.87	0.99	0.2	0.13
Western Dry	1.14	64.71	19.74	0.09	1.43	0	2.58	10.19	0	0.11
Total	**4.31**	**10.9**	**39.47**	**3.54**	**4.56**	**1.91**	**28.69**	**5.26**	**1.1**	**0.26**

Source: NRSA Wastelands Atlas, 2000 and adapted from G.K. Chadha, S. Sen and H.R. Sharma (2004), p. 366.

Land Development Policies

Land must be protected against degradation because there are many areas in the country where land resources are insufficient to meet the food needs of the present and future populations. Land degradation has to be arrested and land must be restored to its original state or reclaimed to a new use after mining. Land development is an important element in land management as it includes change or transformation in the character of land to suit the needs of the society.

Land development is the process for regaining land productivity which provides an efficient way of developing the available land resources and of improving the standard of living for the vast rural population. The main purpose of land development is to prepare the rudimentary land to a situation where land use types are matched with

Table 12.8: Land Developed Under Various Wasteland Development Programmes, Since Inception Upto Tenth Five Year Plan

(Area in Lakh ha. and Expenditure in Rs. Crore)

S. No.	Ministry/Scheme And Year of Start	Progress since inception upto IX Plan		Progress during X Plan		Progress since inception upto X Plan	
		Area	Expr.	Area	Expr.	Area	Expr.
Ministry of Agriculture, (Department of Agriculture and Cooperation							
1	NWDPRA (1990-91)	69.79	1877.74	23.3	1147.82	93.9	3.25.56
2	RVP & FPR (1962 & 81)	54.88	15156.26	9.98	727.98	64.86	2244.24
3	WDPSCA (1974-75)	2.58	166.27	1.35	129.31	3.93	295.58
4	RAS (1985-86)	5.81	76.36	1.3	45.35	7.11	121.74
5	WDF (1999-00)	0	0	0.59	26.02	0.59	26.02
6	EAPs	13.35	2039.81	4.8	1927.54	18.15	3967.35
	Sub-Total	**146.41**	**5676.47**	41.32	**4004.02**	**187.73**	**9680.49**
Ministry of Rural Development (Department of Land Resources)							
1	DPAP (1973-74)	68.95	3284.74	68.32	1557.76	13.27	4842.5
2	DDP (1977-78)	33.56	797.38	45.17	1152.5	78.73	19+49.88
3	IWDP (1988-89)	37.34	616.51	62.22	1821.64	99.56	2438.15
4	EAPs	1.4	18.39	3.6	274.28	5	292.67
	Sub Total	141.25	4717.02	179.31	4806.18	3210.56	9523.2
•		287.66	10393.49	220.63	8810.2	508.29	19203.69

*Includes tentative achievement of 2006-7.

Abbreviations

NWDPRA	-	National Watershed Development Project for Rainfed Area
RVP & FPR	-	River Valley Project and Flood River
WDPSCA	-	Watershed Development Project for Shifting Cultivation Area
RAS	-	Reclamation of Alkali Soil
WDF	-	Watershed Development Fund
EAPs	-	Externally Aided Projects
DPAP	-	Drought Prone Area Programme
DDP	-	Desert Development Programme
IWDP	-	Integrated Wasteland Development Programme

Source: Working Group Report of Natural Resource Management for Eleventh Five Year Plan, constituted by Planning Commission, New Delhi, Feb., 2007.

existing land quality. Development potential is the process in which the regenerated land is maintained or developed in a sustainable manner in which the society gets maximum economic gain without affecting the quality of land (Patra, 2004, 177). Land development is concerned with reclamation of degraded lands, restoring damaged lands, restoration of mined land, wasteland development, desertification control, remedial measures of waterlogging and salinity problem, rehabilitation of land, regeneration of different types

Table 12.9: Meta-Analysis[3] Summary of the Watershed Studies (311 cases)

Indicator	*Particulars*	*Unit*	*No. of Studies*	*Mean*
Efficiency	B:C ratio	Ratio	12	2.14
	IRR	Percent	840	22.04
Equity	Employment	Mandays/ha/year	39	181.5
Sustainability	Irrigated area	Percent	97	33.56
	Cropping Intensity	Percent	115	63.51
	Rate of Runoff	Percent	36	-13
	Soil Loss	Tons/ha/year	51	-0.82

Source: S.S. Singh (2008).

of degraded lands, regeneration measures for gullied lands, rocky/stony outcrops, sandy area and mining wasteland and industrial wasteland. Since the causes are tied to varied mixes of physical, social and economic conditions, the remedied must be diverse and tailored to the problems and need of locales.

Wasteland Reclamation

To support and coordinate a nationwide effort, the Government of India constituted the National Land Use Wasteland Development Council in May 1985. Two Boards were set up to function under the Council: (a) The National Land Use Conservation Board in the Ministry of Agriculture and Rural Development, (b) The National Wasteland Development Board under the Ministry of Environment and Forests. The Council is the highest policy planning and coordinating agency for the management of land resources in India. The functions of National Land Use Conservation Board are (Government of India, 1990, 5-6): (i) To formulate a national policy and respective plan for conservation, management and development of land resources of the country taking into account appropriate land use and soil capability and other factors, (ii) To make an overall review of the progress of implementation of ongoing schemes and programmes connected with conservation and development of land resources, soils and allied matters, (iii) To consider and review proposals concerning soil surveys and general assessment of land resources, (iv) To consider measures for ensuring that good agricultural land is not indiscriminately diverted to non-agricultural purposes, (v) To act in full collaboration with the National Wasteland Board in regard to matters of common interest.

The major focus of the National Wasteland Development Board is to reclaim wastelands in the country through a massive programme of afforestation and other appropriate measures. It also restores the disturbed ecological balance and provides employment opportunities in the rural areas. The key features of the programme are: (i) To formulate within the overall national policy, a perspective plan and programme for the management and development of wastelands, (ii) To identify the wastelands of the country, (iii) To promote, encourage and finance development of wastelands in the country through the active involvement of non-government organizations, voluntary agencies and the public at large, including the landless, (iv) To collaborate with the Central and State Government departments, agencies, local bodies and voluntary agencies with a view to mobilizing manpower, funds and other inputs required for wasteland development programme, (v) To create a reliable data base and documentation centre on related aspects of wasteland development, (vi) To interact with financial institutions for funding wasteland development programmes, (vii) To create general awareness in favour of wasteland development especially through the education system, and (viii) To act in collaboration with the National Land Use and Conservation Board in regard to matters of common concern.

The National Land Use and Wasteland Development Council is headed by the Prime Minister and represented by State Chief Ministers and Heads of the Union Territories. On January 5, 1985, the then Prime Minister Rajiv Gandhi announced ambitious programme of afforesting 5 m hectares per year to enhance the productivity of wastelands (Nagaraja and Gautam, 1996, 145-146). On the basis of normal requirement of 2000 saplings per hectare, the country would require 1000 crore saplings per year. To meet a demand of this magnitude, the programme of decentralized nurseries has been taken up. However, special difficulties have risen because they were not found satisfactory associated with increasing the rate of afforestation, increasing the availability of fuelwood and fodder for the rural poor and securing people's participation in afforestation activities (Thakur, 2007, 24). There was an increasing awareness that policies which may be effective in one part of the country may be ineffective in another. Associated with this trend of evaluation were needs to develop better restructured programme to take concrete action with regard to the emerging concerns for which the Council was set up. As remote sensing technology has become refined, it has emended as an increasingly important tool for monitoring and preparing detailed base level action plan for integrated management of wastelands through soil and moisture conservation, grass and trees cover (Government of India, 1990, 31). Meanwhile, using satellite data, coupled with computer cartography and Geographic information systems, the Agency has displayed unprecedented insight by publishing the maps showing the spatial distribution of wastelands on a scale of 1:3500000 and 1:1000000. The Agency has also completed wasteland mapping at the 1:50000 scale for 237 severely affected districts in 19 states in India and the more is to be done in future (Nagaraja and Gautam, 1996, 145-154). On the other hand, the National Wasteland Development Board has also published district-wise extent of wastelands in the country (National Wasteland Development Board, 1987). The map has been the central tool of the geographer and this will undoubtedly continue. The revolution of harnessing computers to geographic databases, the development of computer mapping technology, plus the refinement of remote sensing, all provide geographers with powerful mapping tools that promise to reveal much about patterns, relations and processes underlying wasteland development and reclamation.

Watershed Programmes[2]

There are many programmes implemented by Ministries of Central Government, namely, Ministry of Agriculture, Ministry of Rural Development and Ministry of Environment and Forests for the development of wastelands and dry land/rainfed areas for enhancing production. With an expenditure of Rs. 19203 crore under Central Sector Programmes, an area of 50.8 m ha have been developed (Singh, 2008, 7). In the process of recamation, watershed plays a very important role because it is the widely used water drains into a stream, river, lake, or other body of water. Watershed management is a process of managing an area in order to protect and rehabilitate land and water and associated resources (Grover, 2007, 1933). The main watershed programmes for enhancing the productivity of wastelands are (Singh, 2008,7): (i) National Wasteland Development Project for Rainfed Areas, (ii) Soil and Water Conservation in the Catchments of Rivers, (iii) Reclamation of Alkali soils, (iv) Wasteland Development Projects in Shifting Cultivation Areas, (v) Drought Prone Areas Programme, (vi) Desert Development Programme, and (vii) Integrated Wasteland Development Programme.

(i) *National Wasteland Development Project for Rainfed Areas:* This is a programme of the Ministry of Agriculture which was launched in the VII Plan (1985-86). Arguably, the most important area of concern of NWDPRA is to improve production in the vast rainfed areas and to rehabilitate ecological balance. It has been extended for operation beyond also. In response, the restructured NWDPRA allows a greater degree of flexibility in choice of technology, decentralization of procedures, provision for sustainability and reemphasizes active participation of the watershed community in the planning and execution of the watershed development projects (Singh, 2008, 8).

(ii) *Soil and Water Conservation for the Treatment of Catchment Areas:* This is a programme of the Ministry of Agriculture for the catchment of River Valley Projects which was launched in the III Plan (1962-63). Another subject that requires additional attention is scheme of Integrated Watershed Management in the Catchments of Flood Prone Rivers which was launched during the VI Plan (1980-81). They have been combined together into a new scheme called "Soil Conservation for Enhancing Productivity of Degraded Lands in the Catchments of River Valley Projects and Flood Prone Rivers" (Singh, 2008, 8). The scheme is being implemented in 53 catchments encompassing 27 states which attempt to treat catchment areas of reservoirs and flood prone rivers on watershed basis to improve the productivity of degraded lands, minimize siltation of reservoirs and moderation of the floods by holding rainwater in catchments (Singh, 2008, 8). It would be apparent from Table 12.8 that an area of 6.50 m ha has been treated with an expenditure of Rs. 2244 crore up to Tenth Plan.

(iii) *Reclamation of Alkali Soils:* The alkali soils are characterized by the presence of sodium carbonate and bicarbonate which are spread over Punjab and Haryana. Its reclamation was launched in 1974-75. Table 12.8 shows that an area of 0.72 m ha has been reclaimed with a central investment of Rs 121 crore.

(iv) *Shifting Cultivation:* It is a way of life marked by low technology characterized by soil erosion and land degradation. It therefore allows relatively low population densities compared to other cultivation systems. It is mostly practiced in northeastern region, Orissa and Andhra Pradesh, where shifting cultivators are mobile in space and time, but with different frequencies, durations and relative locations. The Watershed Development Project in this area was first started during the Fifth Plan with assistance from the central Government. By the end of Tenth Plan, 0.40 m ha of shifting cultivation area has been treated with a total expenditure of Rs. 296 crore (Table 12.8).

(v) *Externally Aided Projects:* There are 8 on-going externally aided projects on watershed management and land reclamation being implemented in states. Under this project about 1.8 m ha area has been developed with an expenditure of Rs. 3697 crore up to Tenth Plan (Table 12.8).

(vi) *Drought Prone Area Programme:* This is a programme of Ministry of Rural Development, launched in 1972-73 to tackle the areas affected by severe drought conditions. The DPAP covers 972 blocks spread over 182 districts in the country. Some of the important features of the programme are soil and moisture conservation measures, social and farm forestry, development of water resources and pasture and range management. The programme is being implemented on watershed basis. In the absence of watershed in sandy areas, cluster of villages is selected. Up to Tenth Plan, 13.7 m ha has been covered with Rs. 4843 crores.

(vii) *Desert Development Programme:* This is a programme of Ministry of Rural Development which covers hot desert as well as cold desert areas, and a few districts of Andhra Pradesh and Karnataka. This was started in 1977-78 and covers 235 blocks of 40 districts in 7 states. The main objectives of the programme are pasture development, plantation of shelter belts, soil moisture conservation and water resource development. Table 12.8 shows that 7 million ha has been covered with expenditure of Rs. 1950 crores up to Tenth Plan.

(viii) *Integrated Wasteland Development Project:* The Integrated Wasteland Development Project (IWDP) deals with total integrated development of wastelands in 374 districts in the country. It was started by Ministry of Environment and Forests in 1988-89 for the development of wastelands based on village/watershed plans and was transferred to Department of Wasteland Development (DWD) in 1992-93, now called Department of Land Resources (DLR) (Chadha, Sen and Sharma, 2004, 276). The main objectives of the programme are the promotion of overall economic development, generation of employment in rural areas and improvement of income conditions of the resource poor (Chadha, Sen and Sharma, 2004, 276). An area of about 10 m ha is estimated to have been covered with an expenditure of Rs. 2438 crores (Table 12.8). The ravines of Chambal in Morena district of Madhya Pradesh are the best examples of operation of Wasteland Development Task Force whose objective was to develop wasteland through afforestation including soil and moisture conservation, plantation and protection (Chadha, Sen and Sharma, 2004, 277).

Impact Assessment of Watersheds

Watershed programmes are explanation and key to agricultural growth and development in rainfed areas. The Government of India has accorded and granted strident priority to the watershed development since the Seventh Five Year Plan and ample investment is made for the success. The rainfed areas (over 50%) in the country, where agriculture is largely dependent on rainfall especially the summer monsoon, face the problem of degrading natural resources, abject poverty and acute unemployment. Contrary to this, it was emphasized that watershed programmes would expedite and quicken the development of a second Green Revolution in these rainfed areas. It has also been recognized in recent years that though watershed programmes have not produced the craving and yearning impact, however, there are gleaming and promising spots also. The functioning of the activity of watershed programmes can be improved through apprehending the drivers of success and putting together suitable policies, institutions and technological interventions and funding scheme based on meticulous and critical evaluation of watershed programmes.

The production and performance of watershed programmes have been appraised by applying meta-analysis. In this regard, based on the exhaustive review of 311 case studies on watershed programmes in India, the actual assessment has been made to account and document efficiency, equity and sustainability benefits (Table 12.9). It should be noted that the mean benefit-cost ratio of watershed programme is quite modest at 2.14. The internal rate of return is 22 per cent, which is comparable with rural development programmes. In overall terms, watershed development programmes generated immense employment opportunities, raised irrigated area and cropping intensity and conserved and sustained soil and water resources. In totality, the performance of watershed development programme was best in rainfall ranging between 70 – 100 cm jointly implemented by state and central government, targeted in low and medium income regions, and had effective people's participation. People's participation is a crucial input for the success and sustainability of watershed development programmes because the participatory approach can play a vital role in adjusting the conservation measures to suit field-specific situations (Chadha, Sen and Sharma, 2004, 272).

Land Reforms Implementation

Land reforms are another approach which have been accorded and granted high priority in land development. Since Independence they have been considered basic in the whole range of rural development activities and reconstruction. They have always been at the top of national agenda and have been reiterated in the successive Five Year Plans. The major objectives of land reforms were as follows: to remove impediments in agricultural production as arising from the agrarian structure inherited from the past, to eliminate all forms of exploitation and social injustice within the agrarian system, to provide security for the tiller of soil and assure equality of status and opportunity to all sections of the rural population. The measures adopted to achieve these objectives were: (i) abolition of intermediaries, (ii) tenancy reforms, and (iii) imposition of ceiling and redistribution of ceiling surplus land among the landless.

India has made enormous strides in land-society relations since it achieved Independence. In the first two plan periods (1951-56 and 1956 -61) legislation was enacted in most states abolishing intermediaries. In 1950-51, it was estimated that 40 per cent of the total cultivable land was under intermediate tenure. By 1960-61, about 20 m tenants of the former intermediaries had come into direct relationship with the state. This was in pursuance with the govt. policy of the 'land to the tiller'. It means that those who actually cultivate land should own it and, conversely, those who do not personally do should be deprived of any ownership right on land. However, the problem still persists, particularly, in some states where absentee landlordism continues to flourish (Nayyar, 1993, 55). The original objective was diluted because some states took long time to bring the law of the statute books; and even after it came into existence, its implementation was blocked by erstwhile landlords through prolonged legal cases (Sahay, 1995, 31). Under the law, intermediaries were also allowed to take with them a huge area of land under the garb of personal cultivation (Sahay, 1995, 31). Surprisingly, there are still a few isolated pockets where intermediary tenures have not yet been abolished. These include zamindari over water rights in Bhagalpur district of Bihar, communidades in Goa, jotedars in Meghalaya, trust estates in Orissa and devasthan inams in Maharashtra (Lingaih, 1997, 2).

Tenancy reforms have been considered one of the key components of land reforms. They provide for conferment of ownership rights from tenants or also allow cultivating tenants to acquire ownership rights on payment of a reasonable compensation to the land-lords. Three important guidelines were laid down in the Five Year Plan for tenancy reforms (Appu, 1995, 6): (i) rent should not exceed one-fifth to the gross produce; (ii) the tenants should be accorded permanent rights in the lands they cultivated subject to a limited right of resumption to be granted to the land owners; (iii) in respect of non-resumable land, the landlord-tenant relationship should be ended by conferring ownership on tenants. Because of this, an estimated 112.92 lakh cultivators have been conferred ownership rights on 153.53 lakh acres of land (Patil, 1995, 13). They were enacted in all states except the northeastern tribal areas. West Bengal, Karnataka and Kerala have achieved more success than the other states. Thus, on the whole, the experience of tenancy reform has been rather limited and the instance of oral or concealed tenancy is very high (Nayyar, 1993, 55).

The distribution of ceiling surplus land is another important programme under land reforms. Imposition of ceiling on the size of single landholding was taken up as a re-distributive measure. This arose from the skewed distribution of land in India, with nearly a quarter of the rural household owning no land at all, and another one-fifth owning less than an acre each; and also provided ample social and economic reason for land ceiling as means of redressing this imbalance (Patil, 1995, 14). This ceiling were 18 acres for wetlands and 54 acres for unirrigated land with family of five as the unit for determining land holdings. The ceiling laws were evolved in 1972. The total area declared surplus up to March, 1995 was 74.09 lakh acres, out of which 65.42 lakh acres have been taken possession (Patil, 1995, 14). Of this, an area of 51.46 lakh acres had been distributed to 49.94 lakh beneficiaries (Patil, 1995, 14). However, it is felt that the release of surplus land as well as its implementation was both slow and indecisive. A large area is, still, covered by litigation. Of the 45.90 lakh acres of land donated as **Bhoodan land** only, 29.42 lakh acres have been distributed (Patil, 1995, 14). Sometimes, the records of the donated land are also not available. It is therefore suggested that ceiling surplus land must be detected and redistributed and there should be a time-bound programme for their distribution.

Consolidation of land holdings brings efficiency and improvement in agriculture at the village level by making ownership holdings compact and facilitating the provision of irrigation and other inputs optimally. However, it is not a re-distributive measure, rather, it is simply a rearrangement of land on the basis of existing rights (Sahay, 1995, 32). Fifteen states have passed laws for land consolidation, but, satisfactory progress has been made only in states like Punjab, Haryana and Uttar Pradesh. Elsewhere, the task has not been accomplished completely. By 1990, scattered land holdings of 56.32 m ha had been consolidated which is one-third of the total cropped area in the country.

Challenges Facing Land Development

Despite improvements in reclamation measures and land reforms implementation, there are several disturbing trends. One of our purpose here is to outline those problems and challenges that may have far reaching consequences. Perhaps the foremost problem is wasteland reclamation which has to be strengthened on the war footing as 55 m ha under wastelands can be brought under productive use. Likewise, there is a serious problem of saline and alkaline soils. Deep gullies occurring along the banks of the rivers in north central India deserve special mention. The mapping of its severity is a challenging and daunting task for cultivation. Ravines are a national problem as they are fast spreading into cultivated lands (Shafi, 2006, 412). Coastal sandy lands along the eastern coast of Orissa and western coast of Kerala and the shifting coastal lands cause serious damage to agriculture (Shafi, 2006, 416). Sandy gravelly lands and lands affected by shifting cultivation and fast spreading sand dunes are another challenges for cultivable lands. This is apparent with respect to issues that the net sown area is stagnant at 140 m ha in the last 20 years in spite of developments. There is increasing evidence that large areas of arable land are inadequately managed for long-term sustainable production.

The impact of new mining technology on land in the mining areas is threatening. It affects the region environmentally and also transforms the entire ecosystems of mineral fields. The challenges are expressed as devegetation, topographic deformation, edaphic loss, drainage derangement, dereliction and air, water and noise pollution. The area becomes biologically unproductive. However, the effective control of the environment in such an area is financially a gigantic task. Because of population pressure on agricultural land resources, depletion has become a common phenomenon. The depletion of land resources is the greatest challenge in this millennium. Pressure on land is rapidly increasing due to ever increasing population competing for finite land resources. Forest lands are being lost to agriculture, agricultural lands are being lost to urbanization and other uses. Around cities and metropolitan cities, good quality agricultural land is scarcely found. Due to growing population, multi-storied buildings are the common features and high-rise dwellings mainly depend on the supply of building materials. Brick kilns prepared from loamy and silty soil are increasingly important in better understanding the process. A fresh look is required to the phenomenon of urban encroachment on good quality agricultural land.

The effective implementation of land reform policies is severely lacking due to many causes like snags in legislation, lack of political will and apathy of bureaucracy in the country. Therefore, need for strong political will to take stern action against those who have been obstructing the program is urgently required. Joshi (1978, A-82) states that the nexus between the politicians, administrator and large land owners has strengthened in different states culminating into rich peasant power. This corridor of power dominates State governments, regional and local administration and acts as an obstacle in the proper implementation of land reforms. Ironically, land to the tiller is a far cry. This has been most visible to the section of agricultural labourers. Resource managers need to work to develop an agenda of this segment of land. It must be linked to a needed concern for the reduction of gap between land supply and demand, and correction of skewed pattern of land concentration.

Providing security to tenants and sharecroppers is the most important issue of agrarian reforms. Likewise, the question of fixing of ceiling on land holding and distribution of surplus land to the rural poor have been half-hearted and ineffective in almost all the states. Another challenge facing land-society relations is redistribution of ceiling limits and consideration of strict definition of 'family' and 'personal cultivation'. Finally, there is the interest in empowering and giving legal status to women in land property. Inequality between men and women has been documented and also indicated that women face constraints that place them at a disadvantage relative to men.

While the trends described above have been broadly concerned with physical problems and land-society relations, another challenge is to learn more about institutional arrangements, functioning and constraints. Institutional arrangements have been identified as a variable needing greater attention (Mitchell, 1989, 242). According to Fernie and Pitkethly (1985), "all resource problems—overpopulation, hunger, poverty, fuel shortages, deforestation are fundamentally institutional problems which warrant institutional solutions. The success or failure of resource management is intrinsically tied up with institution structure – the pattern of agencies, laws and policies which pertain to resource issues." Therefore, a strong and effective institutional framework is needed for monitoring and reviewing the issues. New insights may be gained about the effective implementation of Panchayati Raj Institutions and voluntary agencies. It also requires the ability to challenge constructively and effectively the practices of corruption and malpractices in land resource management. To meet the challenge, the entire administration machinery has to be revamped and missing link in institutional arrangements in order to achieve the most effective management.

Land records are of great importance to contemporary socio-economic imperatives, therefore, relevant to devèlopment process in the country. According to Seventh Plan document, "land records form the base for all land reform measures and, therefore, regular updating of land records is indispensable in all states" (Rai and Bhalla, 2000,38). In other words, planning and maintaining land records is vital before any land reform policies can be progressively and successfully implemented. They provide determined field boundaries, prepared field registers and give the land particulars like ownership, revenue assessment, land classification etc. (Rao and Shekhar, 2000, 35). Equally remarkable is the fact that land records system safeguard the rights of the legal owner of land and form the basis for assignment and settlement of land titles that must stand against legal scrutiny (Rao and Shekhar, 2000, 35). Yet, it tells planners and administrators almost nothing about what they should do for capturing the essentials of changes in social dynamics, where updation and accuracy is always sought. Due to lack of proper survey after Independence, updation of records has suffered because they do not represent the ground realities relating to ownership and possession. Therefore, in many cases, the information about information rights is not complete. The management of land records is still performed manually by Patwaris without computerisation. Patwaris often exercise monopoly in behaviour, seeking bribes and alter data in favour of the party paying bribes (Saxena, 2005). Robbins (2000, 423) is of the view that corruption is an institutionalized system of nature/society interaction forged from state authority and molded around local social power through systems of social capital formation. Therefore, it is powerfully suggested that computerisation of land records saves money by bribes.

Management Strategies

Land management implies strategies, control and direction to develop and conserve land resources so that agricultural production is enhanced in a region. It is a way to maximize lands potential benefits for use which should proceed in a systematic manner. The conservation and management of land resources help in the judicious planning of the use of land, identify the lands of special qualities for special purposes, plan to prevent irrational encroachment on good agricultural lands, suggest after careful survey a better use of village commons for pasture and social forestry and an overall development of the rural landscape and the preservation of the environment (Shafi, 1984).

It is important to mention four aspects of management so that land can be developed in the national interest taking into account its redistribution, reclamation, administration, and information.

Land reform commonly refers to redistribution of land from rich to poor. It is largely achieved on the basis of land ownership reform measures. In an agrarian economy like India with great scarcity and unequal distribution of land, coupled with a large mass of below poverty line rural population, there are compelling arguments for land reform (Ghatak, 2004, 328). During the decades immediately after Independence, India passed a meaningful body of land reform legislation. Access to land, improving tenancy, ceiling on land holdings, agricultural land in favour of less privileged classes are some of the steps towards the development and management of land resources. It is a move towards diminishing the existing land-man relations prevalent in rural society. However, it is quite amazing that despite the acute necessity to change the agrarian structure in many parts of the country, agrarian reforms have not been adequately executed. The total quantum of land declared surplus in the entire country was 73.74 lakh acres up to 1998, out of which about 65.11 lakh acres have been taken. Of a total area of 53.05 lakh acres have been distributed to 55.37 lakh beneficiaries, of whom around 36 per cent belong to the Scheduled Castes and around 14 per cent belong to the Scheduled Tribes. There is a large gap between the annual targets and actual achievements. Unfortunately, such gaps are increasing every year. In addition, an area of 147.44 lakh acres of Government wastelands and Bhoodan land have also been distributed amongst landless rural poor. Legislative provisions have been made in extensive areas of the country providing for conferment to ownership rights on tenants or allowing cultivating tenants to acquire ownership rights on payment of a reasonable compensation to the landlords. Consolidation of agricultural holdings forms an integral part of land reforms policy. In pursuance of this strategy, a number of states have enacted legislation of carrying out consolidation of holdings. So far an area of 1583.45 lakh acres has been consolidated in the country.

Reclamation

This is another name of extensification of land for which our government is mostly concerned. Reclamation measures have been accorded high priority in the process of land management in India. This has received considerable attention from both governments and Earth scientists. There are various programmes implemented by different Ministries of Central Government, namely, Ministry of Agriculture, Ministry of Rural Development and Ministry of Environment and Forests for the management of wastelands. 'Little Rann'—A strategic region along the Indo-Pakistan border in Gujarat has gained appreciable land area during the last century. Remote sensing and LANDSAT data have shown that about 200 sq km of the Gulf of Kachchh had been covered by sedimentation. Of perhaps the greater significance and closely related line of investigation has involved the adoption of watershed of ravines as a planning unit. In 1967, the Government of India constituted a Central Ravine Reclamation Board to regulate and manage ravines and a quick development of the area. The Board suggested measures to restrain further growth of ravines, and to use already reclaimed lands for crops, grasses, horticulture and forestry (Shafi, 2006, 412). Other methods of ravine reclamation are terracing, strip cropping, contour farming and regulatory grazing. To streamline research and implementation, three conservation soil research centers have been established by the Government of India at Kota in Rajasthan, Agra in Uttar Pradesh and Valsad in Gujarat to find out methods of reclamation. To reverse the declining trend in productivity in salt affected areas, a combination of surface and sub-surface drainage, supplemented by improved irrigation management has been adopted. Sand dune stabilization, shelter belt creation, bunding work and the regeneration of rangelands through a variety of earthworks have been affected effectively in the semi-desert and desert areas. Since 1991, a collaborative research and development programme on biological reclamation of wasteland, land spoils and mining over-burdens has been operating. It aims at breaking the dormancy of the rock surface and converting it into the production base for biomass generation through biological reclamation. The derelict and mining landscape is being remodeled by the conservation of top soil by means of sowing, stabilizing plants, first herbaceous plants and then tree or bush seedlings.

Institution

Solutions to the problems of land resource depletion, redistribution and degradation rest with structure and arrangement of institutions. Institutional arrangements include the pattern of government and private agencies and their legislative abilities to tackle issues. The efficiency of institution depends on established laws, customs, behaviour, responsibility and the manner in which they are organized. Institutions for land management are either inadequate or wrought with cumbersome bureaucracy. Because of this drawback, India cannot effectively implement strategies.

Central Land Use Council combines both National Land Use and Conservation Board and the National Wasteland Development Board. These two bodies function as the policy planning, coordinating and monitoring agency for issues concerning scientific management of country's land resources. It appears that state governments are also responsible for policy implementation and the formulation of laws to conserve and manage land resources with encouragement to local communities, Panchayats and district authorities.

The government through several agencies such as Indian Council of Agricultural Research and the National Bureau of Soil Survey and Land Use Planning has been attempting to prepare detailed maps for land resources. The National Bureau of Soil Survey and Land Use Planning at Nagpur and Central Soil Survey Conservation Research and Training Institute at Dehradun conduct training programmes also. The Ministry of Rural Development, Government of India, has recently created a Department of Land Resources of India which is presently under various stages of development. Its real issue is that there is no dearth of land, but of management.

Information

Land information is prime requisite for making decisions related to land investment, development and management (Thakur, Khadanga, Venkatesh and Shukla, 2004). In order to achieve environmentally feasible land use, the site-specific information of the land through systematic survey is basic to such achievement. Land is a precious resource and its information base must be continuously improved by updating land records and digitization of cadastral maps. Modernization of land records forms the base for all reform measures. Thus, regular updating of land records is essential. Land information system is a computerization process which assists in giving land particulars like ownership, revenue assessment and land classification (Rao and Shekhar, 2000, 35). Land information system forms the micro-level land management tool while GIS is a macro-level tool for decision-making. Both are computer-based solution for addressing the requirements of land records management.

The correction and updating of land records are very complex and elaborate as millions of plots are plotted on the cadastral maps. The plot owners keep on changing due to transactions of land among village dwellers. Because of non-operation and non-updation of land parcels since Independence, many states have suffered because they no longer represent the ground realities of ownership and possession. Since the First Plan, planners and administrators have been recommending correct maintenance of land records as the basis of good administration. The challenge of doing this provides a humbling experience of computerization. The conference of Revenue Secretaries of States in December 1988 took apprehension of the poor land records and alluded for immediate action. Subsequently, "the 73rd Amendment to the Constitution refers that computerization of land records is an essential step to accomplish decentralized planning and effective administration (Thakur, Khadanga, Venkatesh and Shukla, 2004, 3). The scheme is being implemented since 1988-89 by Ministry of Rural Development in collaboration with National Informatics Centre extended to 544 districts in the country.

Conclusions and Future Research Directions

The study opens with a summary of major characteristics of physical and social attributes of land. A brief survey of attributes of land addresses the characteristics of mountains, plateaus, plains as well as tenurial systems, insecure tenancies, land concentration and land and caste relations. We set out to investigate the attributes of land drawing on two distinct but connected bodies of literature. The study started with five research questions: (1) To what extent Indian states vary spatially and temporally with respect to agricultural land and land-man ratio? (2) What is the spatial pattern of land degradation? Do the elements of land degradation vary across states? (3) What has been the role of reclamation and reforms in land development and management? (4) What are the challenges facing land development especially physical, social and institutional? (5) Will redistribution, reclamation, institution and information play significant role in land development and management? Using a mixed methodology (cartographic, tabular and statistical) and data from Indian Agricultural Statistics, Directorate of Economics and Statistics, Department of Agriculture and Co-operation, Ministry of Agriculture, and National Bureau of Soil Survey and Land Use Planning, this research has investigated the changing characteristics of depletion and degradation of land resources in India. The paper explores the spatiality of land resource diversity in India and highlights several competitive issues and challenges. Briefly, the findings reveal a diverse pattern.

The answer to the first question is that it is reasonable to speculate about increased demand of agricultural land due to population growth. The study provides a contextual analysis of growth and distribution of agricultural land to show that they vary across space in accordance with agrarian and social environments. With unabated population growth, the pressure of population on arable land is bound to grow. The increase in agricultural land and population has not been similar during the post-Independence period. The growth of agricultural land could not keep up with increasing population. Agricultural land was 0.46 ha per person in 1951 which decreased to 0.19 in 2001. However, there are regional differences in land-man ratio. The high land-man ratios are found in the Thar desert, Chambal badlands, the Malwa plateau, the Deccan lava plateau and the Karnataka plateau. The low land-man ratios are concentrated in well-watered and productive areas of Indo-Gangetic lowlands and eastern fringe of the Deccan peninsula. The number of persons engaged in agricultural activities per hectare of net sown area district-wise range from as low as less than 1 to as high as above 5 persons. This is a reflection of population pressure and a relationship between rural population density and the intensity of land use.

The answer to the second question is that out of a total area as many as 174 m ha suffer from degradation. It means on an average, at least 2 out of every 3 ha of land is in poor health. Half the sickland, one of the total, is completely unproductive. Another one-third is partially productive and it is only the remaining one-third which is in good health. Soil erosion including splash, sheet, rill and ravines and gullies, desertification, waterlogging and salinization, alkalinity, landslides, degraded land due to unregulated mining, deforestation and overgrazing are the basic causes of land degradation. These forces have changed the physical, chemical and biological characteristics of land making it unproductive culturable and non-culturable wasteland. More importantly, the study has assessed the extent of wastelands by National Bureau of Soil Survey and Land Use Planning, National Remote Sensing Agency and Wastelands agro-climatic zone-wise. It is demonstrated that agro-climatic zones are important units for the analysis and comprehension of contemporary patterns and processes of land degradation because the contribution of this unit is to infuse regions individuality with a concern for the construction of spatial inequality.

The answer to the third question is development policies, which is concerned with reclamation of degraded lands, restoring damaged lands, restoration of mined land, wasteland development, desertification control, remedial measures of water-logging and salinity problem, rehabilitation of land, regeneration measures for gullied lands, rocky/stony outcrops, sandy area and mining and industrial wasteland. The Government of India has set up National Wasteland Development Board (NWDB) to reclaim wastelands in the country through a massive programme of afforestation which has published district-wise extent of wastelands in the country. They provide geographers and planners with powerful mapping tools revealing patterns, relations and processes underlying wasteland reclamation. The study has demonstrated the relevance of watershed management in the process of reclamation and enhancing the productivity of wastelands. It is also revealed that though watershed programmes have not produced the craving impact, however, it has potentialities in generating employment opportunities, raising irrigated area and conserving soil and water resources. Land reforms are another approach which have been accorded priority in land development.

With objective to remove impediments in agricultural production, to eliminate all forms of exploitation and social injustice within the agrarian systems, to provide security for the tiller of the soil and assure equality to status to all sections of rural population, the study has provided an overview of the abolition of intermediaries, tenancy reforms and imposition of ceiling and redistribution of ceiling surplus land among the landless.

The answer to the fourth question is that despite improvements in reclamation measures and land reforms implementation, several disturbing trends are discernible. There is a serious problem of saline and alkaline soils, deep gullies in north central India, coastal sandy lands along the eastern and western coasts, spreading sand dunes, lands affected by shifting cultivation, and biologically unproductive area in the mineral fields. The depletion of land resources is the greatest challenge in this millennium because pressure on land is rapidly increasing. Around cities, metropolitan and mega-cities, good quality agricultural land is scarcely found. The effective implementation of land reform policies is severely lacking due to many causes like snags in legislation, lack of political will and apathy of bureaucracy. Providing security to the tenants and sharecroppers is the important issue of agrarian reforms. Redistribution of ceiling limits is another challenge. Women should be given legal status in land property. Likewise, a strong and effective institutional framework is needed for monitoring the issues and to achieve effective management. The management of land records is still performed manually by Patwaris without computerization which has increased corruption in land management.

The answer to the fifth question pertains to four aspects of management strategies, i.e., redistribution, reclamation, institution and information. The study has highlighted that access to land, improving tenancy, ceiling on land holdings, agricultural land in favour of less privileged classes are some of the steps towards land management. Consolidation of agricultural holdings should be given top priority. Reclamation Board to regulate and manage ravines selecting watershed as a planning unit should be streamlined. Institutions for land management are either inadequate or wrought with cumbersome bureaucracy; they should be efficient and responsive. Land information system as well as Geographic information system, powerful tools for decision-making, should be compulsorily introduced to Patwaris and other officials for updating land records and digitization of cadastral maps.

Further research needs to be carried on in the following directions: (1) to identify and analyze the surplus and deficit areas of agricultural land, (2) to analyze patterns, trends and policy implications of the urban encroachment on agricultural land in post-Independent India, (3) to investigate impact of population pressure on land scarcity in densely settled region, (4) to develop a methodology for the measurement of lands carrying capacity, (5) to explore and examine the spatial processes of land resource depletion, (6) to probe and examine the changing land-caste relations in rural India, (7) to analyze the contribution of Acharya Vinoba Bhave to Bhoodan land and its management, (8) to explain the role of reclamation and reforms in the sustainable development of land resources, (9) to analyze the impact of land degradation on future food production, (10) to elucidate the potential impact of future climate change on the nature and extent of land degradation, (11) to probe into the status of region on the basis of depletion and degradation ratio, and (12) to develop clear regional policy to release land for urbanisation, industrialisation, for building roads and opening mines.

Notes

1. In writing attributes of land the author has drawn heavily from his paper on "Emerging trends in India's resource development: Implications for sustainable management", in B. Thakur (ed) (2007). *Perspectives in Resource Management in Developing Countries,* Vol. 2: *Population, Resources and Development,* New Delhi: Concept, 3-87.
2. This section is partly based on an unpublished paper by Shamsher Singh, Additional Commissioner, Department of Agriculture and Cooperation, Ministry of Agriculture, New Delhi (2008) on "Management of degraded lands for sustainable agriculture production: An overview".
3. Meta-analysis summarises results of past studies, suggests potential reasons for inconsistencies in past research findings and direct future investigations. See, for example, E.A. Patall and H. Cooper (2009). "Conducting a meta analysis", in P. Alasuutari, L. Bickman and J. Brannen (eds.) *The Sage Handbook of Social Research Methods,* Thousands Oaks, Cal: Sage, 536-554.

References

1. *Agricultural Statistics At A Glance* (2008). Directorate of Economics and Statistics, Department of Agriculture and Cooperation, Ministry of Agriculture, Government of India, New Delhi.
2. Ahmed, E. (1973). *Soil Erosion in India,* Bombay: Asia Publishing House.
3. Appu, P.S. (1995). "Land reforms: Need for orientation of policies", *Kurukshetra,* 44(1): 6-10.
4. Bandyopadhyay, R. (1993). "Land system in India: A historical review", *Economic and Political Weekly,* December 25: A149-A155.
5. Barlowe, R. (1958). *Land Resource Economics,* New York: Prentice Hall.
6. Bhumbla, D.R. and Khare, A. (1987). "Estimates of wastelands in India", in H.R. Yadav (ed) *Wasteland: Diagnosis and Treatment,* New Delhi: Concept, 27-48.
7. Bowonder, B., Ramana, K.V., Ravi, C. and Srinivas, C. (1987). "Land use, waterlogging and irrigation management", *Land Use Policy,* 4: 331: 341.
8. Brush, J.E. (1966). "India: The land and its resources", *Encyclopedia Americana,* 15, New York: International Edition.
9. Chadha, G.K., Sen, S., and Sharma, H.R. (2004). *Land Resources: State of the Indian Farmer, A Millennium Study,* New Delhi: Academic Foundation.
10. Chisholm, A. and Dumsday, R. (1987). *Land Degradation: Problems and Policies,* Cambridge: Cambridge University Press.
11. Dayal, E. (1984)."Agricultural land and man-land ratio in India: An analysis of change", in O.J. Bee (ed.) *Natural Resources in Tropical Countries,* Singapore: Singapore University Press, 346-67.
12. Dayal, P. (1972). "Agricultural and industrial potentialities", in A.K. Dutt (ed) *India: Resources, Potentialities and Planning,* Dubuque, Iowa: Kendall/Hunt Publishing Company, 38-71.
13. Eden, M.J. and Perry, J.T. (eds) (1996). *Land Degradation in the Tropics: Environmental and Policy Issues,* London: Pinter.
14. Fernie, J. and Pitkethly, A.S. (1985). *Resources: Environment and Policy,* London: Harper and Row.
15. Gautam, N.C. and Narayan, L.R.A. (1988). *Wastelands in India,* Mathura: Pink Publishing House.

16. Geddes, A. (1960). "The alluvial morphology of Indo-Gangetic plain: Its mapping and geographical significance", *Transactions of the Institute of British Geographers*, 28:23-276.
17. Ghatak, M. (2007). "Land reforms", in Kaushik Basu (ed.) *The Oxford Companion to Economics in India*, New Delhi: Oxford University Press, 328-332.
18. Goodall, B. (1987). *The Penguin Dictionary of Human Geography*, Harmondsworth: Penguin Books.
19. Government of India (1990). *Developing India's Wasteland*, New Delhi: Ministry of Environment and Forests.
20. Govind, H. (1989). "Recent developments in environmental protection in India: Pollution Control", *Ambio: A Journal of the Human Environment*, 18(8): 429-33.
21. Haigh, M.J. (1984). "Ravine erosion and reclamation in India", *Geoforum*, 15(4): 543-561.
22. Haque, T. (1995). "Role of land reforms in rural development in the wake of economic liberalisation", *Kurukshetra*, 44(1): 63-69.
23. Houerou, H.N.L. (1994). "Desertification and desertisation: Climatic fatality or human mismanagement", *Environmental Education Dossiers*, No. 7 on Desertification, Barcelone: Centre Unesco de Catalunya.
24. Johnson, D.L. and Lewis, L.A. (1995). *Land Degradation: Creation and Destruction*, Cambridge: Blackwell.
25. Joshi, P.C. (1978). "Land reforms implication and the role of administrators", *Economic and Political Weekly*, 30 September.
26. Kumar, A. and Pandey, R.N. (1989). *Wasteland Management in India*, New Delhi: Ashish Publishing Company.
27. Lingaiah, K. (1997). "Land reforms at a glance", *Kurukshetra*, 45(10): 2-4.
28. Livernash, R. (1995). "India", *World Resources 1994-95*, Delhi: Oxford University Press, 83-106.
29. Ministry of Agriculture and Irrigation (1976). *Report of the National Commissions on Agriculture*, New Delhi.
30. Misra, S.K. and Puri, V.K. (1996). Indian Economy: Its Development Experience, Bombay: Himalaya Publishing House.
31. Mitchell, B. (1989). *Geography and Resource Analysis*, 2nd Edition, London: Longman.
32. Mohammad, N. (1988). "Wastelands in India: Some issues and problems", in S.C. Sharma (ed) *Utilisation of Wasteland for Sustainable Development in India*, New Delhi: Concept. 57-65.
33. Mukerji, A.B. (1963). "Alluvial morphology of the Upper Ganga-Yamuna doab", *The Deccan Geographer*, 2(1): 1-36.
34. Muthiah, S. (1987). *A Social and Economic Atlas of India*, Delhi: Oxford University Press.
35. Nagaraja, R. and Gautam, N.C. (1996). "Role of remote sensing and GIS in wasteland management in India", M.J. Eden and J.T. Parry (eds) *Land Degradation in the Tropics: Environment and Policy Issues*, London: Pinter, 144-154.
36. Nag, P. (ed) (2010). *Atlas of Agricultural Resources of India*, Kolkata: National Atlas and Thematic Mapping Organisation, p. 79.
37. National Wasteland Development Board (1987). *Description, Classification, Identification and Mapping of Wastelands*, New Delhi: Ministry of Environment and Forests.
38. Nayyar, R. (1993). "Land reforms: Still a relevant issue", *Yojna*, 37: 54-56
39. Noble, A.G. (1972). Physical basis of resources", in A.K. Dutt (ed) *India: Resources, Potentialities, and Planning*, Dubuque, Iowa: Kendall/Hunt Publishing Company, 13-37.
40. Pascoe, E.H. (1973). *A Manual of the Geology of India and Burma III*, Delhi: Government of India Publication, 1346-2130.

41. Patil, J. (1995). "Effective implementation of land reforms: Need of the hour", *Kurukshetra,* 44(1): 13-15.
42. Patra, P. (2006). *Land Degradation in the Tropical Highlands: A Case Study of Orissa,* Delhi: Rajesh Publications.
43. Prasad, M. (1983). "Mining and land dereliction in Chotanagpur, India", in O.J. Bee (ed) *Natural Resources in Tropical Countries,* Singapore: Singapore University Press, 141-175.
44. Prasad, M. (1988). "Pattern of mining despoliation in the plateau regions of Bihar", in S.C. Joshi and G. Bhattacharya (eds) *Mining and Environment in India,* Nainital: Himalayan Research Group.
45. Prasad, M. (2007). "Ecological impacts of exploitation of natural resources and their socio-political dimensions in tribal areas of Chotanagpur", in B. Thakur (ed) *Perspectives in Resource Management in Developing Countries,* Vol. 2: *Population, Resources and Development,* New Delhi: Concept, 668-686.
46. Prasad, M. (2008). "Impact of new mining technology on land degradation in coal fields of Chotanagpur", in B. Thakur (ed) *Perspectives in Resource Management in Developing Countries,* Vol. 3: *Ecological Degradation of Land,* New Delhi: Concept, 212-231.
47. Pulsipher, L.M. and Pulsipher, A. (2006). World Regional Geography: Global Patterns and Local Lives, Third Edition, New York: W.H. Freeman and Company.
48. Rai, K. and Bhalla, D.K. (2000). "Computerization of land records in India", *GIS Development,* 4(4): 38-40.
49. Raj, K.N. (1995). "Land reforms in India since 1947: Achievements and failures", *Kurukshetra,* 44(10: 3-5.
50. Rao, T.R. and Shekhar, P.R. (2000). "Land records computerization: Issues and approaches", *GIS Development,* 4(4): 35-37.
51. Raza, M. and Ahmad, A. (1978). *General Geography of India,* New Delhi: National Council of Educational Research and Training.
52. Robbins, P. (2000). "The rotten institution: Corruption in natural resource management", *Political Geography,* 19: 423-443.
53. Sahay, B.N. (1995). "Land reforms: Issues and challenges", *Kurukshetra,* 44(1): 31-33.
54. Saxena, N.C. (2005). "Updating land records: Is computerization sufficient", *Economic and Political Weekly,* January 22, 313-321.
55. Shafi, M. (1984). Geographers and the Society with special reference to India", *Annals of the Association of Geographers,* India, 5.
56. Shafi, M. (2006). Agricultural Geography, Delhi: Pearson Education.
57. Sharma, H.S. (1980). *Ravine Erosion in India,* New Delhi: Concept.
58. Singh, L.R. and Singh, R.P. (1961). "The ravines of the lower Chambal valley: A geographical study", *National Geographical Journal of India,* 7(3): 150-164.
59. Singh, M. (1994). 'Waterlogged Areas in Karnal District: A Geographical Analysis', M. Phil Thesis, Department of Geography, University of Delhi.
60. Singh, R.L. (ed) (1971). *India: A Regional Geography,* Varanasi: National Geographical Society of India.
61. Singh, S.S. (2008). "Management of degraded lands for sustainable agriculture production: An overview", Unpublished paper, Department of Agriculture and Cooperation, Ministry of Agriculture, New Delhi.
62. Sinha, R.R.K. (2009). *Dynamics of Land-Caste Relations in India: A Case Study of Bihar,* New Delhi: Manak Publications.

63. Stamp, L.D. (1962). *The Land of Britain: Its Use and Misuse,* London: Longman Green and Company Ltd.
64. Thakur, B. (2000). "Land distribution and land development in India", in D.K. Singh, R.S. Dube and V.K. Shrivastava (eds) *Geography: Development and Change,* Gorakhpur: Association of Marketing Geographers of India, 225-247.
65. Thakur, B. (2007). "Emerging trends in India's resource development: Implications for sustainable management", in B. Thakur (ed.) *Perspectives in Resource Management in Developing Countries,* Vol. 2: *Population, Resources and Development,* New Delhi: Concept, 3-87.
66. Thakur, B. (2008). "Introduction and Overview", in B. Thakur (ed) *Perspectives in Resource Management in Developing Countries,* Vol. 3: *Ecological Degradation of Land,* New Delhi: Concept, 1-18.
67. Thakur, B. (2008). "Introduction and Overview", in B. Thakur (ed) *Perspectives in Resource Management in Developing Countries,* Vol. 4: Land Appraisal and Development, New Delhi: Concept, 1-37.
68. Thakur, B. and Bhushan, B.S. (1992). "Land-man relationship and distribution of agricultural land in eastern Sikkim", *Transactions of the Institute of Indian Geographers,* 14(1): 13-32.
69. Thakur, B. and Kurl, A. (1996). "Impact of limestone quarrying on water resources in the Tons basin, Dehradun district", *Regional Symbiosis,* 4: 91-106.
70. Thakur, B. and Vir, P. (2003). "Caste structure, patterns of land ownership and crop security in Ramgarh block, Bhabua District", in S. Jha, M.M. Thakur and B. Thakur (eds) *Dimensions of Society and Culture of Bihar,* Allahabad: Academy Press, 288-303.
71. Thakur, B. and Bajpai, M. (2008). "Capability classification and reclamation of wasteland in Etawah district, U.P.", in B. Thakur (ed) *Perspectives in Resource Management in Developing Countries,* Vol. 3: *Ecological Degradation of Land,* New Delhi: Concept, 323-361.
72. Thakur, B. and Sinha, R.R.K. (2008). "Caste and land quality in Bihar: A case study of six villages", in A.K. Dutt, H.N. Misra and Meera Chatterjee (eds) *Explorations in Applied Geography,* New Delhi: Prentice-Hall of India, 491-510.
73. Thakur, B. and Patra, P. (2008). "Biophysical characteristics of land in the Mahanadi delta", in B. Thakur (ed) *Perspectives in Resource Management in Developing Countries,* Vol. 4: *Land Appraisal and Development,* New Delhi: Concept, 158-195.
74. Thakur, B. and Das, A. (2011). "People, resources and space: Changing land and caste relations in rural Bihar", *National Geographical Journal of India,* 51.
75. Thakur, V., Khadanga, Venkatesh, D.S., and Shukla, D.R. (2004). "Land management systems in India: Past, present and future", www.gisdevelopment.net/application/is/overview/ lisrp0004.htm.
76. Tirtha, R. and Krishan, G. (2002). *Geography of India,* Jaipur: Rawat Publications.
77. Tiwari, R.C. (2006). *Geography of India,* Allahabad: Prayag Pustak Bhawan.
78. Troeh, F.R., Hobbes, J.A. and Donahue, R.L. (1980). *Soil and Water Conservation for Productivity and Environmental Protection,* Englewood Cliffs, N.J. Prentice-Hall, p. 686.
79. UNEP (1993). *Environment Data Report, 1993-94,* Oxford: Blackwell Publishers.
80. Valdiya, K.S. (1987). *Environmental Geology: Indian Context,* New Delhi: Tata McGraw-Hill Publishing Company Ltd.
81. Vohra, B.B. (1981). *A Policy for Land and Water,* New Delhi: Ministry of Information and Broadcasting, Publication Division.
82. Young, A. (1998). *Land Resources: Now and for the Future,* Cambridge: Cambridge University Press.

13
Monitoring Desertification in Rajasthan: Status, Processes and Causes

H.S. Sharma

Introduction

Some 64 years ago, in March 1952, National Institute of Sciences, India in collaboration with Government of India organized a National Symposium at Jodhpur to discuss the origin of Indian desert. This was the first academic event on desert and proceeding of this symposium was also published in 1952. The major conclusion of the symposium was that Rajathan desert is a man-made desert. But it was feared that this desert is marching towards eastern side of Rajasthan (Planning Commission, 1951).

Desert and Desertification: Changing Paradigm

Deserts are areas of scanty rainfall where vegetation is scarce or absent because of deficient rainfall or edafic aridity. Thus delimitation of desert boundary is on the basis of aridity indices when the amount of water gained from rainfall and lost by evaporation and transpiration. Generally in deserts evaporation exceeds the amount of rainfall.

Desertification is a term, has been in use since at least 1949 when Auberville, a perspective and well informed botanist and ecologist published a book on Climates, Forests, desertification de L Afrique Tropicale. He defined desertification as the change of productive land into a wasteland as the result of ruination by man-induced soil erosion. He associated it with the humid and sub-humid tropics where he worked. Desertification is a term employed rather loosely specially with reference to the deterioration of arid and semi-arid lands that are used beyond their capacity for sustained production by cultivators and herdsmen.

Despite the fact that a world conference has been held on the subject, there is no generally accepted definition of 'desertification'. In the International Conference held at Nairobi in 1977, desertification has been defined as the 'diminution or destruction of the biological potential of the land, which leads ultimately to desert-like conditions and is an aspect of the widespread deterioration of eco-system under the combined pressures of adverse and fluctuating climate and excessive exploitation.'

Unfortunately, there has been much confusion over the meaning of *desertification.* The largely invalid concept of expanding deserts and advancing sand dunes has become a more permanent image in the public eye than the less visible and much more serious phenomenon of land degradation in drylands. Desertification, as defined by UNEP, is land degradation in arid, semi-arid and dry sub-humid areas resulting mainly from human activities. This definition was modified in the 1992 during the UN Conference on Environment and Development (UNCED), to read as follows:

> "Desertification is land degradation in arid, semi-arid and dry sub-humid areas resulting from various factors, including climatic variations and human activities."

This definition has been internationally negotiated and approved a UNCED as the operational standard for Agenda 21. Unfortunately, however, it does include the three separate elements of short-term drought, long-term climate fluctuations and land degradation induced by human beings. Each of these is different and needs to be addressed in different ways, although there are interactions.

Droughts are natural phenomena which recur periodically. Severe and prolonged drought in degraded dry lands results in increased demands on the production of humid lands. In case of prolonged drought in certain key areas, global food security is already at a precarious state and this situation will become critical if desertification continues unabated.

Desertification and Droughts

A common misapprehension about desertification is that it spreads from a desert core, like a ripple on a pond. The truth is that land degradation can and does occur far from any climatic desert; the presence or absence of nearby desert has no direct relation to desertification. The process being as a patch on the landscape where land abuse has become excessive. From that patch which might be around a watering point or in a cultivated field, land degradation spreads outward if the abuse continues. Ultimately, the patch may merge into homogeneous area, but that is unusual on a large scale.

The second misconception is that droughts are responsible for desertification. Ironically, association of the Sahelian drought with desertification had led to the common misconception that drought causes desertification. In truth drought do not cause desertification, but it only exaggerates the harmful effects of improper land management.

The researches all over the world indicate that desertification is caused by climate change, overgrazing, excessive woodcutting, land abuse, improper soil and water management and land disturbance; its effects appear as reduced productivity of land and environmental degradation. Desertification is thus the continuous degradation of land under the influence of natural and anthropogenic causes in arid, semi-arid and dry sub-hummid climate resulting from various factors including climatic variation and human activities (UNCC, 1994).

Magnitude of the Problem

Desertification affects two third countries of the earth surface, on which about one billion people live (1/6 of the world population). The processes of desertification and land degradation are observed to have accelerated during recent years.

The desertification and environmental degradation are the major problems in Rajasthan. The state embraces a huge area of approximately 3,42,261 square kilometers and population of 6,85,48,437 (2011). It is the first largest state in India. The Aravalli range divides the state in almost two equal halves. The north-western part consists of Thar desert comprising various types of sand dunes as high as 20-25 metres and rocky plateau. The state is frequently drought prone. Drought is a regular feature and its recurrence has rendered the problems very acute. The ecological imbalance of arid part of the state is marked by low annual rainfall and its great variability. High wind speed (10-20 km. per hour) aided by the lack of vegetation and intense solar radiation (450-550 calories per day per cm) cause high evapo-transpiration (6 mm-per day). The result is high aridity index-ranging from 76 to 78 per cent.

The problem of desertification is not a new one, for already the Planning Commission of India (1951) described the magnitude of the problem that 'the great Indian desert of Rajasthan has been spreading outwards in a great convex arc through Ferozepur, Patiala and Agra towards Aligarh and Keshganj at a rate of about half a mile (0.8 km) per year for the last fifty years and is encroaching upon approximately fifty square miles (130 sq. km.) of fertile land every year.'

So far as the genesis of Indian desert is concerned, it may be said that several theories have been proposed by scientists on this issue. However, five theories are important to mention such as Blandfords theory of drifted sand particles from Runn of Kutch (1876), La Touche's theory of encroachment of blown sand and silting of river valleys (1902), Wadia's theory of long continued aridity (1939-1960), Krishnan's theory of increasing aridity and climate change (1956) and Rode's theory of sheet movement (1964).

The problem of desert expansion was thoroughly discussed in various seminars and symposium (1952, 1964, 1975). The studies of Central Arid Zone Research Institute (Roy, B.B. and Pandey, S., 1971; Amal Kar, 2012, 2014; Singh, S., 1977; Dhir, R.P., 2003) have also indicated that except much hue and cry the desert is not expanding towards north-east beyond Aravallis. Though in the past the north-eastern part of Rajasthan has also witnessed arid phase in pre-Middle stone Age or early Holocene (more than 10,000 B.P.) during which major dune formation took place (Goudie *et al.* 1973) (Table 13.1). Alarming views are, however, still held about the continuous march of the desert in the form of sand movement to the north and north-east along with the prevalent south-west wind through various Aravalli gaps (Dhabria, 1986; 1988).

There are four views prevalent on the problem of desert expansion and desertification in Rajasthan. These are as follows:

Stationary Hypothesis: The scientists of CAZRI held the view that Great Indian Desert is stationary since its formation some times in Mid-Miocene period (Roy and Pandey, 1971). Some temporary modifications in the climate may, however, been experienced during the glacial period but this did not appreciably affect the desert condition. According to them, the climate boundary of the desert is neither expanding nor deteriorating but has been occupying essentially the same position. However, green revolution, introduction of irrigation, new tehnologies and better infrastructure have made drastic changes in agricultural landuse in post-independence era as a result farmers economy has improved. Contrary to it, environmental issues like increased sand movement and atmospheric dust load have increased due to deep ploughing and apprehended that in future (mid 21st Century) there is likelyhood of extension of aridity beyond the present desert boundary (Amal Kar, 2012; 2014).

Table 13.1: Geomorphic evidences of desertification/dedesertification during the late Quaternary period in Western Rajasthan

Period in Years B.P (After Singh et al. 1974)*	*Major gemorphic events*	*Geomorphic processes*	*Climatic phases*	*Desertification/ Dedesertification*
0- Early Centuries A.D.	Increasing fluvial activities but increasing biotic stress on the land has resulted as salinity hazard along period drainage channels and more deflation and deposition of sediments in the sandy plains and duny tracts	Biotice activities	Slight amelior-ation of climate	Small scale amelioration of climate but biotic desertification is present
Early centuries A.D. 3,800	Formation of low active dunes like barchans. Shrubscoppice etc. reactivation of the flanks and crests of the stabilized dunes. Inland basins turned into saline depression	Aeolian	Arid	Climatic deserti-fication
3,800-10,000	Major dunes stabilised, drainage channels revived, sandy inland basins filed with fresh water and fluvial sediments	Fluvial	Humid	Amelioration of climate leading to dedesertification
10,000 - ?	Formation of major dunes, disintegration of drainage systems of aeolian sand, creation of sandy inland basins.	Aeolian	Arid	Climatic desertification
?	Well integrated drainage systems, Vast alluvial plains	Fluvial	Humid	No evidence of desertification

*B.P. Means before present.
Compiled by H.S. Sharma (1987).

Sand Spread Hypothesis: Another alarming view is that desert has been spreading in the form of sand movement towards the north-eastern part of Rajasthan through the twelve Aravalli gaps (Dhabria, 1984).

Man-induced Sand Reactivation Hypothesis: The third view holds that the existence of fossilized dunes and sand sheet are not result of current aeolian activity. They were deposited in the former arid phase. As a result of the ameliorating climate, increasing human and animal pressure, land use changes in eastern part of the Aravallis and urbanization in Jaipur and other urban centres sand dunes have been reactivised causing initiation and intensification of desertification hazard in this part (Sharma, 1988; 2008). This group also attributes desertification to the processes of modernization, as the expanding horizon of new science and technology aimed of maximum output from land or returns from these investments (Hooja, 1985).

Similar to other parts of the world, the process of desertification in Rajasthan is the result of a combination of several factors, some of which are natural and completely beyond human control (such as droughts), while others are the result of human and animal interaction with the environment.

Desertification Characteristics: Historical Perspective

Man first placed his imprint on the landscape of Rajasthan thousands years ago.

Historical records and evidences make it clear that the destructive impacts of man was already apparent five thousand years ago in the Ghaggar and lost Saraswati plains. It was then what has been called the first period of accelerated man-made erosion began. Wood cutting and cultivation of slopping lands of the Aravallis and sand dunes were responsible for most of the land degradation that occurred in the Aravallis and around former fertile plains of the Ghaggar, Chambal, Luni and Banas rivers of Rajasthan. Sedimentation of canals, dams, waterlogging and salinization combined to raise havoc in most of the river valleys of Rajasthan.

In recent years grazing pressure seems to have become a much greater problem than it was before. There has been an expansion of human and livestock population between 1971-1981 having higher growth rate than 2011. Cultivation has been extended into pasture lands. Wells have been constructed in arid and semi-arid parts and improved transportation facilities have made formerly remote areas accessible to livestock. As a consequence, cropping has become a high risk land use in the climatically marginal areas that were formerly the best pasture lands. Destruction of vegetations on the Aravallis, sandy lands and upland has led to well spread soil erosion and depletion of soil fertility.

Nature of Dunes of North Eastern Rajasthan

In order to resolve the problem of desert encroachment, it is very essential to analyse the nature of sand dunes and sand sheets of north-eastern Rajasthan which is climatically a part of semi-arid region.

The north-eastern Rajasthan is characterized by widespread sand sheets and obstacle sand dunes on the windward side of the isolated hogbacks and hills of Aravalli mountain. A particularly large sand field occurs in the Sambhar area, west of Jaipur and extends almost to Jaipur. The dunes extend a considerable distance east and north of Jaipur. A large number of obstacle dunes are found at Banada, Chaksu, Jhir, Dausa, Lalsot, Karauli, Gangapur, Newai and Tonk. This would appear to mark the eastern-most extension of the aeolian features in Rajasthan.

The existence of sand features in eastern Rajasthan has been misinterpreted as a product of the present aeolian process. Contrary to it, geomorphological, archaeological, pedological and botanical evidences suggest that the sand dunes of this part are fossile dunes and relict features of former arid phase of climate (later upper paleolithic, Allchin *et al.* 1978). There are quite a number of evidences that indicate the relict nature of sand sheets in eastern Rajasthan.

It has been found that most of the dunes are extensively moulded by fluvial action. In the foothill dunes of Jaipur, Bhairana, Chaksu, Newai, Samod, Boanli, Lalsot and Karauli gullying have developed, the phenomena indicating the aeolian phase was followed by fluvial erosion.

One consequence of the fact that the dunes are essentially fossil is that they have developed upon them extensive soil or weathering profile. The sand is highly oxydised and kankarized in the form of root tubes locally called 'Diikaka'. Such root tubes were exposed in abundance during 1981 flood at Jaipur in the course of Amanisa Nala near Shastri Nagar.

The oxydized red phase of sand still preserve archaeological material over their crest. In eastern Rajasthan many dune surfaces are covered by large number of microliths (Allchin *et al.*, 1978).

The Sand dunes of the Sambhar area do not indicate any sand drift except local movement. The Jaipur City itself was planned on stable dunes about 289 years ago.

Reactivation of Stabilized Dunes

The phenomenon of wind formed sand ripples provide a clear indication of desertification processes on the stabilized dunes in northeastern Rajasthan. Loosening of sand through farming and large scale leveling sets the soil moving. As a result sand ripples appear, first on the crests of the so far stabilized dunes and then on the slopes and lower parts. The fact that these sand ripples appear in overused areas in the vicinity of Jaipur city and other settlements. Less used areas still preserve a part of the rich, original plant cover which proves that ripple formation is anthropogenetically conditioned (rapid urbanization and expansion of city). After the formation of sand ripples the formation of fresh dunes on or along side the old one initiates. Then, the reactivation of the old dunes is complete. Recent dunes formation provides a sure evidence of desertification. But it is not easy to convince that this phenomenon is not simply a naturally controlled one, but, is a consequence of human impact, or that it is a process which is triggered off and enhanced by land misuse in the sand field. The most dramatic indicator of desertification is the encroachment of sand sheets upon buildings, roads, railways and cultivated areas. This process is quite natural in western Rajasthan, but its occurrence in semi-arid north-eastern Rajasthan, however, indicates that ecological degradation has been imposed by the impact of man and animals. Thus, because of rapid urbanization in Jaipur and expansion of agricultural activities during last few decades, the old stabilized dunes have been reactivized and thus the loose sand moves locally and seasonally.

Status of Desertification in Rajasthan

No scientific data was available about the status of desertification in India and Rajasthan till 2002-2003 and 2005-06. In the year 2000, the Regional Action Programme (RAP) to combat desertification was formulated by UN Convention to Combat Desertification (UNCCD) and accordingly RAP for Asian region had been initiated as collective effort by the member countries. Under RAP Thematic Programme Network (TPN-I) is on desertification Monitoring and Assessment. In India Space Application Centre (ISRO), Ahmedabad has been identified as the national focal institution to coordinate TPN-I activities and establish the national network for desertification monitoring and assessment. To implement TPN-I activities in India, a working group was constituted by SAC in which 17 institutions were involved. Accordingly the mapping of status of desertification was worked out on the scale of RF.I:500,000 and an Atlas on Desertification/land degradation has been published in 2007 by Government of India (SAC, 2007). A separate Atlas has also been published for pilot projects at district level on 1:50,000 scale.

As per earliar analysis done by SAC on the basis of AWiFS data 2003-05 about 67 per cent of area was affected due to desertification/land degradation in Rajasthan. Out of this figure the most significant process of desertification in Rajasthan is wind erosion which accounts for 44.42 per cent of area, followed by water erosion (11.22%), vegetal degradation (6.25%), and salinization (1.07%) in canal command areas, waterlogging

(0.01%), man-made (0.09%) and others (4.36%) while only 32.58 per cent area is unaffected by the process of desertification in the state (SAC, 2007).

However, in 2013 a fresh study on spatio-temporal status of desertification was initiated by Space Application Centre, Ahmedabad under the heading Second Cycle of desertification mapping. The study is based on multi-temporal digital IRS AWiFS data (2003-05 and 2011-13), ancillary information and aided data of Forest Survey of India to bring out the changes in desertification and land degradation status of country including Rajasthan (SAC, 2016).

The present study conducted by SAC, Ahmedabad reveals that Rajasthan is the state with highest area under desertification/land degradation with respect to country's total geographical area and second highest area under desertification/land degradation with respect to state TGA. The state is observed with 62.90 per cent of the total geographical area under desertification/land degradation for the period of 2011-13. The desertification/land degradation area in Rajasthan has decreased about 0.29 per cent since 2003-05.

Desertification Processes in Rajasthan

The most significant process of desertification/land degradation in Rajasthan as per second cycle DSM is Wind Erosion (44.41% in 2011-13 and 44.80% in 2003-05) followed by Vegetation Degradation (7.62% in 2011-13 and 7.59% in 2003-05) and Water Erosion (6.18% in both 2011-13 and 2003-05). The Table 12.2 and Fig. 13.1 clearly brings out the details of various processes during 2003-05 and 2011-13.

Table 12.2: Desertification Processes in Rajasthan (2003-05 to 2011-13)

Process of Desertification/Land Degradation	2011-13		2003-05		Change (ha)
	Area (ha)	Area (%)	Area (ha)	Area (%)	(2011-13) - (203-05)
Vegetation Degradation	26062121	7.62	2596003	7.59	10218
Water Erosion	2116314	6.18	2116082	6.18	233
Wind Erosion	15197874	44.41	15332054	44.80	-134180
Salinity	363768	1.06	365666	1.07	-1898
Waterlogging	18421	0.05	18421	0.05	0
Manmade	53058	0.16	50865	0.15	2193
Barren/Rocky	1052374	3.07	1047818	3.06	4556
Settlement	118482	0.35	98696	0.29	19786
Total Area under Desertification	**21526512**	**62.90**	**21625604**	**63.19**	**-99092**
No Apparent Degradation	**12546925**	**36.66**	**12448140**	**36.37**	**98785**
Total Geographical Area (ha)			**34223900**		

Source: Desertification and Land Degradation Atlas of India, Space Application Centre, Indian Space Research Organisation, Government of India, 2016.

The above analysis and table 13.2 reveal a difference in data under desertification in earlier study of 2007 and recent study of 2016 conducted by Space Application Centre, Ahmedabad. As per earlier study the total area under desertification in Rajasthan in

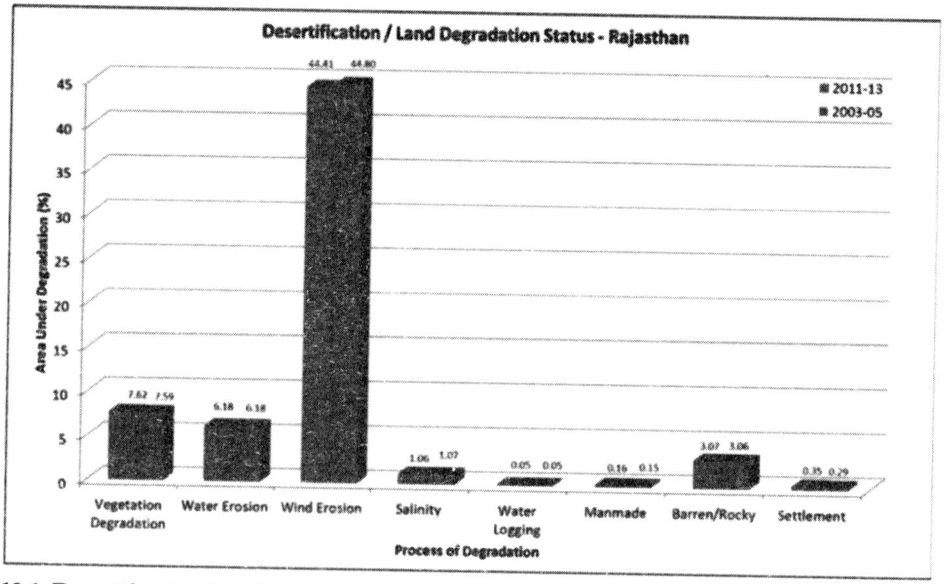

Fig. 13.1: Desertification/Land Degradation Status—Rajasthan

Source: Desertification and Land Degradation Atlas of India, Space Application Centre, ISRO, Govt. of India, 2016.

2003-05 was 67.10 per cent while in the recent study it was reported as 63.19 per cent. The difference of 3.81 per cent is a big difference at the state level which warrants attention and justification.

The spatio-temporal analysis of desertification indicates that the area under desertification in Rajasthan has reduced by 0.29 per cent or 9992 hectares. This reduction in desertification and land degradation or improvement in the non-degraded area, is as a result of combating desertification programmes of the Central and State Governments such as DDP, DPAP, CDP and watershed schemes. The increase in water bodies is because of increasing trend of rainfall after 2003. The Table 13.2 further reveals that the areas under vegetal degradation, man-made activities and settlements have increased. However, the wind erosion and problem of salinity have shown a declining trend in contradiction to the study conducted by Amal Kar (2012) where he stated that "there is likelyhood of aridity spreading beyond the present eastern boundary. Wind erosivity is expected to increase. destabilizing the sandy landscapes contributing more of the atmospheric dust load". (Fig. 13.2).

Causes of Desertification in Rajasthan

At the U.N. Conference on Desertification (Nairobi, 1977) an attempt was made to review on a global scale the nature and causes of desertification. At UNCOD the data were reviewed and three possible causes of desertification have been defined (Baker, 1982) as:

Climate change resulting from changes in global atmospheric circulation.

Mismanagement resulting from an inexorable build up of pressure on resources and destructive misuse of resources reducing large areas to desert—like conditions, i.e., the same effect as first but cause differently.

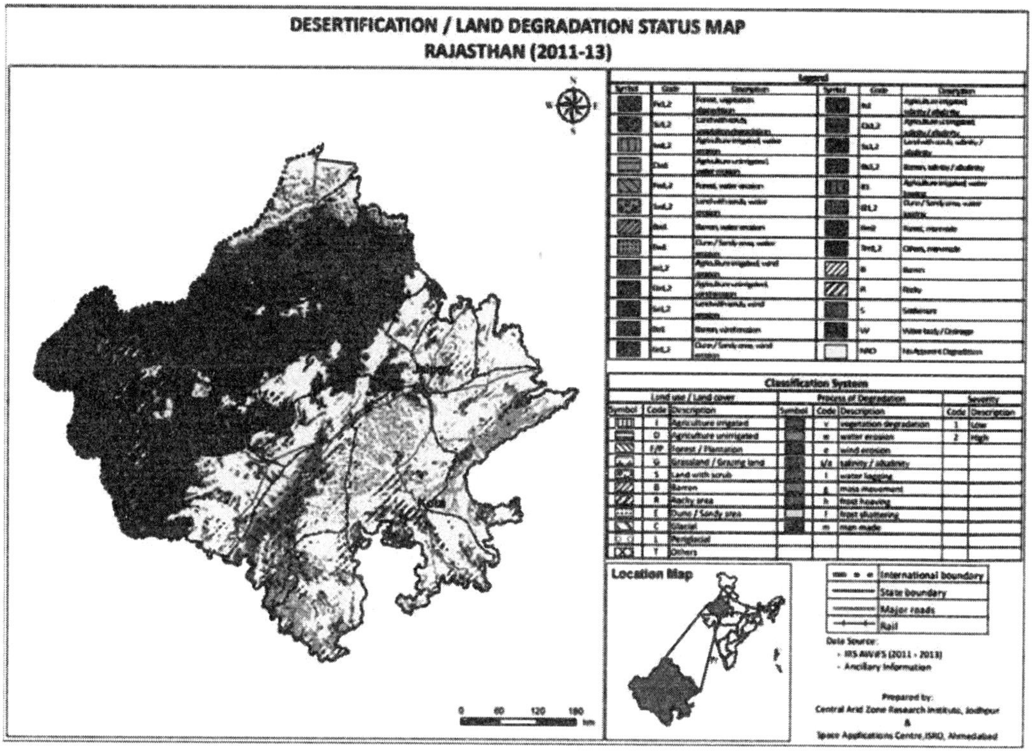

Fig. 13.2

Source: Desertification and Land Degradation Atlas of India, Space Application Centre, ISRO, Govt. of India, 2016.

Biogeophysical feedback relating (second) to (first) by which means, it is hypothesised that changes in land use are able to effect changes in climate. It is very difficult to prove the presence of climatic change from local data of Rajasthan. However, the impact of mismanagement of resources is often locally and regionally very clear and on an alarming scale. This is, however, something which unlike global climatic change, can be tackled at source as it is the work of man rather than nature.

Climate Change in Rajasthan

Over one hundred and twelve years data of rainfall and temperature in Rajasthan now give an indication of departure from the normal. Besides, there are various studies expressing the view that, by and large, India is undergone significant climatic change since the beginning of last century (Parthasarathy and Dhar, 1978; Singh, 1984 more).

The study of rainfall trends of the last 112 years (1901-2012) and more in Rajasthan shows that certain stations have recently experienced apparent downward trend in rainfall, whereas others have shown a high variability from year to year, but no constant trend has been noticed. Now, the question arises, is there any widespread evidence that rainfall in the Indian desert and adjoining areas is decreasing or increasing? If so, is a fluctuation in progress, or is the decrease a lasting change? On a very long time scale there have certainly been great changes in climate specially within the Quaternary period. During the past 10,000 years Rajasthan has been alternating between arid and humid phases resulting in accumulation of sand dunes especially during the arid phase (Table 13.1).

There is thus plenty of evidence of past desiccation of climate that led to the spread of the desert surface in the northeastern part of the present desert boundary. Is there evidence of a similar desiccation in progress today? Is the widespread desertification now in progress due to such a downward trend of rainfall? There is strong evidence to suspect that more and more humid and sub-humid zones of the country, which had earlier experienced no significant moisture stress, are now under the intensification of man-induced desertification process.

The unprecedented drought situation that has occurred during 1987, 2003 and 2009 years in Rajasthan is not a new thing. There are plenty of references of such a situation in the Vedic literature and in the recorded history of India. The droughts of 1877, 1900, 1918, 1940, 1965, 1972, 1979, 1984-85, 1986-87 and 2009 are well-known because of their disastrous consequences. The most notable characteristics of such droughts are a random distribution in time, spatial coherence over large and in some regions, persistence for years at an end.

Climate Desertification Linkages

The prolonged drought for four consecutive years has prompted the question whether the climate is being affected by human action? There are two mechanisms whereby this might happen. First overgrazing, unwise use of land and stripping of forest cover, all tend to make the surface more reflective to solar radiation in this part of the country. Dynamic modeling indicates that such raised albedos tend to diminish rainfall and further potential disastrous positive feedback. Second, in regions like Rajasthan, far from the ocean, much of the rainfall is caused by re-evaporated soil moisture. Faulty land use practices have reduced the storage capacity of enhanced evaporation.

Moisture Index (P/Pe) and Desertification

Gore, *et al.* (2011) using rainfall normal for the periods from 1901 to 1950 and 1941-1990 based on moisture Index classification of the moisture index over India have calculated for arid, semi-arid and dry subhumid as P/PE=0.05-0.20, 0.21.05, and 0.51-0.65 respectively. In the state of Rajasthan, the districts falling in arid, semiarid and dry subhumid having moisture indexes in different regions is given in Table 13.3.

The analysis of the Table 13.3 reveals that in Rajasthan the districts like Ganganagar, Hanumangarh, Bikaner, Jodhpur and Jaisalmer (Arid region), Jhunjhunu, Sikar, Jaipur, Ajmer, Tonk, Sirohi, Pali, Jalore and Churu (Semi-arid region) and Sawai Madhopur and Bundi (Dry sub-humid region) show increase in P/PE ratio. The increase is significant in Sirohi, Jaipur and Sawai Madhopur districts. The districts like Kota and Chittorgarh show decrease in P/PE ratio. The districts like Barmer, Alwar, Bharatpur, Bhilwara and Udaipur show no change. However, the stations having significant change in P/PE ratio from 1901-1950 to 1941-1990 are Sirohi, Jaipur and Sawai Madhopur.

Table 13.3: Moisture Index in Different Region of Rajasthan

Arid (P/PE=0.05-0.20)			*Semi-arid (P/PE=0.21-0.50)*			*Dry Sub-humid (P/PE=0.51-0.65)*		
	1901-1950	*1941-1990*		*1901-1950*	*1941-1990*		*1901-1950*	*1941-1990*
Rajasthan			*Rajasthan*			*Rajasthan*		
District	–	–	*District*	–	–	*District*	–	–
Ganganagar	0.15	0.16	Jhunjhunu	0.27	0.28	Sawai Madhopur	0.51	0.57
Hanuman Garh	0.11	0.14	Alwar	0.44	0.44	Bundi	0.50	0.53
Bikaner	0.15	0.17	Sikar	0.31	0.33	Kota	0.58	0.59
Barmer	0.15	0.15	Bharatpur	0.39	0.39	Chittograrh	0.54	0.53
Jodhpur	0.17	0.20	Jaipur	0.31	0.38			
Jaisalmer	0.08	0.09	Ajmer	0.34	0.37			
			Tonk	0.44	0.46			
			Sirohi	0.26	0.37			
			Bhilawara	0.45	0.45			
			Pali	0.23	0.24			
			Jalore	0.22	0.23			
			Churu	0.21	0.24			
			Udaipur	0.45	0.45			

Source: Impact of climate change of Land degradation over India: National Climate Centre, RR No. 1/2011, May 2011 (NCC Research Report by P.G. Gore, B.A. Roy and H.R. Hatwar).

The decrease in P/PE ratio values for the two periods as above indicates land degradation to less soil moisture availability. There are 23 districts in Rajasthan from three climatic regions show land degradation of which three districts (Sirohi, Jaipur and Sawai Madhopur) show significant land degradation.

Amal Kar (2012) on the basis of GCM predited that there is a strong possibility of a prolonged drought phases during 2021-2025, which may continue with some respite till the mid century. Further this study also shows the likelyhood of aridity spreading beyond the present eastern boundary of the region during the same period.

Recently, Ratnapriya and Padmini Pani (2014) understanding causes of land degradation in relation to climate and deforestation have indicated that in Rajasthan monsoon rainfall is moderately correlated with land degradation having 0.54, therefore, deforestation or less vegetative cover and population pressure are the main cause of land degradation.

Population Pressure and Mismanagement of Resources

Successive census reports have revealed that population of Rajasthan has been growing at an alarming rate. It has registered more than double over the past thirty years, having risen from 1.59 crores in 1951 to 3.41 crores in 1981, 68.62 crores in 2011. The highest growth rate, however, has been recorded during 1971-81. In absolute terms, population has increased by 238 lakhs during last 80 years. In 1981, Rajasthan has recorded a growth rate of 32.97 per cent as compared to the national growth rate of 24.66 per cent which has declined to 21.44 per cent (2011), but still, higher than the national average (17.64%). In particular, the western arid districts have registered an abnormal growth. The alarming growth rate led to an adverse environmental situation enumerated as follows:

1. Reduction in man-land ratio which is normally devoid of adequate natural sustenance. There is a corresponding shrinkage of cultivated land.
2. Rise in net sown area at the cost of grazing land and marginal land.
3. Over-dependence on agriculture has led to land degradation due to non-availability of adequate cultivated lands.
4. Urbanization and industrialization in Kota, Jaipur, Banswara and Dungarpur districts have further aggravated the man-land situation and reduction of forests for fuel and housing material.

Land Use Changes

Due to increased human and livestock population, vast area of Rajasthan have been brought under cultivation often extending to marginal lands, piedmont slopes of the Aravallis and other fringe areas, where the physical conditions do not warrant farming.

The official data shows an increase of forest area from 6.50 per cent in 1985-86 to 7.98 in 2009-10 but the satellite data 2008 reveals that the forest area in Rajasthan is 4.70 per cent (India State Forest Report, 2011). The field study reveals that degradation of forests of the Aravalli and other parts of Rajasthan is taking place at an alarming rate. The total forest area in the state is 32,639 sq. km. which is 9.54 per cent (2011) of the total geographical area. The Aravallis of Udaipur, Dungarpur, Banswara, Ajmer and Jaipur are completely devegetated. The Bundi-hills which were well-vegetated twenty eight years ago, now have become rocky wastes. The thin soil cover has also been completely washed away. Therefore, the possibility of regeneration is dim. If this rate of depletion will continue the Aravallis will be barren by the end of twentyfirst century.

Massive investments on various irrigation projects in Rajasthan have been made but the returns are not commensurate. Introduction of irrigation farming without adequate drainage measures has given rise to accumulation of minerals and salts in the soil and near surface formation in a tract having naturally impeded drainage and heavily charged with minerals and salts. All these factors have led to diminishing returns from land and intensification of man-induced desertification processes.

Development of new settlements on stabilized dunes, urbanization and construction of roads without regard to drainage channels (as in case of Jaipur city), canal without regard to physical constraints have led to rapid mobilization of stable sand belts and consequent desertification by removal of vegetation cover and wind action.

Animal Pressure

Similar to increase in human population, there has been a rise in number of livestock population (Table 13.4). This has led to overgrazing, deforestation and exploitation of the woody bio-mass of the region.

Table 13.4: Growth of Livestock in Rajasthan

S. No.	Type of Animal	1956	1961	1966	1972	1977	1983	Variation in 2007 over 2003 % in- crease (+) or decrease (-)	Variation in 2012 (P) over 2007 % increase (+) or de- crease (-)
1.	Cows	+12	+09	-09	-05	+03	+05	-	9.44
2.	Buffaloes	+13	+17	+05	+07	+11	+19	(+) 6.018	16.99
3.	Sheep	+37	-16	+20	-03	+16	+35	(+)11.111	(-)18.86
4.	Goat	+58	-08	+28	+18	+02	+25	(+)27.794	0.76
5.	Camel	+33	+25	+20	+16	+01	-53	(-)15.643	(-)22.79
6.	Pig	+32	-02	+16	+40	+11	+37	(-)38.120	(-)13.96
7.	Poultry	+39	+75	+14	+50	+25	+46	(-)20.123	
	Total	+26	+02	+11	+04	+06	+23	(+)15.068	1.89

Source: Some Facts About Rajasthan, Department of Planning Rajasthan, Jaipur, 2015.

The table 13.4 indicates that number of livestock in Rajasthan is showing an increasing trend except camel, pigs and poultry. It is evident that the number of camels are declining since 1977 but drastic reduction is noticed after 1983 (-53%). However, the variation in 2007 over 2003 is -15.643. The total livestock population in 2007 has been reported as 56663183 and the variation in 2007 over 2003 is 15.068 per cent. The percentage of goats and sheep are highest in the order 37.89 per cent and 20.26 per cent respectively. Cattle and buffaloes are 21.05 per cent and 19.57 per cent respectively (Census, 2007) in rural areas. The number of sheep and camels have declined in 2012 over 2007 (Table 13.4).

The squeeze of pasture land has created huge stock of hungry, malnourished animals not only unable to provide economic yield, but adding to environmental destruction by creating soil erosion and destroying forest regeneration.

Ecological Crisis in the Aravalli Range

The Aravalli range of Rajasthan is facing serious ecological crisis. The diminishing mountain cover has adversely affected the climate leading to failure of rainfall resulting into severe famines and drought conditions. If the ruthless and indiscriminate cutting of trees, killing of wildlife and mutilation of its topography is not stopped by now, a situation will be more serious than the Himalayan valleys like Alaknanda. The mis-management of forest resources of Jaipur hills and their extension in the northeast has made the surrounding hills barren. The removal of vegetation and large scale leveling of older fossilized dunes have aggravated gullying and choaking of natural water flows. With increased rainoff, reduced vegetal base and loosened sand there is a greater erosion than even before. This was seen earlier in 1978 monsoon and more spectacular so in 1981. In the event of 1981 large quantities of sand from dunal areas spread out on the entire fertile agricultural fields of southern Jaipur and adjoining areas of Tonk district (Sharma, 1984).

Conclusions

From the above discussion, it can be concluded that the environment of Rajasthan has been deteriorating as a consequence of human interference and mismanagement of land resources. The thesis of desert expansion towards east through various gaps of the Aravallis in the form of sand movement appears inadequate. However, overexploitation of resources, over-grazing, large-scale deforestation in the Aravalli range by man and animals have accentuated the reactivation of sand dunes and accelerated sheet erosion in north-eastern and south-eastern Rajasthan, causing a serious ecological imbalance. The continued degeneration of the Aravalli forests would have serious repercussions and jeopardize the future of agriculture in Rajasthan.

Though Rajasthan has exerted some efforts partly by central assistance to control desertification in the form of Desert Development Programme and Drought Prone Area Programme, the results have not been remarkable. The main reason is the largeness of the affected zone. Added to this reason, there prevails among decision-makers the idea that desert control is not an issue of high priority. Whatever development schemes have been envisaged are more of a pilot projects than of development programmes. However, despite deteriorating environmental conditions in the state, it is a ray of hope that an improvement in the condition and desertification is partially halted (DSM 2016; ISRO). A successful plan to combat desertification in Rajasthan has to be all comprehensive. All recommended measures should aim at a reorganization of land-use so as to retrieve the ecological balance between the land exploitation and the ability of plants and soil to regenerate. It is a matter of satisfaction that the decision-makers in the centre have now realized the magnitude of desertification and the ecological crisis in the Aravalli range. As a result the allocation of funds for desert control by the Central Government after Seventh Five Year Plan onward have been enhanched for Aravalli Hill Area Development constituting one hundred twenty nine development blocks. Despite it, much has to be done to combat desertification.

Solutions to desertification in region under study can be implemented readily if resources are available to do so. An exploding population will continue to build pressure on land resources. This would further hasten the destruction of vegetal degradation by the ever increasing demand for firewood and other requirements. The future droughts and Climate Change will have greater damaging effects on all walks of life.

Combating desertification can be done successfully using technology already known if financial resources are available and the political will to act is present.

[**Acknowledgement:** The author is thankful to space Application Centre (ISRO), Ahmedabad for financial support and associating with the National Project on Status Mapping of Desertification in India in Ist and IInd Cycle (2001 to 2017).]

References

Allchin, B., Goudie, A. and Hegde, K.T.M. (1978). *The Pre-history and Palaeogeography of the Great Indian Desert,* Academic Press, London.

Amal, Kar (2012). GCM-Derived Future Climate of Arid Western India and Implications for Land Degradation, *Annals of Arid Zone,* Vol. 51 (3&4), 147-169.

Amal, Kar (2014). Agricultural Land Use in Arid Western Rajasthan; Resource Exploitation and Emerging Issues, *Aeropedology,* Vol. 24 (02), 179-196.

Amal, Kar (2014). The Thar or the Great Indian Sand Desert in Landscapes and Landforms of India (Ed. V.S. Kala) World Geomorphological Landscapes Series, Springer, Dordrecht, 79-90.

Aubreville, A. (1949). *Climate, Forest et Desertification de l' Afrique Torpicale,* Societe d' Editions Geographiques, *Maritimes et coloniales,* Paris, 255.

Baker, R. (1982). Desertification: Cause and control, A study of the U.N. Plan of Action and its Application, *Scientific Review and Arid Zone Research,* 1., 285-313, Scientific Publishers, Jodhpur.

Blandford, W.T., (1877). Geological Notes on the Great Indian Desert between Sind and Rajputana, *Rec. Geol. Surv. India* 10; 1-2.

Desertification and Land Degradation Atlas of India, Space Application Centre, Indian Space Research Organisation, Government of India, 2016

Dhabria, S.S. (1984). Intensification of Desertification Hazard Through Sambhar gap, A problem of Environmental Degradation in Indian Arid Zone. Paper presented 18th International Symposium on Remotesensing of Environment, Paris (France) 1-5 Oct., 1984.

Dhabriya, S.S. (1986). Assessment and Mapping of desertification in Rajasthan using satellite productes, *Proceedings of the 20th International symposium on remote sensing of environment,* vol, II, 4-10 Dec. 1986. Nairobi, Kenya, ERIM, Ann Arbor, Michigan, USA, 859-868.

Dhabriya, S.S., (1988). Desert Spread and Desertification, *Environmentalist,* Malviya Nagar, Jaipur, 1, 2, 14 and 15.

Dhabriya, S.S., (1988). Eco Crisis in the Araballi Hill Region, *Environmentalist,* Malviya Nagar, Jaipur, 1 to 54.

Dhir, R.P. (2003). Thar Desert in Restrospect and Prospect, Proceedings, Indian National Science Academy 69A (2), 167-184.

Gore, P.G., Roy, B.A. and Hatwar, (2011). National Climate Centre Report, R.R. 1, Pune, 1-12.

Goudie, A.S., Allchin, B. and Hegde, K.T.M. (1973). The Former Extensions of the Great Indian Sand Desert, *Geographical Journal,* Vol. 139(V), 243-57.

Hooja, B. (1985). Reforming Desert Development, Administration and Strategies in Rajasthan, An Extensive Approach for Desert Development and Evvironmental Preservation, *Indian Council of Public Administration,* Vol. XXXI, 910-32.

Krishnan, M.S., (1956). *Geology of India and Burma,* Higginbothams, Madras, 513-14 and 52.

La Touche, T.D. (1902). Geology of Western Rajasthan, *Mem. Geol. Surv. India* 35: 1-115.

Parthasarathy, B., Dhar, O.N. (1978). Climatic Fluctuations over Indian Region-Rainfall: A Review, Research Report, RR-025, Indian Institute of Tropical Meteorology, Poona.

Planning Commission, Govt. of India, New Delhi, (1951).

Ratnapriya and Padmini Pani (2014). Understanding Cause to Land Degradation: Is it Climate or Deforestation? *Abstract Volume,* 27th IGI Conference (14-17 November), Midnapur, W.B.

Rhode, K.P. (1964). Geomorphology and Evolution of Rajasthan Desert, Symp. Problems of Indian Arid Zone, Govt. of India, New Delhi, 69-75.

Roy, BB and Pandey, S (1971). Expansion or Contraction of the great Indian Desert, Proceedings of the Indian National Science Academy, Vol. 36 (6), 331-334.

SAC, (2016). Desertification and Land Degradation Atlas of India (Second Cycle), Space Application Centre, ISRO, Ahmedabad (Govt. of India).

SAC, (2007a). Desertification and Land Degradation Atlas of India, Space Application Centre, ISRO, Ahmedabad (Govt. of India).

SAC (2007b). Pilot Project, Desertification and Land Degradation Atlas of India, Space Application Centre, ISRO, Ahmedabad (Govt. of India).

Sharma, H.S. (1987). *Tropical Geomorphology: Morphogenetic Regionalization of Rajasthan,* Concept Publishing Company, New Delhi.

Sharma, H.S. (1988). The Problem of Desertification in Rajasthan, In *Development of India's Natural Resource Base Patterns, Problems and Prospects* (eds.) V. Vidyanath & R.R. Rao, Gyan Publishers, New Delhi, 358-366).

Sharma, H.S. (2008). Problem of Desertification and Strategy for Combating Desertification with reference to Rajasthan, In B. Thakur (Ed.) *Perspectives in Resource Management in Developing Countries,* Vol. 3 Concept Publishing Company, New Delhi, pp. 149-164.

Singh, N. (1984). Fluctuations of Different Moisture Regimes in India, *Arch et. Geogph. Biocl., Ser. B.* 35, 239-256.

Singh, S. (1977). Geomorphological Investigations of Rajasthan Desert. Monograph No. 7, Central Arid Zone Research Institute, Jodhpur.

Some Facts About Rajasthan (2015). Directorate of Economics and Statistics, Department of Planning, Rajasthan, Jaipur.

India State Forest Report, (2011).

U.N.C.O.D. (1977). *Plan of Action,* Nairobi.

UNCED (1992). *UN Conference on Environment and Development,* Rio-de-Janeiro.

UNCOD (1977). Round-up, Plan of Action and Resolutions, United Nations Conference on Desertification, Nairobi, 43.

UNEP (1997). *World Atlas of Desertification,* Second Edition, United Nations Environment Programme, Nairobi and Edward Arnold, London.

Wadia, D.N. (1939). *Geology of India,* London, Repr. 1953, Repr. 1978, 4 and 373-74.

Wadia, D.N. (1955). Desert of Asia – Their Origin and Growth in the late Pleistocere time, Second Sir Albert Charles Seeward Memorial Lecture, Birbal Sahani Institute of Palaeobotany, Lucknow.

Wadia, D.N. (1957). *Geology of India,* Macmillan & Co. Ltd., New York.

14
Implications of Land Degradation for The Environment: A Field Based Perspective

Padmini Pani

Introduction

Land degradation is the key environmental challenges before mankind. The Sustainable Development Goals cannot be meaningfully achieved without taking care of the problem of land degradation. In many developing and less developed countries land degradation problem is a major concern for sustainable development.

Degradation is, by description, 'a process of change over time' (Scherr and Yadav, 1996). When a land has lost its biological productivity due to the complex biophysical processes the land is considered as a degraded land. Environmental degradation is the deterioration of the environmental resources through reduction of some important life support resources like air, water and soil etc. Land degradation and loss of biodiversity creates an immense pressure for the world's ecosystems and their capacity to support the important services is decreasing at the point of time when the need for these services are even more. (MEA, 2005a). Land degradation is ascertained to be one of the severe global environmental challenges (Eswaran *et al.*, 2001; Lal, 2001; Scherr and Yadav, 1996). Land degradation is having impact on billions of people around the world and may lead to the overexploitation of soil resources, decline of groundwater resources, loss of ecosystem productivity, shifts in vegetation composition, and loss of rural livelihoods (Shroder *et al.*, 2016). The impact of land degradation and the degradation of environment are related to each other and change of one component of it change the balance of environment thereby creating problem or threat to the land and its sustainability.

Land degradation is process in which, biological productivity of land is reduced that has varied social, economic and ecological consequences (Pani & Carling, 2013). Water resources, forest and air, when negatively impacted by the natural processes and human activities, might result in various types of land degradation. The various components of land degradation such as soil loss, loss of soil fertility, salinity, waterlogging, wind and water erosion etc. are closely associated with water and forest resources. Judicious management of all these components is vital for achieving overall sustainable development.

India being agrarian economy the land is the primary asset for national development. The impact of land degradation in respect of productivity was estimated to be 4.5 per cent of GDP in 1992 in India at the cost of land degradation (TERI, 1998). Several case studies have stated land degradation in Indian context is regarded in terms of decreasing land use intensity, high input (fertilizers) use (in terms of cost) and declining profits (in terms of productivity) (Reddy, 2003).

In context of the Indian scenario, the NRSA (National Remote Sensing Agency) has given estimates of five broad categories of degradation namely, water erosion, wind erosion, water-logging, salinity and nutrient depletion. Various agencies and organizations provide data regarding land degradation. As per one study by National Soil Bureau of Soil Survey and Land Use Planning in 1994 it is estimated that up to 187.8 million hectares of land are under soil degradation. The recent estimation has been done by NRSA in 2010 and 2016 at the country level. As per recent assessments by Indian Space Research Organisation, in 2007, 32.07 per cent (105.48 mha) of the total geographical area of the country was disturbed by different types of land degradation. About 10.21 per cent (33.56 mha) of the total degraded area was because of water erosion. State-level approximations from the same study recommend that in Madhya Pradesh, of the total 3.5 mha of degraded land, nearly 37.92 per cent was caused by water erosion [SAC (Space Application Centre)-ISRO 2007]. An assessed 3.98 million ha of ravines appear along the rivers Yamuna, Chambal, Mahi, Sabarmati and others (Ministry of Agriculture, 1984). Recently remote sensing data has been extensively used to monitor changes in spatial extent of ravines. Pani *et al.* (2001) observed an increase of ravine affected area and the villages have constantly been shifting from their original locations in response to soil degradation and erosion. The evaluation of different landform units and the implication of neo-tectonic activity in the area using remote sensing techniques have been carried out by Pani (2017a,) along with the erosion intensity assessment in a part of the Lower Chambal Valley. The study of farmers' perception reveals that farmers are deeply aware of the problems of land degradation in the region and have adopted various coping strategies that include changes in farming practices, reliance on multiple livelihoods and spatial reallocation of household labour (Pani *et al.*, 2011; Pani *et al.*, 2013, Kumar and Pani, 2013). Numerous works on ravine reclamation have been done by various boards under U.P., M.P., Gujarat and Rajasthan state governments and ministries of Government of India. The work done by Soil Conservation Department, Ministry of Agriculture, Forest Department and Planning Commission, Government of India are to name a few. In terms of severity of the erosion there are five major ravine erosion zone had been identified in India (Sharma 1980). The major affected ravine zones are Yamuna-Chambal Ravine Zone, arid Gujarat zone consist of Tapti, Watrak, Sabarmati, Mahi, Narmada and their tributaries and parts of Chotanagpur Plateau. The tract of Ganga plain, the valleys of Ramganga and Sai, and some part of Godavari valley in Maharashtra and Andhra Pradesh and Kathiawar come under the moderately ravine affected area. The less severe affected areas are the whole of Deccan, south of Godavari, east of Varanasi, and western Rajasthan.

Objective of the Study

The objective of this study is to understand the dynamics, extent and degree of land degradation and evaluation of its impact of the natural and socio-economic environment.

Materials and Method

The following sources and materials have been used to fulfil the objectives of the study. (i) The Base map prepared from topographical map of Survey of India 1984; (ii) Village boundary map prepared from District Census Village Boundary Map, (iii) Land use/land cover maps of 1974 and 2014 have been prepared using LANDSAT 1 MSS (Multispectral Scanner sensor) image, and LANDSAT 8 (Operational Land Imager Sensor OLI) images, for the year 1974 and 2014 respectively; (iv) Ground truth verifications and field observations, FGD, have been done in 2014, 2015, 2016, 2017, respectively; (v) Field photographs of different seasons and years have been used to understand the ravine formations and its implications.

Study Area

The area has been chosen on the basis of its severity, and its long history of land degradation. The Yamuna-Chambal Badlands is the most severe and largest ravine area of the world. The study area has been taken one severely ravine affected district, Morena in Chambal Region. The study area is located between the geographical coordinates of 26°17′15″ N to 26°52′22″N latitude, and 76°28′30″E to 78°32′55″E longitude. The geographical area covers near about 4989 km². The region is well-drained by the river Chambal and its tributary Kunwar (Figure 14.1). Most part of the Chambal valley is highly dissected and inaccessible. Steep ridges, low-sloping hills, deep trenches and broad meanders are the common features of the study area (Pani, 2017a). Soils of the area are broadly of two types: one category is sandy loam to loam with low phosphorous and salt content and the other category is clayey loam with low phosphorous and salt content (GoMP, 1996).

The study area is lacking in healthy vegetation cover and most of the area is covered by thorny bushes and sparse vegetation. The district has a forest cover of 14.63 per cent out of the state average of 25.21 per cent. There is no area under dense forest in the district (as per the State of Forest Report 2011). Moderately dense forest cover is nearly 13.42 per cent and open forest account for 90.29 per cent of the total forest area. A further 8.22 per cent of the total geographical area is under scrub (Forest Survey of India, 2011).

Results and Discussion

Overview of Land Degradation Dynamics

Due to the water erosion the huge amount of topsoil loss and Ravine formation is a major concern for the region which have developed a major physical and economic implications (Carling and Pani, 2013). The gullies and ravines are widely spread and form extensive network and develop a wide range of Badlands. The Badlands are locally called *beehad.* These are known in scientific world is as Ravine, means hunger or hungry land, these are consisting of deep gullies, which are running parallel to sub-parallel. The severely eroded deep ravines are a narrow steep sided V-shaped valley which are larger than a gully.

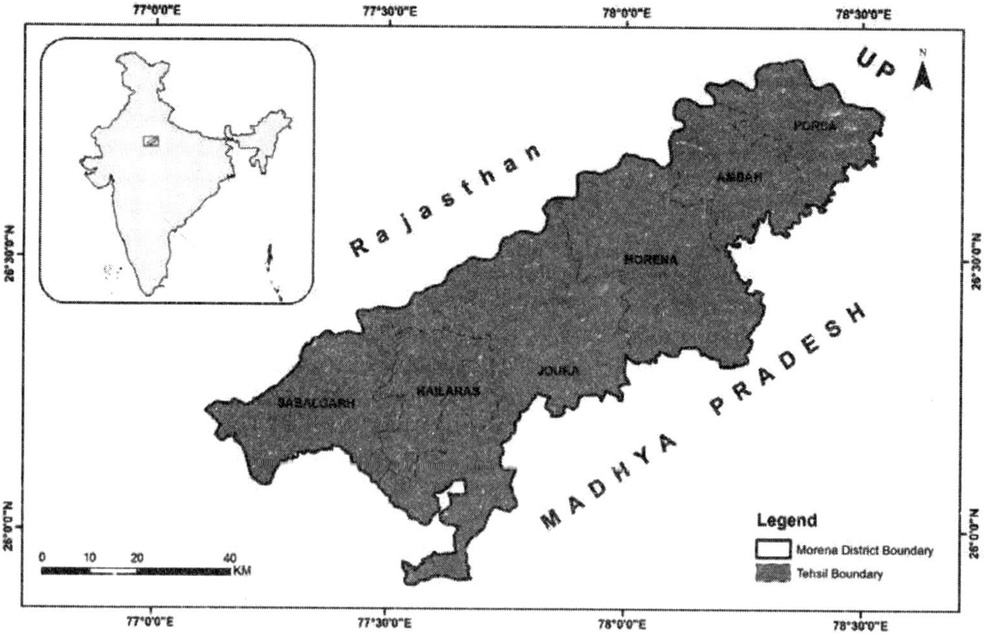

Fig. 14.1: Study Area: Morena District

The ravine formation is an important geomorphic process in many parts of the world in a range of landscapes. Natural causes of ravine formation incorporate climate change, catastrophic storms and isostatic rebound, tectonic uplift or base level lowering, removal of vegetation and deforestation, overgrazing, population pressure, unsuitable agricultural practices, institutional settings and public policy among the significant anthropogenic factors causing land degradation (Pani and Carling, 2013, Sharma 1980). The relations between social and natural processes impacting on land degradation are usually complex and locale specific (Blaikie, 1989).

The implications of gully erosion are not only limited to the loss of agricultural land, in numerous localities it has also caused for shifting of villages either in a village level or in household level (Pani and Mohapatra, 2001). The ravine development and expansion are having significant impact on environment of the surrounding region, which leads to large volume of soil loss after every rainy season. It has been observed that topsoil erosion has resulted in loss of productive capacity of agricultural land in long run. This semi-arid ravine affected district Morena is a comparatively less developed part of India. Around 80 per cent of the rural people of this region are dependent on primitive occupation like agriculture. Ravine erosion in India is a very isolated topic in world academic forum. It is still considered as a region-specific problem in India (Haigh, 1984).

Land-Use/Land-Cover Changes

Land use practice is one of the key factors of land degradation of any area. Decadal remote sensing data have been used to analyse the land use pattern of the study area. It has been observed that the study area has been witnessing a rapid change of land use practice which is one of the key factors of land degradation in the study region. The land use of the area mainly encompasses abundant alluvial plains and ravines and some parts of the district are occupied by barren lands which exist in a scattered manner mainly in south-western parts of the investigated area. The barren lands are basically structural and residual hills. The southern part of the Sabalgarh area comes under dissected plateau. The other common land use land covers are mainly rivers, sand bars and water bodies.

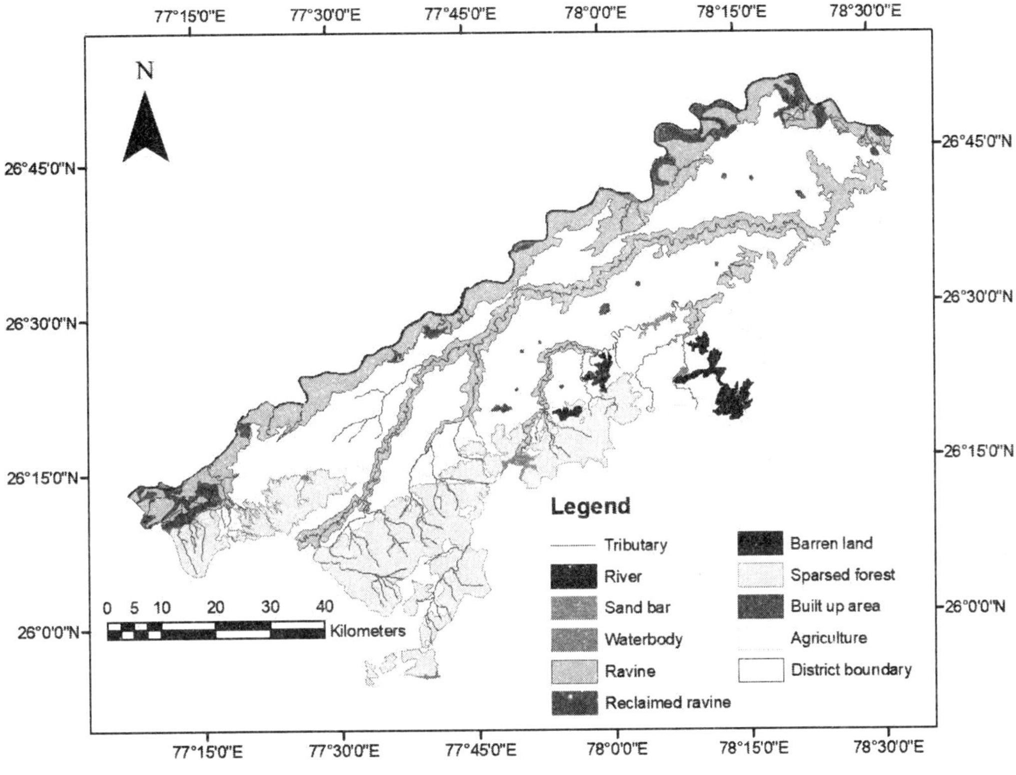

Fig. 14.2: Land use Land Cover Map of The Morena District (1974)

So far as changes in land use classes are concerned, the major changes are: an increase in built-up area and agricultural land and decline in forest cover, however the forest cover was always sparse in nature in the study district. Forty years of remote sensing analysis along with field survey by the author brought enough evidences of rapid change of land use practice in the study region. Loss of natural vegetation cover, change of natural topography and acceleration inland-levelling and increase in agricultural land are shown in the trend of rapid change of land use/land cover of the study region. To estimate the extent and severity of the area and its degradation the forty years of remote sensing data have been proved to be useful (Figure 14.2 and 14.3). A vast area of this region is comprised by highly dissected ravine and the most striking geomorphic unit of the area (Figure 14.2 and 14.3). The ravine lands extend on both flanks of Chambal River and its tributaries (river Kunwari). Ravines are characterized by a very fine drainage network and steep head and side slopes with narrow gully divides. Ravine has been formed in nearly 2 to 2.5 km area in the both banks of the river Chambal and its tributary in the region. The major part of the area, mainly doab area between Chambal and Kunwari, is used for the agricultural land. Apart from it the substantial area has been reclaimed since 1974. Around 2.92 per cent ravine area had been reclaimed during 1974 and in 2014 the 15.67 ravine area has been reclaimed (Table 14.1), subsequently the total Ravine area from 1082.71 km² in 1974 has been reduced to 415.98 km². The Kunwari river bank and its ravine land is almost completely reclaimed as per the 2014 data analysis. The river Kunwari has witnessed a long history of land levelling. Most of the reclaimed levelled ravine area have been used as an agricultural land in the entire district. Thus, the areal contraction of ravines has been in the magnitude of around 8.35 per cent over the last 40 years (1974–2014) (Table 14.1). The area under reclaimed land has increased from 145.48 km² to 781.09 km² during the same period.

However, this reduction could not stop soil loss or able to brought any control over erosion for the region. During field survey it has been found that there are many fresh initiations of headward erosions on the unaffected fertile land and as well as in levelled land along with some fresh ravine formations (Plate 1a and Plate 1b). Natural gully

Plate 1a: The photograph is showing a newly formed gully head erosion more than 2-meter depth approaching towards the road.

Source: Field photographs.

Plate 1.b: Showing how coalition of gully head-ward erosion engulfed an agricultural land near Bilpur Village in Morena District.

Source: Field photographs

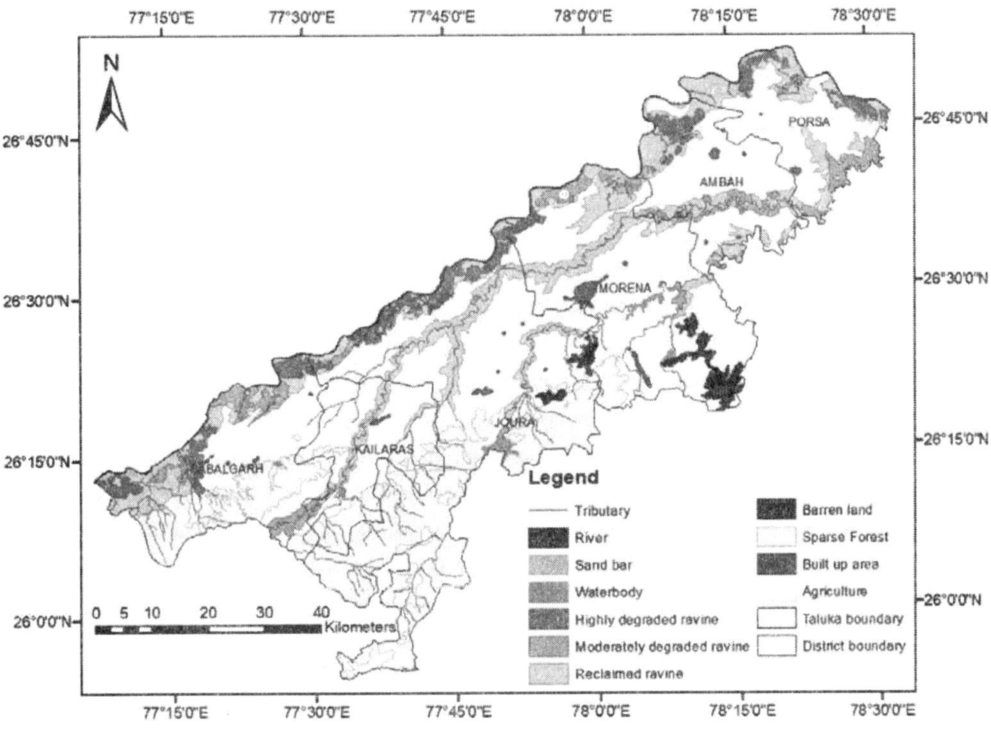

Fig. 14.3: Land use/Land Cover Map of Morena District (2014)

Table 14.1: Land Use/Land Cover Change Statistics of Morena District: 1974-2014

Land Use Land Cover Classes	*1974*	*2014*	*Percentages 1974*	*Percentages 2014*
Built-up	9.24	33.70	0.19	0.68
Agriculture	2615.79	2612.14	52.49	52.42
Sparse Forest	961.11	973.64	19.28	19.54
Ravine	1082.71	415.98	21.72	8.35
Reclamed Ravine	145.48	781.09	2.92	15.67
Barren land	78.29	78.29	1.57	1.57
Waterbody	24.74	24.75	0.50	0.50
Sandbar	32.98	31.63	0.66	0.63
River	33.49	32.14	0.67	0.64
Total	**4983.83**	**4983.35**		

formation has got accelerated by the intensification of the farming systems. The depletion of the soil organic matter, the vulnerable soil structure and levelling the deep ravine intensify the soil loss and increase by the runoff process. It has been observed that apart from loosening of the soil by land levelling, irrigation channels in the levelled land can also favour gully erosion or gully initiation. Field-survey investigations suggest that the levelled ravine in Bagchini village, Ambah tehsil on the bank of Kunwari river, which has a long history of land levelling, is also experiencing the gully formation as a result of irrigation (Plate 2a and 2b). Other researchers have also reported that irrigation can also accelerate gully erosion (e.g., Vanacker *et al.*, 2003; Nyssen *et al.*, 2004b; Poesen and Vandekerckhove, 2003).

Plate 2a. Irrigation initiates gully formation in a wheat field levelled 30 years back: Bagchini, Ambah Tehsil

Source: Field photographs.

Plate 2b. In the same field a closer view of the gully formation

Source: Field photographs.

The ravine lands are invariably levelled down for agriculture practices for the last few years in a rapid manner without following any scientific approach **(Plate 3a)**. These levelled lands are the most vulnerable towards soil erosion and it is difficult to check the flow during rainy season due to high runoff. The earthen bunds built to check the erosion are also not sustainable in such levelled lands (Plate 3b). There are many off-site impacts of land degradation as well, as there is huge amount of siltation in the nearby water bodies due to high soil erosion (Plate 4).

Plate 3a. Massive land levelling is going on and Bund have been built up to check runoff

Source: Field photographs.

Plate 3b. Constructed earthen bund is no longer sustainable, a closer view of an eroded bund (Near village Esah, in Morena).

Source: Field photographs.

Plate 4: Off-site impact, siltation in the river Kunwari, Near Bagchini, Ambah Tehsil, due to land levelling and the soil loss of the area.

Source: Field photographs.

Coping Mechanisms and Implications

Farmers adopt a number of coping mechanisms to cope up with and mitigate the impacts of land degradation. These include agriculture and land-specific coping mechanisms and those which are outside agriculture. Among the standard responses to gully formation in the Chambal region are to build earthen bunds, check dams, gully diversion, ravine mounds removal, terrace levelling, plantations, especially bamboo plantation, fitting pipes in the farmland and in the road to drain extra accumulated water during rains and reduce soil erosion (Plate 5). Use of *andi* (castor oil) plant is also a common practice to minimise gully erosion (Pani 2017a, Pani 2017b). Globally, reclamation of land has been one of the coping mechanisms to counter land degradation. It is an important measure because soil is a non-renewable natural resource. But the way through which land is reclaimed has important implications for the sustainability of land reclamation (Pani 2017a, 2017b). It also has important consequences of socio-economic environment (Pani 2016, Pani 2017a, 2017b).

Plate 5: Pipe fitted to drain the overflowing water during rainy season on the road and prevent the road from further damaging

Source: Field photographs.

Primarily the impact of ravine formation on agriculture is manifested through decline in soil productivity (Pani *et al.* 2011; Kumar and Pani, 2013; Pani, 2018). The shifting of villages and households as a result of ravines, that has been a persistent feature of severely affected villages, has led to disintegration of the villages and loss of social capital (Pani and Mohapatra, 2001; Pani, 2017b). There are several implications for the livelihoods of people in the affected areas. As a result of inadequate earnings from agriculture, many young males have started to move out seasonally. Non-farm employment in nearby towns and cities has also become a part of the livelihoods diversification processes for the poor (Pani 2018). Farmers respond to the loss of productivity through changes in cropping patterns. The land use/land cover investigation is based on the idea that skewed allocation of farmland has caused unsustainable land use behaviour that is creating land degradation, although there are many other dimensions to land degradation.

Land reclamation is a cost-intensive activity. In the recent years, with the use of heavy machinery like earth movers and tractors in land levelling the cost of levelling land has increased manifold. Few years back it used to be done through manual labour using rudimentary tools and techniques. While the better-off households could mobile wage labour for levelling land, the small and marginal farmers were dependent on household labour and exchange labour of friends, relatives and other villagers. The rising per unit cost of levelling land has changed this scenario. Now all kinds of farmers are dependent on heavy machinery for levelling land. Those who cannot affort such expenditures either borrow money from others or forego land levelling. The scale of land levelling has increased as well because of mechanisation (Pani 2017b). An important aspect of levelled land is its legal status. Land reclamation can be done in privately owned land that has been affected by the ravines, but more often than not, it is not confined to legally owned land. In many villages, land levelling has emerged as a mechanism for privatisation of commons (Pani, 2017 b). Like in many other regions, the impact of such privatisation has been unequal. Women and marginalised sections are more severely affected as a result of such reallocation of rights over land (Pani 2017 b).

Land levelling is known to have created various social problems. The reason of these conflicts could be varied. Some of these reasons are related with ill-defined property rights over land. But some other conflicts arise due to externality problems. Land levelling by one farmer might create problems for others, such as loss of grazing or forest land, denial of passage, restrictions on movement of livestock, increased gully erosion, soil loss and siltation in nearby water bodies etc (Plate 4). Apart from the loss of natural resources and reduced natural habitat for the wildlife, the human-animal conflicts are more frequent now-a-days. Land levelling or encroachment of ravine land for private use is the major reason for loss of local trees like *tenti* (wild berr), *parbati babul*, neem, kher etc. which were in many ways among the alternate sources of livelihoods for the poor. Not having access to the forest land, people are forced to use the crops residue for cooking, which is the cause of respiratory health problems for the women and children of the region[1] (Pani 2017 b).

Problems and Failure of Land Reclamation

On the basis of field survey and discussions with grassroots level government functionaries, it was noticed that land reclamation initiatives in the study region suffer from various problems related to faulty conceptualisation and poor implementation[2].

1. Some of the young women in the villages of Eshah, Bagchini have shared their concern regarding health problems they are facing these days, during survey.
2. These findings have been discussed in greater detail in Pani 2016; 2017a; 2017b; Marzolff and Pani, 2017.

First, there are alack of serious scientific studies and cost-benefit analysis of the specific reclamation strategies in specific micro-contexts. A great deal of ad hoc conceptualisation of land reclamation has been noticed on the ground[1]. According to several studies the best way ravine and gully lands are need to leave as a fallow lands for sometimes but in most of the cases farmers are not ready to leave the land due to insufficient land and land insecurity. Ranga *et al.* (2015) have argued that levelling practices are manifestations of growing land hunger.

Secondly, there has been no attempt to involve the local people or their practice-based knowledge to the land reclamation programmes. Several studies show that incorporation of indigenous knowledge helps in local ownership and management of the programmes. It is not necessary that the practice followed by people on the ground are scientific and sustainable in the long run. In fact, on the basis of their own initiatives, 'local farmers are using the interfluves, gully valley sides, and gully channel beds for cropping after sloping and removal of natural vegetal covers such as grass, wild bushes, and dwarf trees. Such cropping has adverse effects on soil erosion, and there has hardly been any attempt to create a framework for internalisation of such externalities' (Pani 2016). The reliance on a top-down approach to land reclamation has resulted in parallel existence of programmes and individual coping strategies. There are hardly any conversation and mutual learning that is taking place on the ground.

Thirdly, the livelihoods based initiatives suffer from a similar mismatch with the requirements of the local environment. Livelihood support programmes, if successful, programmes aiming at support to livelihoods indirectly help in reducing the pressure on fragile environments. However, some 'livelihood-enhancing efforts such as those based on livelihood support have an ambiguous affect, as while more productive and sustainable livelihoods based on animal husbandry would enhance employment and earnings, they might lead to overgrazing and destruction of *bihads* as well. However, at the implementation level these interfaces are neither being integrated into project formulation nor are being coherently operationalised' (Pani 2016). Fourthly, the lack of horizontal coordination among various state governments involved in land reclamation in the region (i.e. Rajasthan, Madhya Pradesh, and Uttar Pradesh), has limited the scope for scaling-up of successful interventions.

Conclusion

The geologic factors are beyond human control. Therefore, gully and ravine formation are inevitable in this region. But other factors like geomorphology, slope and several human induced factors can be minimised and managed to reduce ravine formation. It is possible with suitable agricultural practice, proper land management and sustainable land reclamation policy. But the rapid change of land use practice is one of the key factors causing land degradation. Utilising available data from different sources, such as remote sensing data, published evidences or information on land use, a deeper

1. On several occasions, farmers expressed dismay over the thrust of the programmes implemented by the government. Similarly, the ground-level staff expressed their lack of comprehension regarding the conceptualization of the various components of the programmes they were involved in implementing.

understanding of the nature of land degradation in the region, would certainly help to minimise land degradation of the region. Unequal access to land resources create a social, economic and environmental threat to the people of the region (Pani 2017 a, 2017b). Erosion and unscientific land use practices, mainly land levelling, not only deteriorate the soil quality on-site, but also results in significant sediment-related problems off-site (Field photo Plate 4) for segmentation problem due to uncheck soil erosion in the level land which leads to deposition of huge sediment in the river system). The approach towards the land degradation is also another noticeable factor for the region. Land degradation in the form of gully erosion is a serious problem in the study area.

The problem of land degradation is faced acutely by most of the agriculture dependent population. The villagers have started to adopt various coping mechanisms to overcome the problems of gully erosion. These include construction of bunds, change of cropping pattern, planting of trees and cultivation on slopes. These coping mechanisms have important implications for cost of production. The process of land levelling was also investigated through this survey. The results show the significance of land levelling as well as various problems associated with it. Thus, while farmers have been trying to cope up with the challenges of gully erosion in this region, there is yet to be a sustainable and equitable solution to their problems associated with land degradation.

[**Acknowledgement:** The author gratefully acknowledges financial support to carry out field work by the Indian Council of Social Science Research (ICSSR), New Delhi and the University Grants Commission (UGC), New Delhi, India, in different phases while conducting the research. Financial assistance from JNU-UPE II (Project ID53) is gratefully acknowledge by the author.]

References

Ahmad, E., (1968). Distribution and causes of gully erosion in India. *Selected papers Seminar, 21.*

Blaikie, P. and Brookfield, H., (2015). *Land degradation and society.* Routledge.

Blaikie, P., (1989). "Explanation and policy in land degradation and rehabilitation for developing countries". *Land Degradation and Development, 1*(1): 23-37.

Eswaran, H, and P Reich. (1998). "Desertification: A Global Assessment and Risks to Sustainability." In *Proceedings of the 16th International Congress of Soil Science,* Montpellier, France.

Eswaran, Hari, Rattan Lal, and P F Reich. (2001). "Land Degradation: An Overview." In E.M Bridges (ed.) *Responses to Land Degradation,* 20–35. New Delhi: Oxford & IBH Publishing.

FAO (1996). *The State of Food and Agriculture,* 1996, FAO, Rome.

Forest Survey of India (2011). *India: State of Forest Report 2011,* Dehradun. Available at: http://fsi.nic.in/details.php?pgID=sb_16. (accessed on 5 July 2016).

Government of Madhya Pradesh, (1996). *Madhya Pradesh District Gazetteers. Morena,* Bhopal: Government of Madhya Pradesh.

Haigh, M.J., (1984). Ravine erosion and reclamation in India. *Geoforum, 15*(4), pp. 543-561.

Kumar, H., and P. Pani. (2013). "Effects of Soil Erosion on Agricultural Productivity in Semi-Arid Regions: The Case of Lower Chambal Valley." *Journal of Rural Development* 32 (2).

Lal, R. (2001). "Soil Degradation by Erosion." *Land Degradation and Development* 12 (6): 519–39. doi:10.1002/ldr.472.

Marzolff, Irene, and Padmini Pani. (2017). *Dynamics and Patterns of Land Levelling for Agricultural*

Reclamation of Erosional Badlands in Chambal Valley (Madhya Pradesh, India). Earth Surface Processes and Landforms. doi:10.1002/esp.4266.

MEA (Millennium Ecosystem Assessment) (2005). *Ecosystems and Human Well-Being.* Washington D. C.: Island Press.

Ministry of Agriculture (1984). "Report of the Working Group on Reclamation and Development of Ravines for Formulation of the Five Year Plan." New Delhi.

NCA (1976), Report, National Commission on Agriculture, Government of India, Ministry of Agriculture and Irrigation, New Delhi

NRSA (1995). Wasteland Atlas of India. National Remote Sensing Agency

NRSA (2011). http://bhuvan.nrsc.gov.in/gis/thematic/tools/document/wl/WL0809.pdf

NRSA (2008-2009). https://nrsc.gov.in/Atlas

Nyssen, J., Poesen, J., Moeyersons, J., Deckers, J., Haile, M. and Lang, A. (2004). Human impact on the environment in the Ethiopian and Eritrean highlands—a state of the art. *Earth-science Reviews, 64*(3-4), pp. 273-320.

Pani, P., (2016). Controlling gully erosion: an analysis of land reclamation processes in Chambal Valley, India. *Development in Practice, 26*(8), pp. 1047-1059.

Pani, P., (2017a). Ravine erosion and livelihoods in semi-arid India: implications for socioeconomic development. *Journal of Asian and African Studies, 53*(3), pp. 437-454.

Pani, P., (2017b). 'Chambal Without Ravines'. *Down to Earth,* 26(8): 80-81.

Pani, P., (2018). *Socio-Economic Factors Affecting Land Degradation and Its Impact on Sustainable Rural Development: A Case Study of Chambal Region, India,* Project Report Submitted to the ICSSR, New Delhi.

Pani, P., and Paul Carling (2013). "Land Degradation and Spatial Vulnerabilities: A Study of Inter-Village Differences in Chambal Valley, India." *Asian Geographer* 30 (1). Routledge: 65–79. doi:10.1080/10225706.2012.754775.

Pani, P., and S.N. Mohapatra (2001). "Delineation and Monitoring of Gullied and Ravinous Lands in a Part of Lower Chambal Valley, India, Using Remote Sensing and GIS." *22nd Asian ConferenceonRemot Sensing,* no. 54: 5–9.

Pani, P., Mishra D.K. and S.N. Mohapatra (2011). "Land degradation and livelihoods in semiarid India: A study of farmers' perceptions in Chambal valley". Asian Profile 39 (1): 505–519.

Poesen, J., Nachtergaele, J., Verstraeten, G. and Valentin, C., (2003). Gully erosion and environmental change: importance and research needs. *Catena, 50*(2-4), pp. 91-133.

Priya, Ratan, and Padmini Pani. (2017). "Land Degradation and Agricultural Productivity: A District Level Analysis, India." *Journal of Rural Development* 36 (4): 557–68.

Ranga, Vikram, Anton Van Rompaey, Jean Poesen, S N Mohapatra, and P Pani. (2015). "Semi-Automatic Delineation of Badlands Using Contrast in Vegetation Activity: A Case Study in the Lower Chambal Valley, India." *Geocarto International* 30 (8). Taylor & Francis: 919–36.

Reddy, V. Ratna. (2003). "Land Degradation in India: Extent, Costs and Determinants." *Economic And Political Weekly* 38 (44): 4700–4713.

SAC, ISRO. (2007). *Desertification and Land Degradation Atlas of India.* Ahmedabad: Space Application Centre, ISRP.

Scherr, Sara J., and Satya Yadav (1996). "Land Degredation in the Developing World Implications for Food, Agriculture, and the Environment to 2020."

Schröder, J J, R P O Schulte, R E Creamer, A Delgado, J Van Leeuwen, T Lehtinen, M Rutgers, H Spiegel, J Staes, and G Tóth. (2016). "The Elusive Role of Soil Quality in Nutrient Cycling: A

Review." *Soil Use and Management* 32 (4). Wiley Online Library: 476–86.

Sharma, H.S. (1980). *Ravine Erosion in India*. New Delhi: Concept Publishing Company.

Sharma, H.S., (1968). Genesis of ravines of the lower Chambal Valley, India. In *Proceedings of* (Vol. 21, pp. 18-19).

Sharma, H.S., (1979). *The Physiography of the Lower Chambal Valley and its agricultural development: a study in applied geomorphology*. New Delhi: Concept Publishing Company.

TERI (1998). *Looking Back to Think Ahead: GREEN (Growth with Resource Enhancement of Environment and Nature) India- 2047*. New Delhi: TERI.

Vanacker, V., Govers, G., Barros, S., Poesen, J. and Deckers, J., (2003). The effect of short-term socio-economic and demographic change on landuse dynamics and its corresponding geomorphic response with relation to water erosion in a tropical mountainous catchment, Ecuador. *Landscape Ecology, 18*(1), pp. 1-15.

15

Monitoring and Combating Desertification using Geoinformatics in Bellary District, Karnataka

P.S. Dhinwa, S. Bandopadhya, M.A. Paul, A Dasgupta and *Ajai*

1. Introduction

Land degradation occurs all over the world, but it is only referred as desertification when it takes place in drylands. This is because these areas are especially prone to more permanent damage as different areas of degraded land spread and merge together to form desert like conditions. 70 per cent of these dry lands are affected by degradation, which support over 1 billion people in more than 110 countries. The process of desertification, in which land is degraded by decreased rainfall resulting from various factors such as climatic variations and human activities, has become a major focus of scientists to understand the earths' climatic system (Grotinger, *et al.* 2003).

The 1977 Nairobi conference of UNCOD (United Nations Conference on Desertification) describes desertification as the diminution or destruction of the biological potential of the land, which can lead to desert like conditions. Today about 33 per cent of the earth's surface exhibits desert characteristics to some degree or other. Desertification is initiated when the vegetative cover is reduced by tree cutting, over-grazing, over-cultivation, shifting cultivation, mining, road construction, mismanagement of land and water practices, urbanization and other activities that disturb the natural ecosystem. There are two types of parameters that contribute to the process of land degradation. These are socio-economic conditions of the local population (poverty and illiteracy) and pressure of population on marginal lands. Both are interrelated. Basically, it is the increasing population (human and cattle) pressure which is responsible for disturbing the fragile ecosystem and exerts undue pressure on natural resources which activates the process of desertification and thus the fertile piece of land is converted into a desert. There are five types of indicators—climatic, hydrological, physical, biological and socio-economic which directly or indirectly contribute to desertification processes. For combating the processes of desertification in a dry sub-humid region, not much work has been reported in the literature. Spatial inventory on desertification/land degradation status, a prerequisite for preparation of combating plan, has not been available for our country. Moreover, it has also been one of the major requirements towards fulfilling our commitment to TPN-1 UNCCD (Anon, 2007). Arable land per capita is shrinking throughout the world, threatening food security, particularly in poor rural areas and triggering humanitarian and economic crises. The per capita availability of cultivable land is declined from 0.48 ha in 1951 to 0.15 ha in 2001 and is projected to decrease to 0.08 ha by 2035 (Dwivedi *et al.*, 2006). As per UNCCD estimates, arable land per person has declined from 0.32 ha. per person in 1961-63 to 0.21 ha in 1997-1999 and is expected to drop further to 0.16 ha by 2030 (Kofi Annan, 2003). Because the poor often farm degraded land that is increasingly unable to meet their needs, desertification is both a cause and consequence of poverty. Fighting desertification must, therefore, be an integral part of our wider efforts to eradicate poverty and ensure long term food security. Let us today recommit ourselves to the goals of convention, and to achieving sustainable development for all, including in the dry land rural areas where the world's poorest people live (Kofi Annan, 2003).

Land and water are the basic natural resources for biomass production and constitute the core of ecological systems. The ecological systems, the very base of production is under great stress in many parts of the country due to growing pressure of human and livestock population and ever intensifying development demands. Forests are being denuded at alarming rates. Marginal lands are being cultivated and grasslands are being overgrazed increasingly. There is soil erosion in the upstream and sedimentation and flood in downstream part of a stream. Desertification is on the increase and the renewable resources are getting depleted. Large land area is classified as degraded land. The very resource base of production is thus under threat. The challenge is to make ecological resources more productive and their exploitation less hazardous. A development strategy based on integrated management of soil, water and other resources coupled with appropriate cropping and other agro techniques should be adopted for sustained production (Anon. 1995). Watershed constitutes the basic unit of development of rainfed areas. Unlike the earlier isolated efforts through soil and water conservation or crop production or other agro-techniques in a disjointed manner, the watershed approach aims at optimizing the use of land, water and vegetation in an area in an integrated way and thus helps alleviating drought, moderating floods, preventing soil erosion, improving water availability and increasing fule, fodder and agricultural production on a sustained basis (Anon. 1995).

Remote sensing data along with GIS has been useful for desertification monitoring and assessment. The indicators of desertification amenable to remote sensing include salinity, erosion and sand sheets etc. **(Navalgund, 2006)**. As per the inventory of Desertification and Land Degradation Atlas of India, 105.48 mha. i.e., 32 per cent of the geographical area of the country is undergoing the process of land degradation (Anon., 2007). The indicators used for Desertification Monitoring and Assessment can be categorized into four types.

(A) **Pressure Indicators** characterize driving forces both natural and man-made, affecting the status of natural resources and leading to desertification. Pressure indicators are used to assess desertification trends and make an early warning for desertification. Natural indicators describe natural factors, mainly climatic conditions, natural disasters, which promote the occurrence and development of desertification. Non-natural indicators describe the pressure on land leading to land degradation from human activities **(Ajai *et al.*, 2003)**.

(B) **Status Indicators** characterize the status of natural resources including land. The physical and biological features of desertified land ecosystems are the main factors to be considered. Physical indicators describe the land characteristics, physical and chemical properties of soil and hydrological features of the land ecosystem. Biological indicators are used to describe biological characteristics of the land ecosystem.

(C) **Impact Indicators** are used to evaluate the effects of desertification on human beings and environment.

(D) **Implementation Indicators** are used to assess the action taken for combating desertification and to assess its impact on natural resources and human beings. Such impacts refer to improvements of socio-economic and natural conditions. The positive changes in the land cover conditions owing to mitigation measures aimed at combating desertification. These are also sometimes referred to as response indicators.

2. Objective

The present study aims to monitor the decadal changes in the degree of severity of a desertification process in that land use/land cover type and generate locale specific action plans for combating desertification through the use of geospatial data and GIS techniques.

3. Study Area

Bellary district is located in the central-eastern part of the Karnataka state with an area of 8463 km^2 and lies between 14^0 33′ and 15^0 15′ N latitude and 75^0 40′ and 77^0 10′ E longitude. It is bounded on the north by the Raichur and Koppal districts, on the west by the Gadag and Haveri districts, on the south by the Davangere and Chitradurga districts and the east by the state of Andhra Pradesh. Tungabhadra and Hagari are the major rivers of Bellary district. River Tungabhadra flows through the west and northern side boundary of the district. The area receives an annual average rainfall of 545 mm with a wide spatial variation, ranging from 815 mm in the hilly tracts of Sandur Schist belt to 243 mm in the immediate east of the belt. The rest of the district has an average of about 425 mm rainfall. The temperature of the district varies from 23 to 42°C. The major rock groups of Bellary district are Dharwars, Peninsular Gneisses and Granites. Granite outcrops as in selbergs are prevalent in southeastern and eastern parts, whereas knobs are present everywhere.

Red soil covers the entire peninsular gneisses granite complex; whereas black soil along with a minor extent of characteristic ferruginous red soils cover the Dharwars. Red soils cover about 55 per cent and black soils about 45 per cent of the district. Both areas have distinctly different land scrapes mainly arising out of differences in water infiltration and runoff of rainwater, that often lashes the district torrentially whenever few times it occurs over the district. Since the black soils are the most productive agricultural lands, they are extensively and intensively cultivated and inhabited. Whereas as the red soils with occasional rock outcrops are less extensively and intensively cultivated, except in valleys, all the reserved forests and scrublands are seen in the red soils region of the district. Among them, the forests covering the Sandur Schist belt are still preserved, the rest of the forests are variously degraded.

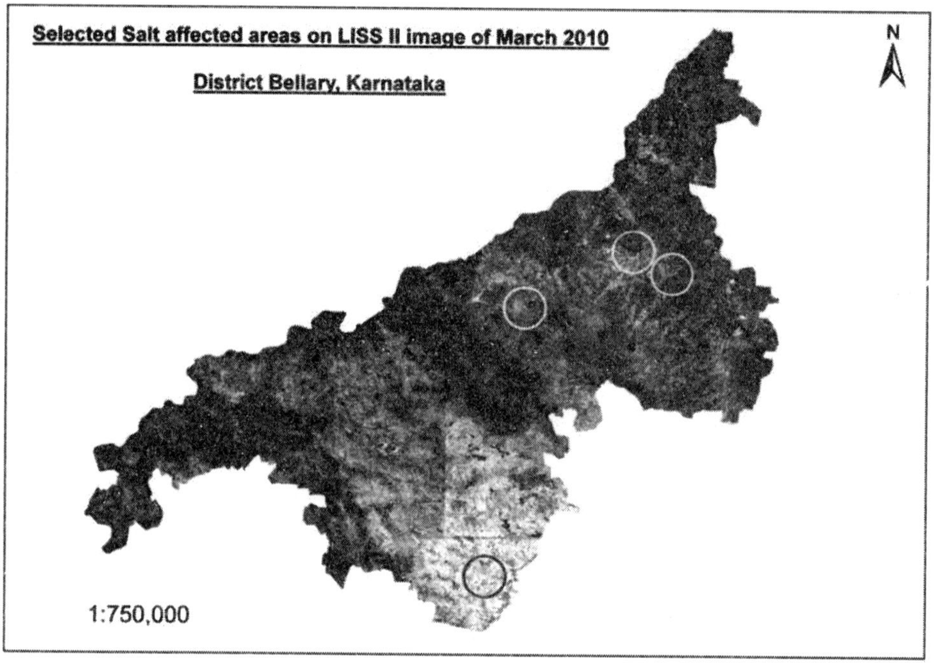

Map: Study Area

4. Data Used

The IRS-1C/1D multi-date satellite data, ancillary information and collateral data has been used to prepare various thematic maps, to generate locale specific action plan maps for Bellary District, Karnataka and its details of the data are given in Table 15.1.

Table 15.1: Data Used

Satellite Data	*Satellite ID*	*Sensor ID*	*Date of Acquisition*
	IRS-1C/1D	LISS- 3	October 2001 February 2002
Ancillary Information	48M/16 48N/9,10,13,14 57A/4,7,8,11,12,13,14,15,16 57B/1,2,5,6,9,10 57E/1,2,3,4 (1:50,000) Scale	-	-
Collateral Data	1. Population 2. Climate 3. Geology 4. Soils etc.	-	-

5. Methodology

By using LISS-2 satellite data of 2001 and LISS-3 satellite data of 2011 and employing

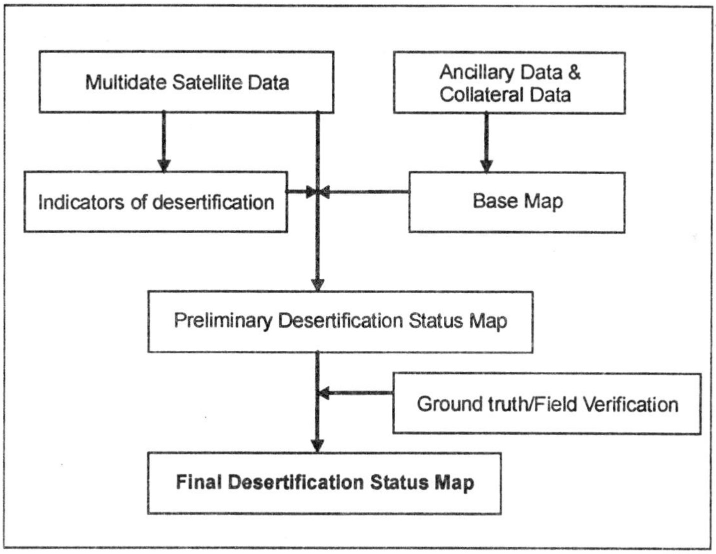

Fig. 15.1: Methodology for Desertification Status Mapping

Comprehensive Classification System for Desertification Status Mapping (Table 15.2), desertification status maps were prepared, and later on, as a follow up of desertification status mapping phase-1, a study was carried out to assess decadal changes in desertification by using LISS-3 satellite data of 2011 as shown in Fig. 15.1. Visual analysis of multi-date satellite data along with ancillary information and collateral data were used to prepare various thematic maps, e.g. land use/land cover, groundwater prospect, land capability and slope. Realising the importance of adopting an integrated approach, and recognizing the mutual interdependence of natural resources, thematic information is integrated in GIS environment. In order to integrate various themes, firstly, land use/land cover layer is integrated with groundwater prospect layer. The resultant of these two themes is unionized with slope and finally with land capability. The resultant coverage has the basic information of all the four themes—land use/land cover, groundwater prospect, land capability and slope and referred in resource data base (Fig. 15.2). Various map units known as composite land development units (CLDU) are created in this composite layer. Based on the CLDU characteristics various measures are suggested for conservation and protection of natural resources. The decision rules, given in Table 15.3 are used to suggest the appropriate action plan for each of the CLDUs (Anon. 1995).

Table 15.2: Comprehensive Classification System for Desertification Status Mapping

Level 1: Land Use/Land Cover		
Agriculture-Unirrigated	D	
Agriculture – Irrigated	I	
Forest/Plantation	F/P	
Grassland/Grazing land	G	
Land with scrub	S	
Barren/Rocky area	B/R	B(sc) indicating scree areas in cold deserts

Dune/Sandy area	E
Waterbody/Drainage	W
Glacial/Peri-glacial (In cold region)	C/L
Others (Urban, Man-made)	T

- (Rocky areas within forest annotated as only V3-R in the map)
- (Vegetal degradation in Land with Scrub around periphery of notified forests delineated as Sv_)
- (Encroachment in forest area especially agricultural practices, is Fv3)
- (Barren and Rocky areas delineated separately as B or R and shown in others category of the legend)

(All settlements are hatched)

Level 2: Processes of Degradation

Type of processes resulting in degradation:

Vegetal Degradation	V
Water Erosion	w
Wind Erosion	e
Waterlogging	l
Salinization/Alkalinization	s/a
Mass Movement (in cold areas)	g
Frost heaving (in cold areas)	h
Frost shattering (in cold areas)	f
Man made (Mining/Quarrying, Brick Kiln, Industrial Effluents, City Waste, Urban Agglomeration etc.)	m

(Gully/ravines are shown as Xw3, where X is the Land use/cover class of surrounding area).

(Salinization or Alkalinization shown as 's or 'a' separately. Where both occur, are shown together i.e. sxay, where x and y are respective degree of severities)

Level 3: Severity of Degradation

This level represents the degree of severity of degradation. It is coded as '1' for slight, '2' for moderate and '3' for severe degradation.

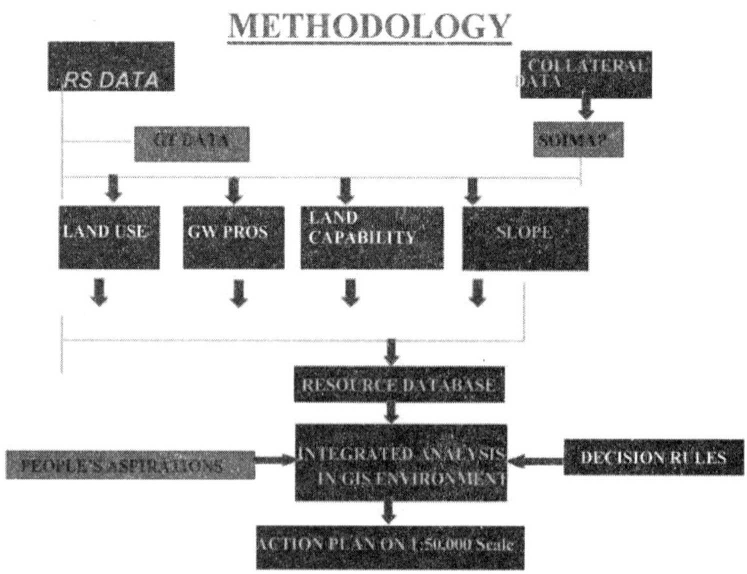

Fig. 15.2: Flow Chart for generation of Land Resources Action Plan for Combating Desertification

Table 15.3: Decision rule for generation of action plan for combating desertification

Sr. No.	*Land Use/ Land Cover*	*Groundwater Prospect*	*Slope In per cent*	*Land Capability*	*Measures*
1.	Scrub Forest	Poor	5-10	Shallow/Moderate	Afforestation with proper soil and water conservation
2.	Kharif+Rabi, Dense Forest	Good	3-5	Shallow/Moderate	No action
3.	Forest	Moderate	5-10	Shallow/Moderate	Protection and Gap Filling
4.	Forest	Moderate	>35	Very Shallow	Afforestation with staggered trenches
5.	Crop land in Forest	Moderate	0-3	Shallow/Moderate	Agro-forestry with Contour bunding
6.	Kharif Crop	Moderate	0-3	Shallow/Moderate	Double cropping with groundwater exploration
7.	Kharif Crop	Moderate	0-3	Shallow/Moderate	Agro-horticulture with groundwater irrigation
8.	Kharif Crop	Good	0-3	Moderate Deep	Minimal action -Farm ponds
9.	Kharif+Rabi, (Salt Affected)	Good	0-1	Moderate Deep	Drainage Treatment
10.					

5. Results and Discussion

Bellary district has a geographical area of 8463 sq. km. The analysis of multi-date satellite data (2001/2002) reveals that vegetal degradation, salinity/alkalinity, wind erosion, water erosion and manmade are the significant desertification processes observed in this district which comprises of 27 per cent of area of the district. The remaining area of 6,10,376 ha, that is around 73 per cent is under No Apparent Degradation (NAD). The analysis of satellite data reveals that out of total geographical area of Bellary district, about 1,45,812 ha that is 17.23 per cent of area is under vegetal degradation. It is observed that 370.35 ha (4.53 per cent) of area is affected by water erosion and 298.94 ha (3.83 per cent) of area is under salinity/alkalinity. The analysis of multi-date satellite data (2001/2002) also reveals that 43.4a ha (2 per cent) of area is under man made activities and the remaining 73 per cent is under No Apparent Degradation (NAD) (Table 15.1). The satellite data also reveals that in 2001 the total area under desertification is 181439 ha (21.50 per cent), of which 4341 ha. (2 per cent), is under manmade, 110170 ha (61 per cent) is under vegetal degradation, 37035 ha (20 per cent) is under water erosion and 29894 ha (17 per cent) is under salinity/alkalinity. The analysis of multi-date satellite data (2011) reveals that 113222 ha. (24 per cent) of area is under vegetal degradation representing various categories of severity of desertification in forest and land with scrub classes. It is observed that 4.53 per cent of area is affected by water erosion and 3.83 per cent of area is under salinity/alkalinity. 2 per cent of area is under man-made activities as shown in Fig. 15.3. The comparision of desertification status maps prepared by using 2001 and 2011 satellite data reveals that there is an overall increase in area under different processes of desertification. It denotes an increase of 02.50 per cent, that is from 21.50 per cent (2001) to 24 per cent (2011). The area under man-made activities has increased from 43.41 ha to 81.80 ha, while area under vegetal degradation has increased from 1101.70 ha to 1132.22 ha. Similarly area under water erosion has increased from 370.35 ha to 497.33 ha and area under salinity/alkalinity has increased from 298.94 ha to 314.21 ha. as shown in Fig. 15.4.

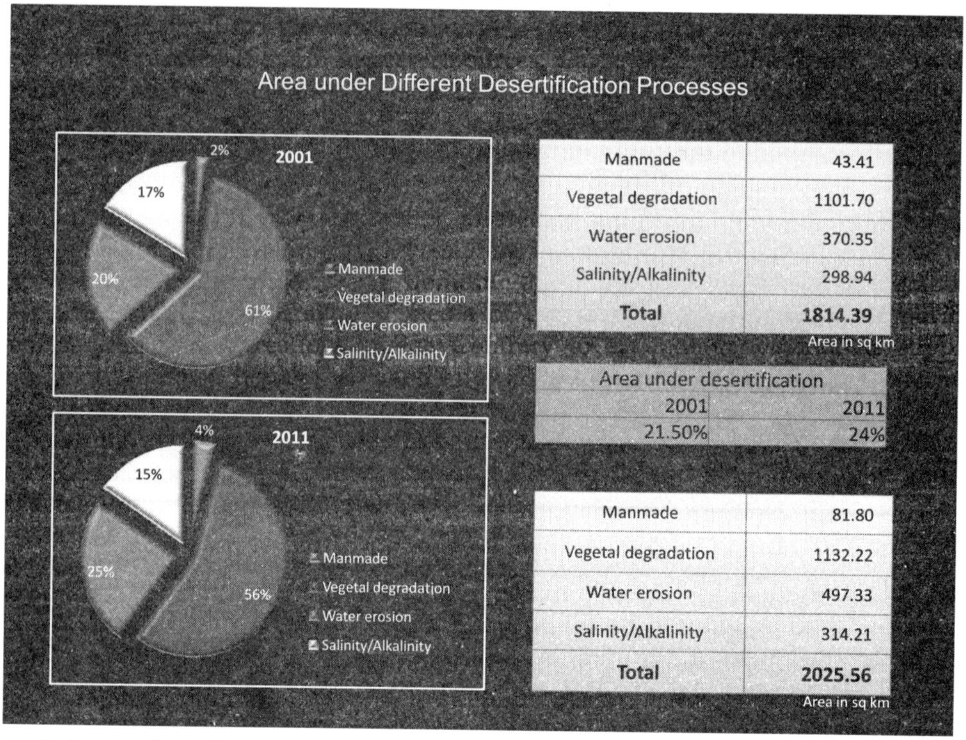

Fig. 15.3: Area under different desertification processes

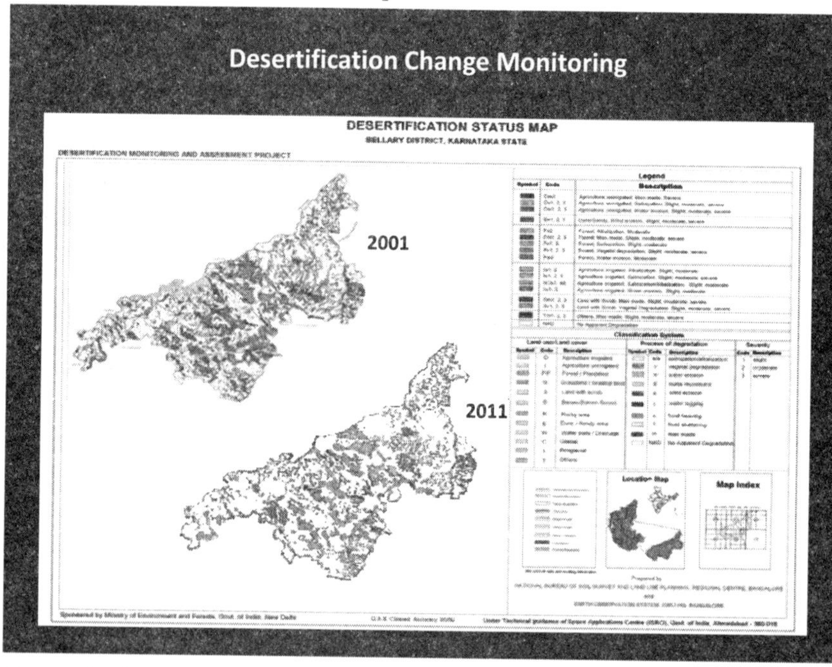

Fig. 15.4: Desertification Change Monitoring

5.1 Land Use/Land Cover Map

Land use/land cover map is prepared by visual analysis of multi-date satellite data and through the use of ancillary information and collateral data. In Bellary district, Karnataka, there are twenty seven categories of level-III land use/land cover types, namely, town/cities, habitation with vegetation, industrial area, villages, rabi crop, kharif crop, double crop agricultural plantation, fallow land, aquaculture pond, moist and dry deciduous open forest, degraded forest, mixed vegetation, forest plantation, prosopis juliflora, scrub forest, tree groves, barren rocky/stony waste, sheet rock area, gullied/ravinous land, land with scrub, land without scrub, mining/industrial wasteland, salt affected land, sandy area, lake/pond, lake/tank (dry), reservoir, river/stream, river island (Fig. 15.5). Land use/land cover map gives an account of spatial distribution and a real extent of various categories of land use/land cover. When it is integrated with other requisite themes, e.g., groundwater prospect, slope and land capability, it gives a clear picture of availability of natural resources at a particular place and thus very helpful in suggesting measures for combating desertification at a specific locale.

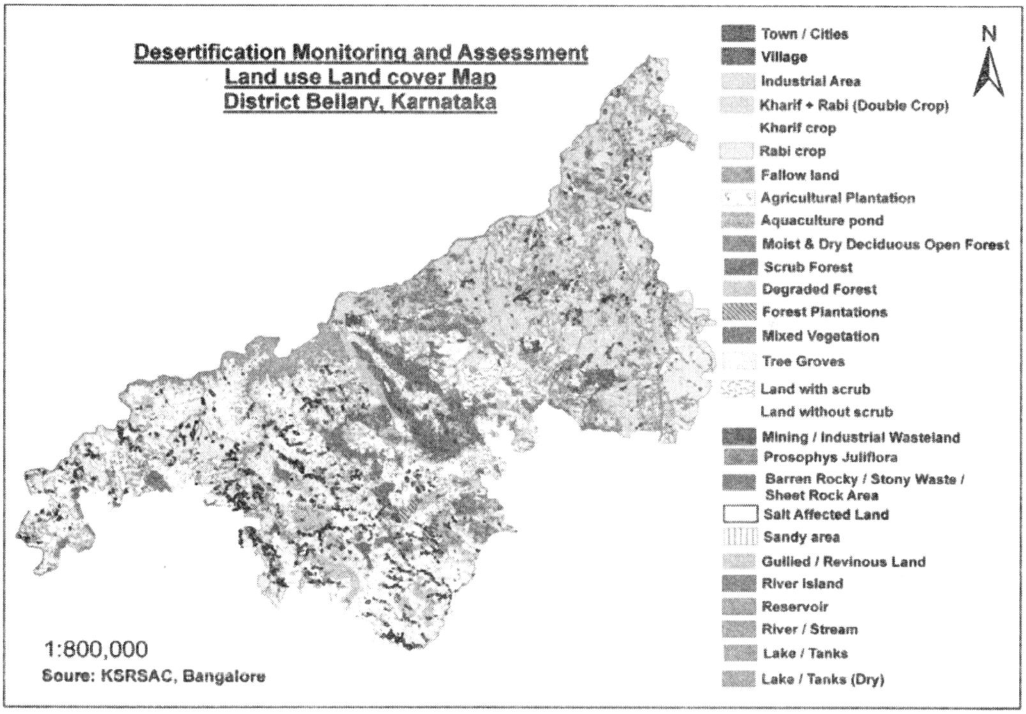

Fig. 15.5: Land Use/Land Cover Map

5.2 Slope Map

Slope map is prepared by using Survey of India (SOI) topographical maps on 1:50,000 scale. It gives an idea terrain and topography of the area. In Bellary district, there are seven categories of slope. The area in the eastern and western part of Bellary district is under nearly level (0-1 per cent) and very gently sloping (1-3 per cent) slope category and it constitutes eighty per cent of area of the district. Gently sloping (3-5 per cent) slope category lies in central and southern part of Bellary district. Moderately sloping (5-10 per cent), strongly sloping (10-15 per cent), moderately steep to steep sloping (15-35 per cent) and very steep sloping (>35 per cent) slope category is present in central part of the district where iron and manganese ores are available (Fig. 15.6).

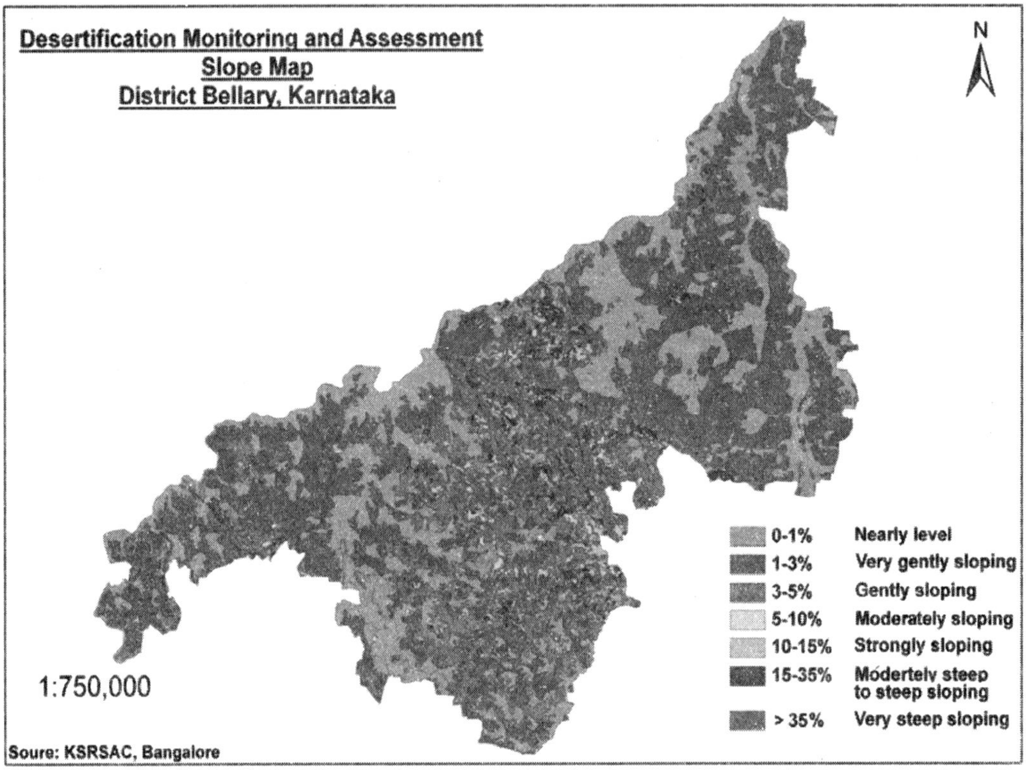

Fig. 15.6: Slope Map

5.3 Land Capability Map

In Bellary district, five classes of land capability, i.e., II, III, IV, VI and VIII are reported. Classes II and III of land capability are present in eastern and western part of the district. Classes IV of land capability is mainly present in southern part of the district. Classes VI of land capability is reported in central part of the district whereas class VIII lies in northern part (Fig. 15.7). For suggesting measures for combating desertification, information about soil type and land capability is essential (Anon.,1997).

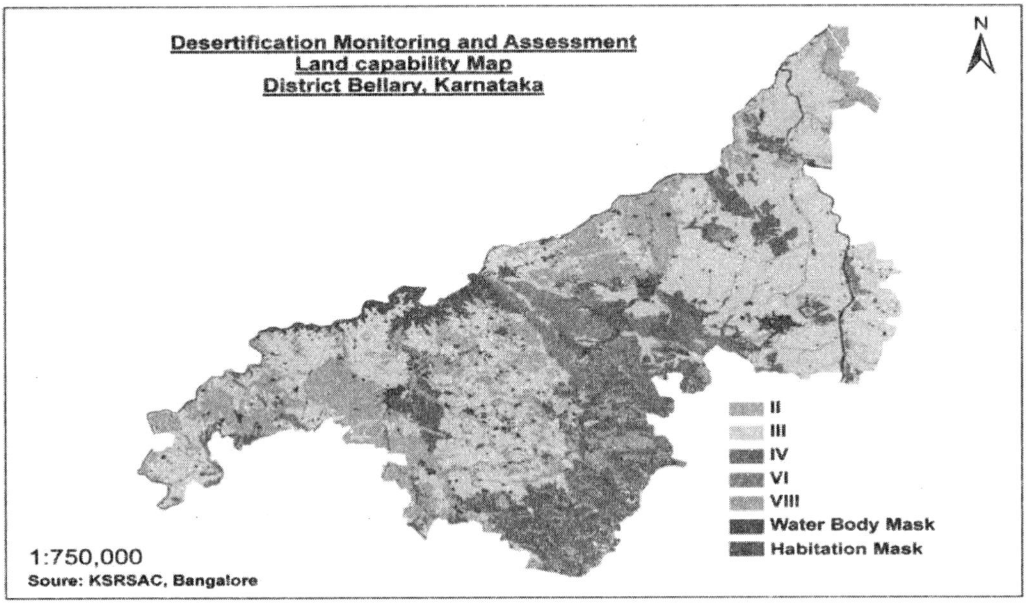

Fig. 15.7: Land Capability Map

5.4 Groundwater Prospect Map

Groundwater prospect map is prepared using multi-date satellite data, ancillary information and collateral data and by employing visual analysis technique. In Bellary district, there are seven categories of groundwater prospect, ranging from poor to very good. Sixty per cent area in Bellary district is under moderate category of groundwater prospect. The region lying in east of the district is under good to moderate categoty. Good category of groundwater prospect is observed in northern eastern part of the district. Poor and poor to nil category of groundwater prospect is reported in central part of the district which is hilly (Fig. 15.8).

Based on the integrated analysis of above mentioned thematic layers and decision rules (Table 15.3), the land resource action plans have been prepared as shown in Fig. 15.9. The details are given as under:

1. **Soil and water conservation**: Appropriate soil conservation and development practices play a key role in maintaining overall productivity in a watershed and maintaining the ecological balance as well (Anon.,1995). Afforestation with proper soil and water conservation measures is suggested for scrub forest, where groundwater prospect is poor, slope category is moderately sloping (5-10) and land capability is class III. It reveals that in this area anthropogenic forces are very active hence dense forest is reduced to scrub forest which is prone to soil erosion and hence afforestation with proper soil and water conservation measures is suggested.

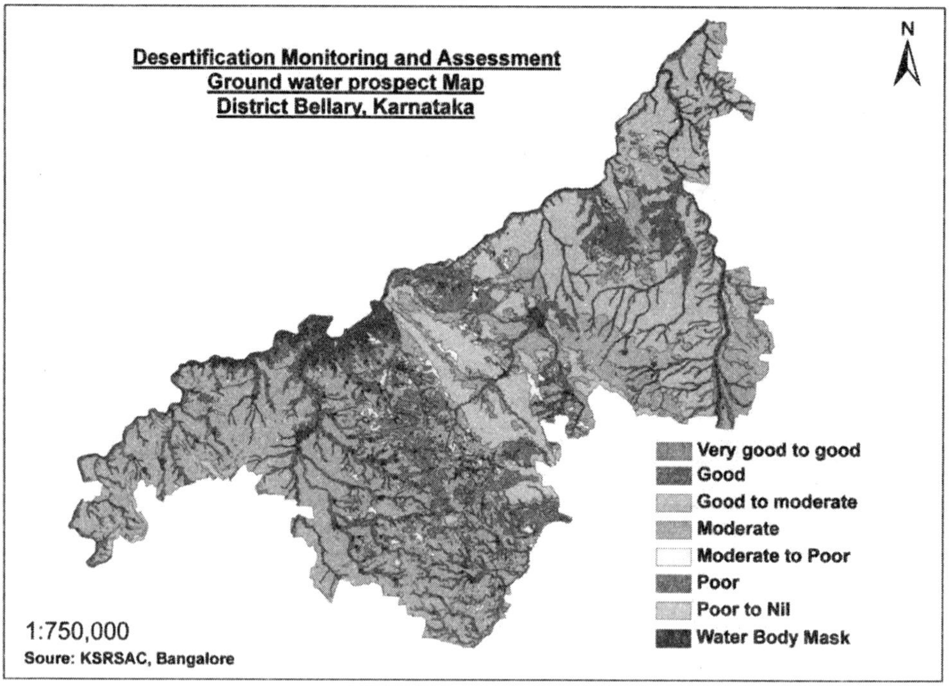

Fig. 15.8: Groundwater Prospect Map

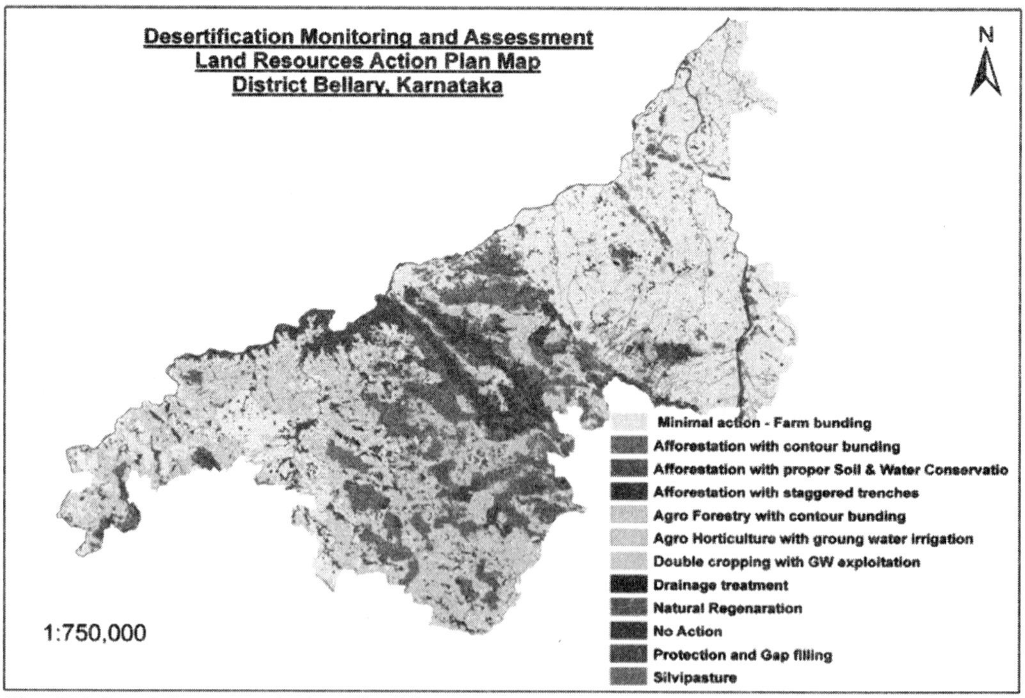

Fig. 15.9: Land Resources Action Plan Map

2. **No action-** is suggested for the areas lying under dense forest, water bodies and habitation.
3. **Protection and gap filling-** is suggested in open forest region of Protected Forest, where slope is moderately sloping category (5-10), groundwater prospect varies from moderate to poor and lies under class VI of land capability.
4. **Afforestation with staggered trenches.** This type of trenches are best suited to plant trees and protect existing plantations from soil erosion and conserve moisture and plant nutrients for growth (Subramanian, S. *et al.*, 1987). Afforestation with staggered trenches is suggested for area under mixed vegetation and open forest, very strong sloping category (>35 per cent), groundwater prospect varies from poor to nil and lies under class VI of land capability.
5. **Agro-forestry with contour bunding** is executed in lands where the slope is less than 6 per cent and which receives an annual rainfall of 1000 mm or below. When contour bunding is combined with conservation farm practices the yield of crop is increased because it helps to prevent soil erosion (Subramanian, S. *et al.*,1987). For imparting stability and providing sustainability to the farming systems in marginal areas, a tree-cum-crop farming system could prove more useful (Anon.,1995). Agro-forestry with contour bunding is suggested for crop land in forest.
6. **Double cropping with groundwater exploration** is suggested for agricultural land presently under Kharif crop, with moderate groundwater prospect., is under gently sloping (3-5) slope category and belong to class III and IV of land capability. This system plays an important role in those regions, where production of annual crops is inefficient. Fruit trees, if suitably integrated, would add significantly to overall agricultural production including food, fuel and fodder, conservation of soil and water and stability to production and income (Anon.,1995). Agro-horticulture with groundwater exploration is suggested for moderately sloping agricultural land presently under Kharif and with moderate groundwater prospect.
7. **Minimal action – Farm ponds** and is suggested for area under double crop, where groundwater prospect varies from good to moderate, is under nearly level (0-1) and very gently sloping (1-3) slope category and it lies class III of land capability.
8. **Silvipasture-** This system essentially consists of top feed tree species and grasses or legumes) preferably perennial) as under-storey crop. This system is most suited to marginal drylands and is preferable where fodder shortages are experienced frequently (Anon.,1995). Silvipasture is suggested for a region under land with scrub and land without scrub, where groundwater prospect varies from moderate to poor, is under strongly sloping (10-15) slope category and it lies class IV of land capability.
9. **Drainage treatment** is suggested for the region which is under salt affected class of land use/land cover, where groundwater prospect is moderate, is under gently sloping (3-5) category of slope and belong to class III of land capability.
10. **Natural Regeneration** is suggested for area under barren rocky/stony waste/sheet rock area class of land use/land cover, where groundwater prospect

is poor, is under moderately sloping (5-10) and strongly sloping (10-15) category of slope and belong to class VIII of land capability.

6. Conclusion

Remote sensing data along with GIS has been useful for monitoring and combating desertification. The analysis of satellite data reveals that vegetal degradation, water erosion, manmade (mining/quarrying), salinity/alkalinity and wind erosion are the significant desertification processes observed in Bellary district, Karnataka. The comparision of desertification status maps of 2001 and 2011 reveal that there is an increase in area under different processes of desertification. Specifically, the area under man-made (mining/quarrying) activities is increased from 43.41 ha. to 81.80 ha. and area under vegetal degradation is increased from 1101.70 ha. to 1132.22 ha. Similarly area under water erosion is increased from 370.35 ha. to 497.33 ha. and area under salinity/alkalinity is increased from 298.94 ha. to 314.21 ha. The results clearly reveal that process of desertification is showing a positive trend which is very harmful and needs to be reversed. Based on decision rule, action plans are generated for land resources development. The measures suggested for combating desertification are: a) afforestation with suitable soil and water conservation measures in forest region, b) shelter belt plantation to prevent wind erosion in sand dune area, d) crop rotation along with growing of pulses and gram for improving fertility of soil in crop land region, e) proper irrigation management and drainage treatment for area affected by salinity/alkalinity and f) judicious use of fertilizers in irrigated agricultural land.

[**Acknowledgements:** The authors are grateful to Dr. Irfan Khan, Director, TALEEM Research Foundation, for his valuable guidance and moral support in preparation of this research paper.]

References

Ajai, Arya, A.S., Dhinwa, P.S. (2003). Desertification: The creeping Disaster, *NNRMS Bulletin* (28), pp. 52-64.

Anon. (1995). *Integrated Mission for Sustainable Development – Technical Guidelines,* National Remote Sensing Agency, Hyderabad.

Anon. (2007). *Desertification and Land Degradation Atlas of India.* Space Applications Centre, (ISRO), Ahmedabad.

Anon. (2007). Desertification Monitoring and Assessment using Remote Sensing & GIS: A pilot Project under TPN-1 UNCCD. SAC/RESIPA/MESG/DMA/2007/01 Space Applications Centre, (ISRO), Ahmedabad.

Dhinwa, P.S., Pathan, S.K. and Ajai (2010), Action Plan for combating desertification in dry sub-humid region of Kathua district (J&K), *Journal of Geomatics,* Vol. 4, No. 1, pp. 43-48.

Grotinger, J., Jordan, T.H., Press, F., Siever, R. (2007), *Understanding the Earth.* W.H. Freeman and Company, New York. pp. 453-471.

Dwivedi *et al.,* (2006). Proc. of Regional Workshop on Deserts and Desertification, at Hyderabad during Dec. 29-30, 2006), p. 31.

Kofi Annan, (2003). Message by UN Secretary General KOFI ANNAN for the World Day to combat desertification and drought, June 17, 2003.

Navalgund, R.R. (2006), Geoinformatics in Disaster Monitoring and Mitigation, *ISG Newsletter*, Vol. 12, No. 1 and 2, pp. 4-23.

Subramanian, S., Kothandaraman, G.V., Natarajan, S. and Ramaswamy, P.P., (ed.) (1987). Soil Survey and Land Use Planning for watershed management, Directorate of Soil and Crop Management Studies, Tamil Nadu Agricultural University, Coimbatore.

16
Mining in Goa: Environmental Issues and Challenges—Post-Ban on the Path of Recovery

Nandkumar Sawant

Introduction

Land is a resource of colossal importance for the survival and sustenance of all the nature's bounty in general, and humankind in specific. Over the ages man has excelled in tapping the immense potential of land resource for his benefit. However, the demands are infinite and the resources on the contrary are limited. Increased pressure on land resources has eventually degraded the quality and quantity of land (FAO, 2012). Land cover has been undergoing changes over the years, but in recent times, the transformation is unrestrained and haphazard.

Tourism and mining have been the major contributors to the Goa's State Domestic Product (SDP) along with agriculture, fishing and other allied activities. The mid land region and the foot hills have rich deposits of minerals. It is endowed with a number of mineral resources like Iron ore, manganese and bauxite that are the major minerals of economic importance in addition to other minerals (including minor minerals) like limestone, clay, basalt, lateritic and river sand. This geographic region has major traversing rivers, namely two major being Zuari and Mandovi and many other small rivers. The eastern side, is flanked by the Western Ghats, accommodating 4 Wildlife Sanctuaries and 1 National Park, truly store house of biodiversity.

Geographically, Goa is divided into three physiographic divisions: The coastal plains, Midland Plateau and the Mountainous region. The coastal plains at the backdrop, have the beaches interspersed with headlands in the west. The midland region, mainly consisting of talukas like Bicholim, Sanguem, Quepem, Sattari and Dharbandora, are the agricultural hearth but also store house of minerals. The Shayadris, the part of Western Ghats in the east, rich in biodiversity are one of the Hotspots of the world.

Objective

Mining has been in prominence during the regime of the Portuguese during the early 1940's, eventually the exploitation accelerated over the decades and reached the pinnacle in first decade of this century. This has led to land use changes that have negatively affected the ecological balance and posed threat to its natural vegetation. In this context, the present research aims to understand mining nexus environment in the State of Goa.

Therefore the objectives are stated as:

a. To understand the mining in Goa.
b. Mining in Goa and its impact on the forest cover.

3.1 Methodology and Data Source

The following research is based on culmination of primary and secondary data—Though the study is largely based on secondary data sources which include Governmental as well as Non-Governmental publications but there is considerable dependence on primary data too.

For assessment of Landuse/Landcover the following satellite data for Sanguem talukas of South Goa District was used. The details are as follows:

Month and Year	*Satellite*	*Sensor*	*Resolution*
January 2005	Landsat 7	Enhanced Thematic Mapper Plus (ETM+)	30 m
January 2016	Landsat 8	Operational Land Imager (OLI)	30 m

Further, ArcGIS 10.1, ERDAS Imagine 9.2 and Microsoft Office 2007 and Microsoft Office Excel 2007 software were used. Both supervised and unsupervised classification was used.

With the reference of primary data, questionnaire based survey and personal interviews were undertaken in the mining village of Codli, Sanguem taluka, Goa. The number of respondents administered was 84. The purpose was to extort perceptions of the local populace towards environment. The sample consist of locals of diverse occupations - mining, agriculture and other sectors. Further, to get finer details, personnel interviews were also conducted.

4.1 Discussion

Overview of Mining in Goa.

Though the presence of mineral ores in Goa was known since ancient times, mining activity on a commercial scale started in Goa at the end of the Second War II. Amongst all the mineral ores found, the iron ore contributes the bulk of the production.

The first export of 100 tonnes was in 1947. The figure rose to a million tonnes by 1954, 10 million tonnes by 1971, and 13-15 million tonnes in the 1980s. Iron ore export in 2012 reached a whopping 52-55 million tonnes per year. The export volume accordingly showed exponentially upward growth over the years, thereby making Goa a major iron ore exporting State, where over 60 per cent of India's iron ore export share was contributed by Goa. In terms of foreign exchange reserves it accounted for nearly Rs. 1000 crore per year. Iron ore mining thereby became the major extractive industry in Goa.

Some of the Salient Features of Mining in Goa

(a) Mining belt stretches 65 kms from south-west to the north-west covering about

700 square kms (Central Pollution Control Board, 2008). All the mines in Goa are concentrated in small area of five midland talukas of the State, namely Bicholim, Sattari, Dharbandora, Sanguem and Quepem. (Fig. 16.1) In 2011, there were 336 mining leases (concessions) in the State, over an area of nearly 24,168 hectares comprising 8 to 9 per cent of the Goa's geographical area (Directorate of Mines, 2011). From the environmental perspective, all these mines are located along the foothills of Shayadris, a part of Western Ghats, the store house of biodiversity.

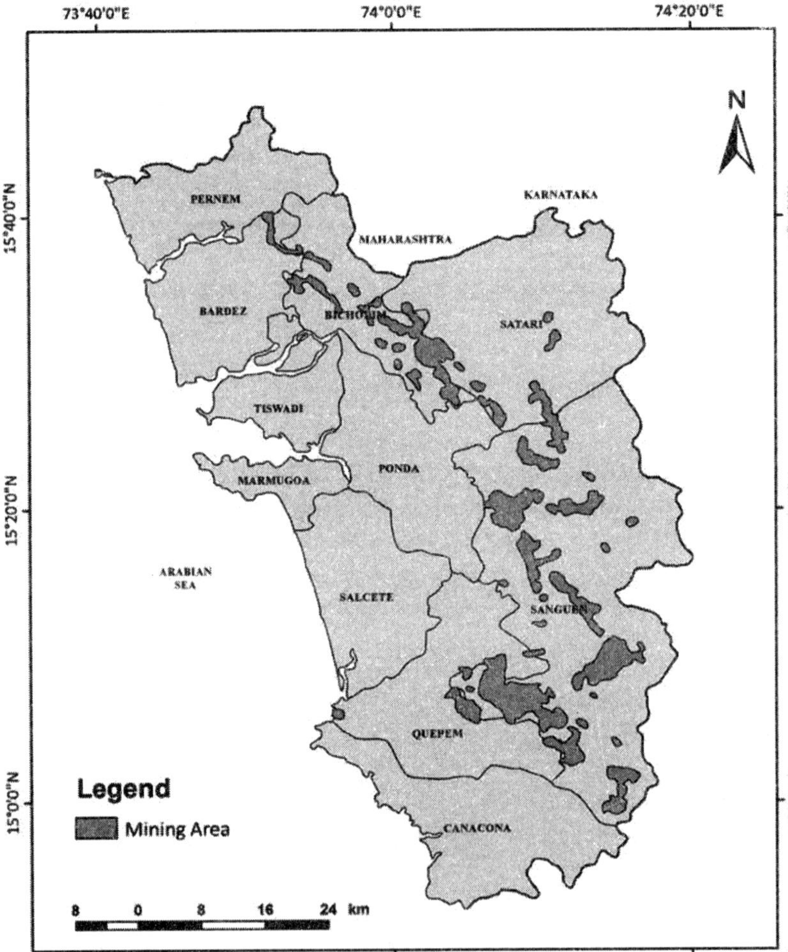

Fig. 16.1: Goa: Mining Area

Source: Indian Bureau of Mines

(b) Another distinguishing feature of Goa mining is that the area of individual mining concessions is limited to a maximum area of 100 hectares; in fact out of the total number of leases, as many as 73 mining leases had each an area of less than 50 hectares

(TERI, 1998). Small size of the mining lease hinders proper development of mines infrastructure, in particular, the arrangements for proper stockpiling of mineral ores and waste materials. In many of the cases these material was piled beyond the mining site or area owned by the mine owner.

(c) A idiosyncratic feature of Goa's mining industry is the high overburden to ore ratio, which requires a large volume of material to be handled, and extensive places for dumping. The excavation ratio in Goa is 1:3 (for every tonne of ore, three tonnes of soil is extracted). Forest areas were used for dumping the rejected ore or the waste.

What Went Wrong?

Undoubtedly, iron ore mining had a positive and significant impact on the economic development of Goa. There have, however, been several adverse effects on environment and local communities living in the vicinity of the mining areas that have attracted considerable attention from several quarters.

Through the 1980s and 1990s, exports averaged 14 million tonnes annually, According to the records, mining in the Goa touched 55 million tonnes in the financial year ending March 2011 as against 45 million tonnes in 2010. The mining business in the state is estimated to be worth Rs 16,000 to Rs 22,000 crore a year. (Department of Mines and Minerals, 2012). Endorsement of "Robber" industry, unplanned and unconcerned, triggered the environmental and social crisis. (D'Mello P, 2018)

Mining Nexus Ecosystem

Goa has fragile ecosystems and mining operations in the Western Ghats have had impacts on forests, agricultural land and water bodies including the two main rivers - Mandovi and Zuari. Some of these problems are due to the unique features of mining in Goa and a few others as the result of inappropriate mining practices and poor environmental management. (Terence Jorge, Hagen De Sa, Pushp Jain, 2013).

A report of "Down to Earth"(2012) put forth that damage to the environment was far greater than loss of revenue. Mining utterly devastated Goa's pristine forests. According to the Ministry of Mines (2008) in the period of 28 years from 1980 to 2008, around 1,453 hectares of forest land has been diverted for mining activities. Large areas of forests in the state are not classified under government records. These are private forests or community lands, and hence forest clearance is not necessary for mining. The Economic Survey of Goa mentioned that around 2, 66,000 sq. km. of government land is being illegally encroached upon by mining companies. The Energy and Resources Institute (TERI) states that absorb 18 per cent of Goa's forests have been lost to mining. In majority of the mines, the overburdens are not properly stored and managed hence hugely affecting the forest cover of the state (Talule and Naik, 2014)

As a whole, 13 mines were found within the precincts of Wildlife Sanctuaries/ National Parks of the State, while 6 were partly within the State and 151 leases were found within 10 km and only 9 were beyond 10 km of WLS/National Parks. (Jorge T, D'Sa H & Jain, 2013)

The heavy rainfall during the monsoons leads to surface runoffs of heavy metals and siltation in fields and rivers thereby causing huge agricultural losses and the pollution of river water. The State of Goa being the region of heavy monsoon the deposition of silt and runoffs of overburden piled up at the various mining sites have come down in the agricultural fields due to monsoon runoff that has created problems for fertility of the soil (Talule, D. and Naik, R.G., 2014).

In short, overall major environmental problems caused due to mining operations are Deforestation, Land degradation, Groundwater pollution, Surface water pollution, Dust pollution and over all landscape. Mining, more than any other industrial activity tends to leave a strong negative impact on environment and society but this has been a view point where environmental impact has been overdriven to social and economic costing.

The Verdict

Making an argument of ban on mining, Goa's mining operations were suspended first by the Goa Government on September 10th, 2012 after the Justice M.B. Shah Commission Report was released. The ban on mining is looked as a process of reclamation in order to avoid any further damage to the State's environment, social well-being and traditional occupations, a successful recovery from about 60 years of continued mining operations. Prior to this, in 2002, the Ministry of Environment and Forests, acting on a direction by the former Indian Board of Wildlife, issued a circular which directed all State Governments to *'demarcate a 10 km buffer zone from the boundaries of all Wildlife Sanctuaries, which would be considered ecologically sensitive.'*

According to Shah Commission Report, the common illegalities reported in Goa were 'mining without license and mining outside the lease area. The report studied the damages to environment, livelihood issues, and rights of tribal and other people living in the mining area. (*The Hindu*, 2013). The operators going beyond their boundaries and unregulated and unrestricted illegal mining have resulted in a loss of Rs 350 billion to the State and also caused environmental destruction.

The report stated that parliament prompted state government to ban mining. On top of that the Ministry of Environment suspended the environmental clearances of mining leases. The Shah Commission stated that about 55 per cent iron ore exported from India comes from Goa. The common illegalities the report points to are mining without license, mining outside lease area and transporting minerals stands illegally. The Bombay High Court bench in Panaji directed the Goa government to file an FIR against the mining lobby. Since the ban, the state has been battling with economic woes as the sector contributes significantly to the State's economy.

However, Ministry of Environment and Forest (MoEF) in 2014 constituted Expert Appraisal Committee to look into the issues relating to environmental clearances of iron ore mining lease cases in the State of Goa with this capping of 20 million Metric tonnes. However, the court has also asked for a mechanism that will monitor the environmental impact of mining on Goa's ecology. By this time there has been awareness and many of the local are looking towards mining in a sustainable model.

4.1 Case Study and Voices: Sanguem, South Goa District, Goa

Sanguem taluka, the largest taluka of Goa, lying in the midland region is marked by highly dissected and undulating topography with hills and mountains running longitudinally in the talukas. A very significant part of Western Ghats traverses the taluka, running from north to south, which is also a protected zone. It contains wide variety of flora and fauna; and hence the forest cover is very high. Vegetation dominates the area with 74 per cent of areal coverage. Followed by vegetation, barren land covers 22 per cent of the area. The western stretch is a mining zone which has left the land barren and devoid of any vegetation.

Sanguem also has the distinction of having the largest area under mining. The maximum area under mining is found in Sanguem taluka. The mining activity prior 2012 ban, affected the Salaulim dam on the Salaulim river in Sanguem, which supplies water to half of the Goa's population. It is also a major source of water for irrigation and industries. Over 20 mines lie in close proximity of the dam. Sesa Goa, Fomento, Vedanta are few mining companies that operate in Sanguem. There are also many mining villages like Codli, Sigao, Sanvordem, Dabal, Camarconda etc (Ashwin, 2011).

Research by Nandini Fernandes (2015) reveals that the fish down stream of Sanguem river showed synergistic effect of change in water quality and increase in heavy metal concentration in the downstream region on account of mining may be the possible cause of impaired growth of fishes, indicative that marine life is also affected due to mining.

Mineral extraction is the primary economic activity in Sanguem. It generates 70 per cent of the taluka's income. Activities related to mining and other associated activities have modified the natural set up in and around the mining areas. Opencast mining technique is used to extract iron ore. It is carried out by forming systematic benches on the hilltop and along hill slopes (India, 2005).

Table 16.2: Change in Landuse and Land Cover in Sanguem Taluka

LU/LC	2005		2016		
	Area (hectares)	% Area covered	Area (hectares)	% Area Covered	% Change
1. Vegetation	65673.4	74.33	60753.8	61.76	**–12.57**
2. Settlement	322.83	0.37	479.16	2.54	**2.17**
3. Mining/Barren Land	19303.1	21.85	24227.73	32.2	**10.35**
4. Agriculture	1092.15	1.24	1229.22	1.39	**0.15**
5. Water Body	1959.84	2.22	1661.445	1.88	**0.34**
Total	**88351.32**	**100**	**88351.36**	**100**	

Source: Landsat Image 7 & 8.

Landuse and land cover in Sanguem taluka has undergone drastic change over the decade (2005-2016). Firstly, natural vegetation which covered almost three-fourth of the area in 2005 is presently around 62 per cent of the total taluka's geographical area. (Table 16.2). Human actions are the main drivers of this change. Vegetative land has been transformed for mining. As a result forests are fragmented. The region has a huge scope for eco-tourism being a world heritage site and a hotspot but is eventually losing its aesthetic value because of faulty economic practices.

Decrease in vegetative cover is coupled by increase in barren land. Mining has been exploiting the area over the years. This problem has now become intense. Some mines are working, some are closed in 2012 due to illegal operations, but the damage they have caused is irreversible. Continuous and colossal patches were bought under ore extraction transforming pristine forest lands to barren, useless and which is as good as waste. From just 22 per cent in 2005 it has risen to 32 per cent (2016). The change is 10 per cent (Table 16.2)

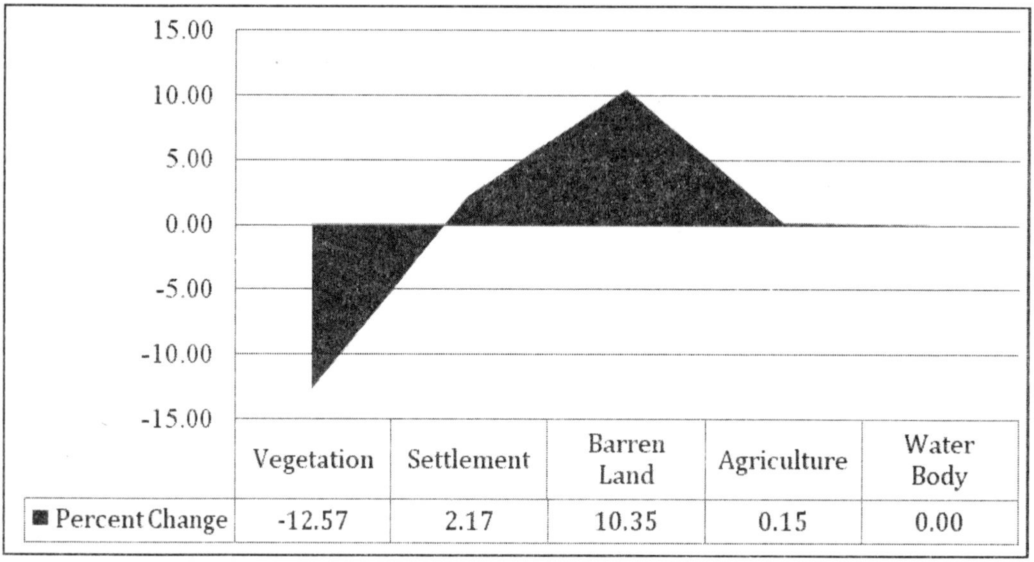

Fig. 16.2: Percentage Change in Land use/Land Cover (2005-2016)

Codli village in Sanguem Taluka of South Goa district, have been primarily mining village with agriculture being secondary. The Codli mine is mainly operated by the mining company, Sesa Goa. The mine covers an area of 40.76 hectares. These ores are predicted to last for 15 years. This is one of the largest mines owned by Sesa Goa and is a multi-pit operation mine. With the ban imposed by the Central Government in 2012, the mines remained abandoned. Research clearly indicates that mining has affected the employment structure. Unemployment has increased and the village economy has deteriorated but definitely agricultural areas have increased and cultivation of crops and production has increased. Apart from paddy cultivation, there has been effort to undertake plantations like coconut, areca nut, cashew and mangoes.

There are a couple of major effects that mining can have on the environment. Onslaught of mining for decades has led to heavy air, water, noise pollution, deforestation, land degradation etc. Eventually the condition is seen to be getting restored leading to positive effects on the natural setup of the village. 47 per cent responded that water quality has improved. Earlier the water would get dirty due to heavy siltation and but ever since the ban the water quality is comparatively better. The villagers also said that the small springs and fresh water sources are getting restored.

A local resident stated "mining had devastated the village. The environment was dusty, noisy and hardly any vegetation bloomed next to the road. But now with the ban the village environment is slowly changing. Though there were no drastic changes but forests are slowly regenerating. Now however the roads are safe and congestion free.

According to her, "ban on mining has been the best decision made and it should prolong in the future."

Table 16.3: Post-Mining Ban Perceptions of the Locals

Sr. No	Statement	Strongly Agree	Agree	Neutral	Disagree	Strongly Disagree
1.	Mining encroached forests area	20	41	21	0	0
2.	People have taken agriculture as a source of employement	8	32	23	18	1
3.	Water quality has improved	15	32	19	10	6
4.	Forests are regenerating	2	8	28	42	2
5.	Air quality has improved	21	43	18	0	0
6.	Noise pollution has decreased	21	48	6	4	3
7.	Govt. is working hard to restore damaged environment	8	12	21	29	12
8.	Health conditions have improved	25	47	6	3	1
9.	Should mining resume in the village in full swing?	6	19	15	26	16

The above table clearly indicates that post-mining ban, the village. There are really a couple of major effects that mining can have on the environment. Onslaught of mining for decades has led to heavy air, water, noise pollution, deforestation, land degradation etc. eventually the condition are seen to be getting restored leading to positive effects on the natural set up of the village. 47 per cent responded that water quality has elevated. Earlier the water would get dirty due to heavy siltation and but ever since the ban the water quality is comparatively better. The villagers also said that the small springs and fresh water sources are reviving. During mining there was dust emission that would get settled on the leaves of the plants along the roads. The air would be laden with harmful oxides and particulates that caused many health issues. Around 60 per cent respondents opined that air quality has improved and is a relief for the residents. The constant movement of trucks had caused heavy noise pollution. Nearly 70 per cent of them informed that noise has drastically decreased. Presently, locals are enjoying noise free environment. Students can study and adults can work peacefully. Apart from this, half of the respondents felt that there is a long way for the forests that were felled to regenerate. However, small patches of greenery blossomed everywhere.

Table 16.4: Air quality in Codli Village

	Air Quality	
Pollutants	**Feb-2011**	**Feb-2016**
SO_x	12.57	5.31
NO_x	25.66	13.9
pm10	97.44	39.54
pm 2.5	28.29	15.85

Source: The Goa State Pollution Control Board, 2016.

The air quality samples collected from Codli station showed positive results after the ban on mining. SO_x refers to sulfur dioxide, NO_x is nitric oxide. Pm10 is particulate matter with 10 microns diameter and pm2.5 is particulate matter with < 2.5 microns. NO_x is caused by combustion where there are heavy motor vehicles. All the above matter causes pollution in mining area due to drilling, blasting, loading and unloading, crushing of ore and transportation (GSPCB, 2016). The air quality in Codli has changed drastically from mining to post mining-situation. All the suspended matters show stark decline indicating improved air and environmental quality, rendering fresh air to the disposal of the locals.

Codli being worst hit by the mining activity can now slowly move towards growth and development. Sanguem taluka's significant area is under wildlife sanctuaries and protected forests. Tourists come here to enjoy the distinct flora, fauna and the natural beauty. Likewise, Codli village has great scope to develop tourism. Nowadays, Eco-tourism is gaining great impetus and in this context the worst hit mining zones in the village can be converted and transformed into a rehabilitated zone.

Post-Ban: On the Path of Environmental Sustainability

Even though 'mining dependents' in the form of truck drivers, barge owners, contractors, mine owners, employees and other ancillary industries have 'claimed' losses, there is a section of the public benefitting economically from the stoppage of mining activity, indicated by improvement in agricultural yield and others exploring new options.

Due to closure of mines, there have been varied and multi-dimensional social and economic impacts on the lives of people. Many people lost their livelihood and also faced economic constraints. This has led to revival of the traditional occupations, agriculture, plantation and fishing.

Mining had caused the wanton destruction of the States environment, which is irreparable and irreversible. As such, the damage caused by mining to the environment is far greater than the loss sustained to the economy since the ban. The ban on mining has allowed for rejuvenation of the environment on the whole. The decrease in the air, water and noise pollution has benefitted not only human being but also animals. Nature's resilience is evident by sightings of rare animals, butterflies and growth of rare and endemic orchids. There are signs of revival in sacred groves and community conserved forest too. (Terence, J *et al.*, 2013).

There has been steady reclamation of forest land and scrub lands, would take years to come to its natural state.

Conclusion

Goa being endowed with a number of mineral resources like iron ore, manganese and bauxite and some minor minerals, mining has developed as one of the major economic activities along with tourism, agriculture, fishing and other allied activities. Due to inappropriate mining practices, there have been several adverse effects on environment and local communities living in the vicinity of the mining areas that have attracted considerable attention from various quarters. To combat this problem, a ban was laid on mining practices in Goa which led to economic and social constraints for the people engaged in mining. The ban on mining has allowed for rejuvenation of the environment on the whole. The decrease in the air, water and noise pollution has benefitted not only human being but also animals. It also helped in the revival of the traditional occupations, agriculture, plantation and fishing.

Taking all these points into consideration, it would be prudent to make a statement that the Goa's environment impacted by the mining is on the recovery path; it would take certain amount of time and effort to make it ecological sound. It is time for Goans to reflect and analyze the factors needed to envision the holistic and sustainable development of Goa.

References

Alvares, Claude (Ed.). (2002). *Fish Curry and Rice: A Source Book on Goa, Its Ecology and Lifestyle,* Fourth Edition, Mapusa: Goa Foundation.

Ashwin, A. (2011, September 30). *Down To Earth.* Retrieved March 15, 2016, from www.downtoearth.org.in: http://www.downtoearth.org.in/news/mining-in-goa-damage-to-the-environment-greater-than-loss-of-revenue--34142

Directorate of Mines and Geology (2011). *Statistical data collected from the Department of Mines,* Government of Goa.

FAO. (2012, August 9). *Land Resources and People: Dependence and Interaction.* Retrieved March 10, 2016, from www.fao.org: http://www.fao.org/docrep/004/x3810e/x3810e04.htm

GMOEA. (2010). "Goan Mineral Ore Exports: Selected Statistics," Goa Mineral Ore Exporters' Association (GMOEA), Panjim, Goa

Government of Goa (2010). *Economic Survey 2009-10,* Directorate of Planning, Statistics and Evaluation, Government of Goa, Panaji, Goa.

Government of Goa (2011). Forest Statistics, Forest Department, Government of Goa, available at www.goaforest.com/goastate, accessed 28th May, 2018.

Jorge, T., De Sa H & Jain, 2013). Survey of Environmental and Socio-Economic Impacts of Interim Ban on Mining in Goa, EIA Resource and Response Centre (ERC), Goa, India.

Nandini Fernandes (2015). Impact of Iron Ore Mining: Comparison of Physicochemical Parameters and Heavy Metals in Upstream and Downstream Sites of a River, *International Journal of Scientific Research,* 7(5) July 2015, pp. 254-257.

Planning Commission (2012). *Sustainable Development: Emerging Issues in India's Mineral Sector,* Institute for Studies in Industrial Development, New Delhi.

Shah Commission report given to Centre", Times of India, March 16, 2012 at http://articles.timesofindia.indiatimes.com/2012-03-16/goa/31201025_1_illegal-mining-shah-commission-report-iron-ore

Talule, D. and Naik, R.G. (2014). Impacts of Indiscriminate Mining on Agriculture and Biodiversity in the State of Goa in India, Universal Journal of Agricultural Research 2(6): 211-215, 2014

http://www.downtoearth.org.in/ *Mining in Goa Damage to the Environment,* Down to Earth 30.09.2012

http://www.indiaenvironmentportal.org.in/content/44798/goas-mining-problems/

http://articles.timesofindia.indiatimes.com/2011-03-23/goa/31201025_1_illegal-mining-shah-commission-report-iron-ore

17

Model for Identifying Areas Vulnerable to Desertification Using Fuzzy Membership Analysis and Geospatial Techniques

A Case Study of Kota District, Rajasthan

P.S. Dhinwa and *Arunima Dasgupta*

1. Introduction

Desertification, the land degradation in drylands, has become an alarming global environmental issue. Areas affected by desertification processes lose progressively their level of biological quality and productivity. This is one of the major challenges that countries are facing in 21st century. The present research is carried out with an objective to identify the areas vulnerable to desertification along with their severity in terms of degradation in natural parameters. The study adopted an integrated model with fuzzy membership analysis and geospatial techniques. The model would help to locate the vulnerable areas along with their risk severity for prioritizing efficient conservation plan of resource allocation and decision-making.

Desertification is "Land degradation in arid, semiarid and dry sub-humid areas resulting from various factors including climatic variations and human activities", where "Land" is defined as terrestrial bio-productive system and "land degradation" is defined as reduction or loss in biological and economic productivity (UNCCD[1], 1994). Very recently, LADA[2] has further developed this definition as "the reduction in the capacity of the land to provide ecosystem goods and services, over a period of time, for its beneficiaries". "Ecosystem goods" are products of land, which have an economic and/or social value, including land availability, animal and plant production, soil health and water quantity and quality. "Ecosystem services" include biodiversity and the maintenance of hydrological, nutrient and carbon cycles (LADA, 2011). In particular, dryland ecosystems, which are the areas having aridity index[3] less than 0.65 (*UNEP, 1997*), are more prone to desertification. Drylands cover around one third of the world's

---

1. United Nations Convention To Combat Desertification
2. Land Degradation Assessment in Drylands
3. Aridity index is the ratio between average annual precipitation and potential evapotranspiration. Aridity index of four Dryland subtypes are; arid: 0.05 to 0.2, semi arid: 0.2 to 0.5, dry sub humid0.5 to 0.65. Aridity index of hyper arid areas is less than 0.05, which are considered as desert (UNEP, 1992).

land area which are extremely vulnerable to over-exploitation and inappropriate land use. Drylands are the home to more than 38 per cent of the total global population (MEA[1], 2005, Reynolds et al, 2007, UNCCD, 2008). Most likely the true level of degradation in drylands lies somewhere between the 10 per cent (MEA, 2005) and 20 per cent (*GLASOD*[2], *1990*), and over 250 million people are directly affected by land degradation (Reynolds *et al.*, 2007, UNCCD, 2008).

The conventional way to classify parameters for environmental or land degradation is by defining classes with certain lower and upper limit value and assign the parameter values to suitable classes. For socio-economic classification this system may suit as most of them are discrete values derived for certain administrative boundaries. But natural parameters are continuous and there remain some representative values in transitional areas between two classes. If observed in GIS environment, a considerable number of polygons can be found out which do not exactly represent a particular class of value, but a transitional zone. These are the most important areas in terms of degradation, as they have the lowest probability to be in a certain class, hence highest probability to be extended or narrowed down in next or previous class respectively. Corrective methods should be applied to these vulnerable areas to make them less vulnerable (Sasikala, 2001). Thus, to embrace the variability of natural parameters like slope, soil depth, pH and NDVI are selected for this study. The use of remote sensing techniques and GIS along with fuzzy logic to evaluate the degree of risk would help an expert in a very efficient planning of resource allocation and decision-making.

In Kota, one key natural cause of land degradation is water erosion. The river banks of both Chambal and Kalisindh River are vigorously eroded by gully erosion. In the southern part of the district, the entire Aravalli ridge had been deforested and exposed where land degradation by sheet erosion is predominant. On the other hand northern part of the district is facing land degradation by vigorous gully and ravine erosion, particularly on the banks of river Chambal and Kalisindh.

2. Study Area

Kota district in Rajasthan state, India, is selected as the study area (Fig 17.1). The district extends from 24°32′ to 25°50′56″ North in latitude and from 75°37′245″ to 76°35′22″ East in longitude. It covers an area of 5217 sq. km, among which 4906.95 sq. km is rural area and 310.05 is urban area (Census, 2011). The district is bounded by Bundi District in North, Bara District in East, Jhalawar District in South and Chittorgarh District in West. According to the census 2011, the total population is 1950491, which is 2.48 per cent of the state of Rajasthan, among which 77.48 per cent are literate (Census, 2011). The district is at present comprised of five sub-divisions namely Kota, Digod, Itawa, Sangod and Ramganjmandi and five tehsils namely Ladpura, Digod, Pipalda, Sangod and Ramganjmandi with 874 villages.

Physiographically, different morphological features can be observed in Kota, starting from mountains of Aravalli Range in south, Lakes or Reservoirs like Kishore Sagar, Sur sagar, Jawahar Sagar and Dheba lake in the middle, Riverines and gullies of rivers

1. Millennium ecosystem assessment
2. Global Assessment of Human-induced Soil degradation

Chambal, Kalisindh and Parwan in north, east and west. In the southern part of the district, there lie Aravalli ranges which are totally deforested and exposed where water erosion in the form of sheet erosion is predominant. Presence of desertification processes is also observed in northern part of the district where water erosion in the form of gullies and ravines on the banks of river Chambal and Kalisindh is rampant.

Kota district lies in hot, semi-arid (moist) with dry summers and mild winters agro-climate sub region. The mean annual temperature ranges from 24 to 25 degree Celsius rising to a maximum of 38-40 degree Celsius. The mean annual rainfall ranges from 800-1000 mm (NBSS & LUP[1], 1999).

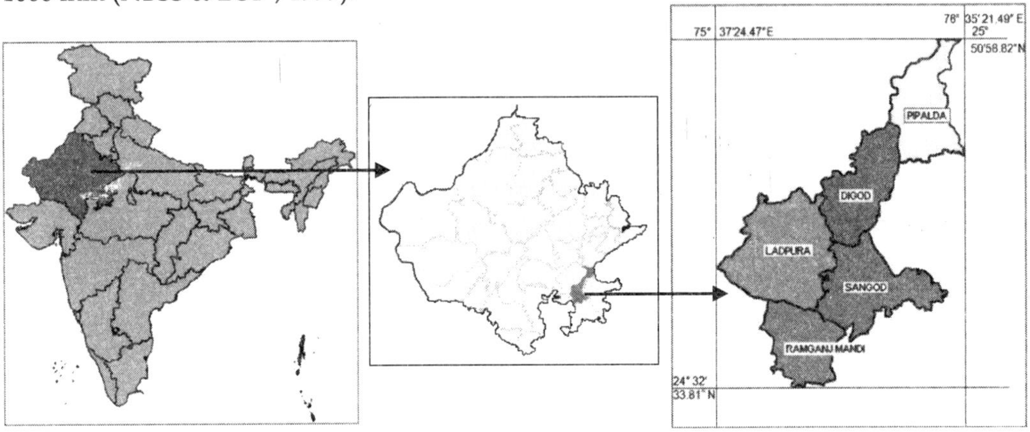

Fig. 17.1: Study Area; District Kota, Rajasthan, India

The city is the trade centre for an area in which cotton, millet, wheat, coriander and oilseeds are grown; industries include cotton and oilseed milling, textile weaving, distilling, dairying, and the manufacture of metal handcrafts. Kota also has an extensive industry of stone-polishing of a particular kind of limestone called "Kota Stone". It is used for the floor and walls of both residential and business buildings. It is a cheap alternative to marble. Kota's economy today is driven by the all-India fame of its coaching classes.

3. Data Used

Satellite images of post-monsoon season (5th September, 2010) along with ancillary information are used for identifying desertification risk areas of Kota district in Rajasthan as shown in Table 17.1. The forests in Kota are deciduous in nature, which shed all the leaves in winter. Therefore to get the true vegetation vigor the images of post-monsoon season were used for the study.

1. National Bureau of Soil Survey and Land Use Planning, Indian Council of Agricultural Research, India

Table 17.1: Data used for the study

Data	*Specifications*	*Year*	*Source*
Toposheet	45-O/12,15,16, 45-P/9,13,14, 54-C/1-11, 54D-1,2,5,6,9,10,11,13,14,15. Scale is 1:50000	1971	SOI
Images	LISS III, spatial resolution 23 meter	2010	IRS P6
Soil data	Depth in meter, pH values	2001	NNRMS

4. Methodology

Natural factors accountable for desertification are broadly of four types, namely climatic factor, terrain characteristics, soil type, and vegetation. Kota district is falling under semi-arid (moist) region (NBSS& LUP, 1999). Assuming the aridity of the whole district as homogeneous, a single aridity value that is 0.35, is taken for the entire study area. 0.35 is the mean value of the total range of aridity of semi-arid region that is from 0.20 to 0.50.

The present study included specifically, slope, soil pH, soil depth and NDVI for terrain, soil and vegetation analysis. First the parameters were classified into three risk classes, according to their deviation from mean. The risk severity was derived for each class of individual parameter using membership analysis. Four different indices were generated, namely Climate Index (*AI*), Terrain Index (*TI*), Vegetation index (*VI*) and soil Index (*SI*). These indices are defined as the normalized values of each individual parameter. The values of individual parameter were normalized in a range from 1 to 10 to obtain corresponding individual indices. The Climate Index, Terrain Index, and Vegetation index were derived from normalized aridity values, slope values and NDVI values respectively. The values of soil depth and pH were combined to generate Soil index. Thus Soil Index is a function of soil depth and pH. $SI = f(SD + SPH)$, where SI is Soil index, SD is Soil depth, SpH is Soil pH. Once the risk indices of variables were evaluated, they were integrated in GIS environment using geostatistical operations to derive the overall Natural Parameter Risk Index (*NPRI*). Thus Natural Parameter Risk Index (*NPRI*) is a function of Climate Index (*AI*), Terrain Index (*TI*), Vegetation index (*VI*) and soil Index (*SI*). $NPRI = f(AI + TI + VI + SI)$. Finally, the total range was classified according to their deviation from mean to locate final severity risk classes. The overall methodology of the study is described in the flowchart (Fig. 17.2).

On the whole, the methodology involves three steps. First the parameters were classified according to their deviation from mean. The total range of values for each individual parameter were classified into three risk classes namely "Low", "Moderate" and "High". Eventually the classes are:

LOW:	$\mu-\sigma < x \leq \mu+\sigma$
MODERATE:	$\mu+\sigma < x \leq \mu+2\sigma$ OR $\mu-2 < x \leq \mu-\sigma$
HIGH:	$x > +2\sigma$ OR $x \leq \mu-2\sigma$

The parameters, which are inversely related to desertification, such as soil depth and NDVI, were classified following the descending order of values. In the contrary, the slope

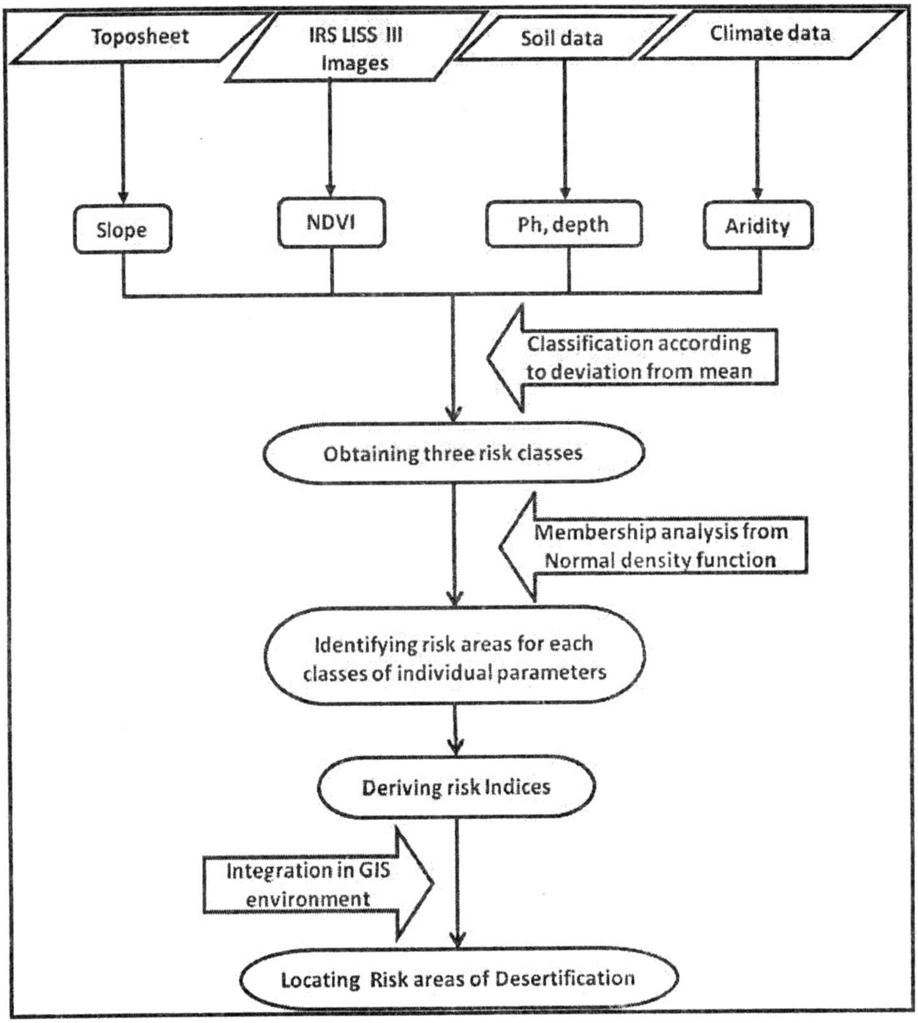

Fig. 17.2: Methodology flowchart

values, being positively related to the risk, were classified following the ascending order. On the other hand, both the lower and higher values of soil pH are unfavourable for cultivation indicating low productivity. pH values are classified in three classes following the ascending order. Then both the class Low and High are classified in High risk and the values near neutral range are classified in Low risk class.

Second the risks of each class of individual parameter were assessed from membership analysis of normal density function. The distribution of each class of a single parameter was assumed to be Gaussian with μ mean and σ standard deviation. Normal probability density function was used to obtain the membership of individual variable to be in a particular class, following the statistical formula as follows:

$$f(x, \mu, \sigma) = \frac{1}{\sigma\sqrt{2\pi}} * e^{-(x-\mu)2/2\sigma^2}$$

Where, x is the value or an individual member of a class of chosen parameter, μ is the arithmetic mean and σ is the standard deviation of the class. Choice of normal probability density function implies that there is infinite distribution tail, represented by typically the values beyond $\mu+2\sigma$ and $\mu-2\sigma$. These are not representing a certain class, but a transitional zone between two successive classes, indicating the most important areas in terms of degradation, as they have the lowest probability to be in a certain class, hence highest probability to be extended or narrowed down in next or previous class respectively. Thus, both the areas of high and low risk could be categorized.

A detailed explanation implied that, for each of the class of a single parameter, there are some values more than $\mu+2\sigma$ and less than $\mu-2\sigma$, not strongly representing that particular class but a transitional zone between two successive classes. If we consider a threshold ranging from $\mu-2\sigma$ of previous class to $\mu+2\sigma$ of the next class, the values falling in this threshold are considered to be in high risk. These values are referring a transitional zone, hence are susceptible to be altered easily to a higher value, if unchecked. In other words, these are highly vulnerable areas for degradation. On the other hand, the values in between $\mu-\sigma$ and $\mu+\sigma$ are considered in a stable condition and not to be altered easily, as they strongly represent their respective class, hence in no risk. The values in between $\mu-2\sigma$ to $\mu-\sigma$ and $\mu+\sigma$ to $\mu+2\sigma$ are on moderate risk. The areas that are under high risk to be degraded more could be checked. Also the vulnerable areas whose degradation can be narrowed down easily could be spotted and proper corrective methods could be planned and implemented.

Finally, the risk indices of individual variables were integrated in GIS environment using geostatistical operations to derive the overall Natural Parameter Risk Index (*NPRI*). This index was classified according to their deviation from mean into three risk classes namely, "Low", "Moderate" and "Severe". Thus, the areas under overall desertification risks in terms of considered natural parameters were located along with their severity in GIS platform. The way to derive overall desertification risk categories in terms of natural parameters is given below:

$$NPI_{VLR} = TI_{VLR} \cup SL_{VLR} \cup VI_{VLR} \cup AI_{VLR}$$

Where, *NPI* is Natural Parameter Risk index, *TI* is terrain Index, *SI* is Soil index, *VI* is Vegetation Index, *AI* is Aridity index, and *VLR* indicates Very Low Risk.

5. Observations and Analysis

The analysis was carried out for four natural parameters namely slope, soil depth, soil pH and NDVI following the method mentioned before. The detail analysis and observations are given below.

5.1 Terrain Index; Slope

For terrain analysis slope was considered. Slope is positively related to land degradation.

Soil erosion by water increases as the slope length increases due to the greater accumulation of runoff. Slope was derived by IDW[1] (Yang & Hodler, 2000) method. The spotheight valus ranges between 200 metres to 450 metres approximately giving a range of slope from about 0.0004 to 9.26 degree with a mean value $\mu = 0.57$ degree and standard deviation $\sigma = 0.55$ degree. There area no values less than μ-σ, i.e. 0.014. Thus three classes are as below:

Low : x < +, i.e. x < 1.125
Moderate : + ≤ x < + 2, i.e. 1.125 ≤ x < 1.68
High : x ≥ + 2, i.e. x ≥ 1.68

For each individual classes the membership has been calculated following the normal probability density function, and accordingly the risk has been derived (Fig 17.3.a, b, c). Observations from normal density funtion of class Low (Fig 17.3.a) shows that values in between μ-σ and μ+σ are strongly belonging to this class. These are the areas with slope values 0.15 to 0.7 degree. Thus these areas are under very low risk of degradation in terms of change in slope. On the other hand, the values in between μ+σ to μ+2σ, are in moderate risk. These values have a tendency to be altered into next class, i.e. class Moderate. These are representing the areas with slope values from 0.7 to about 1.0 degree. These areas may need little attention with easy measures to be in stable condition. In addition there are some values beyond μ+2σ. These are the values not representing its present class strongly, but a transitional zone between class Low and class Moderate. These areas are in high risk to be converted into next higher class. These are the areas with slope value from 1 (μ+2σ of class Low) to 1.2 (μ-2σ of the next class, i.e. class Moderate). These areas are in high risk of desertification in terms of slope unstability and need immediate attention with proper measures to stable and lower down the slope.

Perusal of the normal density function of class Moderate (Fig 17.3.b) also confirms some areas both in moderate risk and high risk to be changed into next higher class, i.e. class High. The values between μ+σ and μ+2σ, are refering areas with slope values from 1.5 to 1.64 degree, are in moderate risk. The values more than μ+2σ, are representing areas with slope values more than1.64 are in high risk to be changed into next class.

The areas falling in class High (Fig 17.3.c) are already in high risk of desertification because of higher slope vlaue. The normal density function of class High, indicates some areas beyond μ+2σ, referring the areas with slope value more than 4.5. These areas are in very high risk with a tendency to attain higher value of slope.

5.2 Soil Index

Soil Depth

Soil depth is inversely related to soil degradation. The soil having high depth shows differentiation between distinct horizontal bands, known as soil horizon, indicating a

---

1. Inverse distance weighted method.

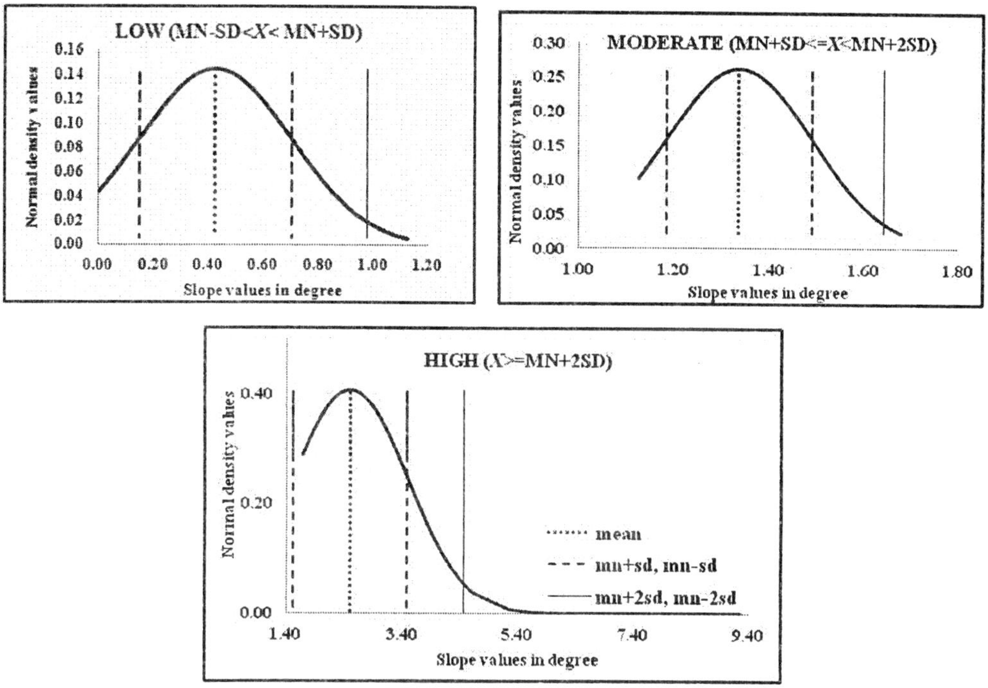

Fig. 17.3. Normal density function of individual classes of slope; i.e. 3.a Low, 3.b Moderate and 3.c High, showing membership of values in each respective class.

well developed mature soil. Less deep soils are easily erodable. In Kota, the depth of soil ranges between 7 to 30 metres, with μ=15.40 and σ=5.68. It was classified in descending order, as higher the depth value lower is the risk. There are no values less than $\mu - 2\sigma$ but some values more than $\mu + 2\sigma$, i.e. 26.75. Thus the classes are:

High	:	$x \leq -$, i.e. $x \leq 9.71$
Moderate	:	$- x < +$, i.e. $9.71 < x < 21.07$
Low	:	$x \geq +$, i.e. $x \geq 21.07$

where x is a single depth value in meter. Different risk categories had been identified from the normal distributions of each indiviual classes (Figure 17.4.a, b, c). The graph of normal density function of class Low (Fig. 17.4.a), and High (Fig 17.4.c) show a steep slope in the both their sides due to less amount of data.

One notable observation from the normal density function of the class High (Fig 17.4.a) and class Moderate (Fig 17.4.b), is that there are some values in between $\mu+\sigma$ of higher class and $\mu-\sigma$ of lower class, i.e. areas having depth values from 8.5 to 10.6 meter. These are representing the transitional zone in between High and moderate class, mostly shown in the right tail end part of the graph of class High. These areas having tendency to move towards lower depth values. Thus these areas High risk zone, can be lowered down from class High to class Moderate with easy measures.

The normal density function of class Moderate (Fig. 17.4.b) also demonstrates some values in between $\mu+\sigma$ and $\mu+2\sigma$, i.e. areas with depth values from 17.2 to 20.5 meter with a propensity to shift towards higher values hence in Class low. These areas are referring area under moderate risk.

Furthermore a considerable amount of values are observed in between $\mu+2\sigma$ of class Moderate and $\mu-\sigma$ of class Low (Fig. 17.4.c), i.e. areas with depth values from 20.5 to 23. These values are in Low risk zone, but are not strongly belonging to a single class, but a transitional zone. Therefore, these areas also can be recognized for quick measures to stable down the depth value.

Thus areas having vulnerability towards positive change are also recognized in the model.

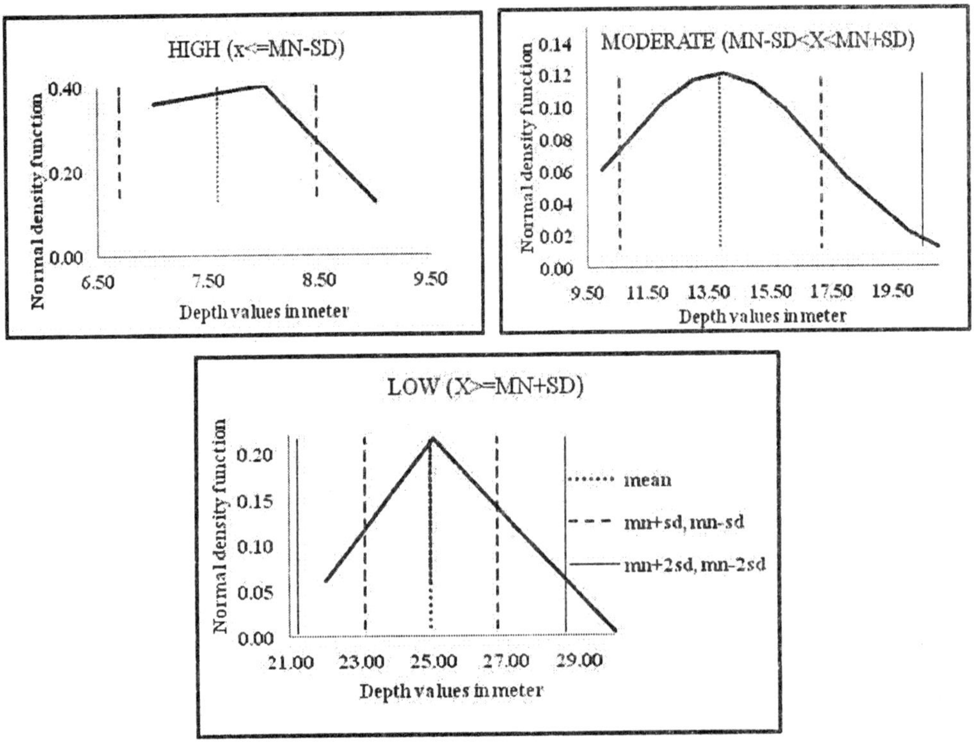

Fig. 17.4: Normal density function of individual classes of soil depth; i.e. 4.a Low, 4.b Moderate and 4.c High, showing membership of values in each respective class.

Soil pH

Soil pH is a measure of the acidity or alkalinity in soils, which is commonly called soil reaction. pH is defined as the negative logarithm (base 10) of the activity of hydrogen ions (H^+) in solution. It ranges from 0 to 14, with 7 being neutral. A pH below 7 is acidic and above 7 is basic. Soil pH is considered as a key variable in soils as it controls many chemical processes that take place. It specifically affects plant nutrient availability by controlling the chemical forms of the nutrients. The optimum pH range for most plants is between 6 and 7.5, however many plants have adapted to thrive at pH's outside this range (*Townsend, 1973*). Classes of soil based on pH factor (USDA[1], 1999) are given in Table 17.2.

Table 17.2: Soil reaction classification

Denomination	*pH range*
Ultra acid	<3.5
Extremely acid	3.5-4.4
Very strongly acid	4.5-5.0
Strongly acid	5.1-5.5
Moderately acid	5.5-6.0
Slightly acid	6.1-6.5
Neutral	6.6-7.3
Slightly alkaline	7.4-7.8
Moderately alkaline	7.9-8.4
Strongly alkaline	8.5-9.0
Very strongly alkaline	>9

Both the higher and lower pH values indicate incompatibility of land for productivity hence degradation. pH values are classified in three successive classes, namely Low, Moderate and High, according to their Deviation from mean in ascending order. It is to be noticed that in this case both the class Low and Class High are vulnerable and to be considered in high risk for desertification. In Kota, the pH values of soil ranges between 6.45 to 8.72, with μ=8 and σ=0.3. Thus, the classes are (Fig. 17.5. a, b, c):

Low : $x \leq 7.7$

Moderate : $7.7 < x \leq 8.3$

High : $x > 8.3$

The membership analysis of class Low (Fig 17.5.a) and class Moderate (Fig 17.5.b) of soil-pH shows a considerable number of values in between $\mu+\sigma$ of the previous class and $\mu-2\sigma$ of the later class, i.e. areas with soil pH values from 7.4 to 7.8. These values are referring a intermediary area betweeen Low and Moderate class and could be lowered down to strongly belong to the previous class.

Class Moderate is designated as in very Low risk, as mostly the pH values are ranging around the neutral value 7. There are some values in between $\mu+\sigma$ nd $\mu+2\sigma$, representing the area with pH values ranging from 8.2 to 8.4. The pH values of these areas have a inclination to be increased and need attention to be stabled down.

The areas represented by the pH values falling in Class High are already in High risk of desertification. Additionally the normal density graph of this class shows a considerable amount of values more than $\mu+\sigma$, i.e. areas with pH value more than 8.5. The pH values of these areas are very much susceptible to be incresed, indicating very high risk of desertification, hence need immediate attention to lower down the values.

1. United States Department of Agriculture

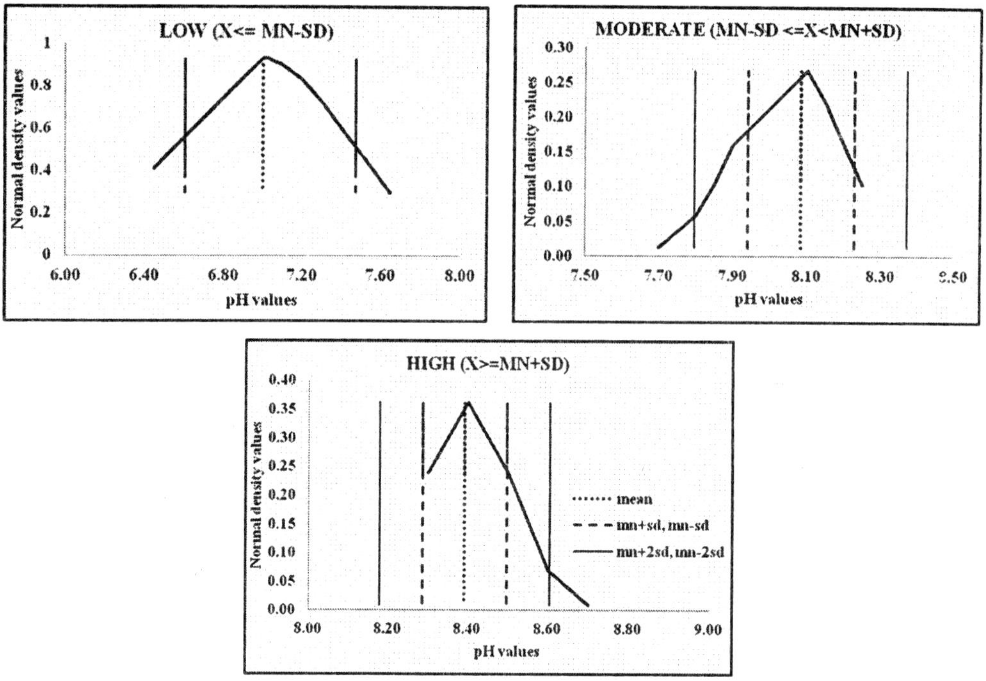

Fig. 17.5: Normal density function of individual classes of soil pH; i.e. 5.a Low, 5.b Moderate and 5.c High, showing membership of values in each respective class.

5.3 Vegetation Index; NDVI

The most important single factor in the protection of soil fertility, hence productivity, is vegetation. Destruction of vegetation, most often by human activities accelerates soil degradation, hence coupling with other responsible factors, leading desertification. Vegetation vigor can be portrayed in NDVI (normalized difference vegetation index), which can be obtained from the NDVI image, following the formula **NDVI= (IR-R)/(IR+R)**, where, IR and R are the reflectance value of pixels in infra-red and red band respectively. NDVI values range between -1 to +1, representing different chlorophyll content, indicating the health status of vegetation. From the whole image of Kota district, only the area covered by forest and scrubland were taken under consideration to analyze the NDVI of natural vegetation. The DN value images were converted to spectral radiance image, and further to apparent reflectance image (Lillesand and Kiefer, 2000). NDVI values ranges from about 0.37 to 0.67 with 0.15 mean and 0.3 standard deviation, and are classified into following three classes, x being a single variable;

High	:	$x \leq 0.15$
Moderate	:	$0.15 < x < 0.45$
Low	:	$x \geq 0.45$

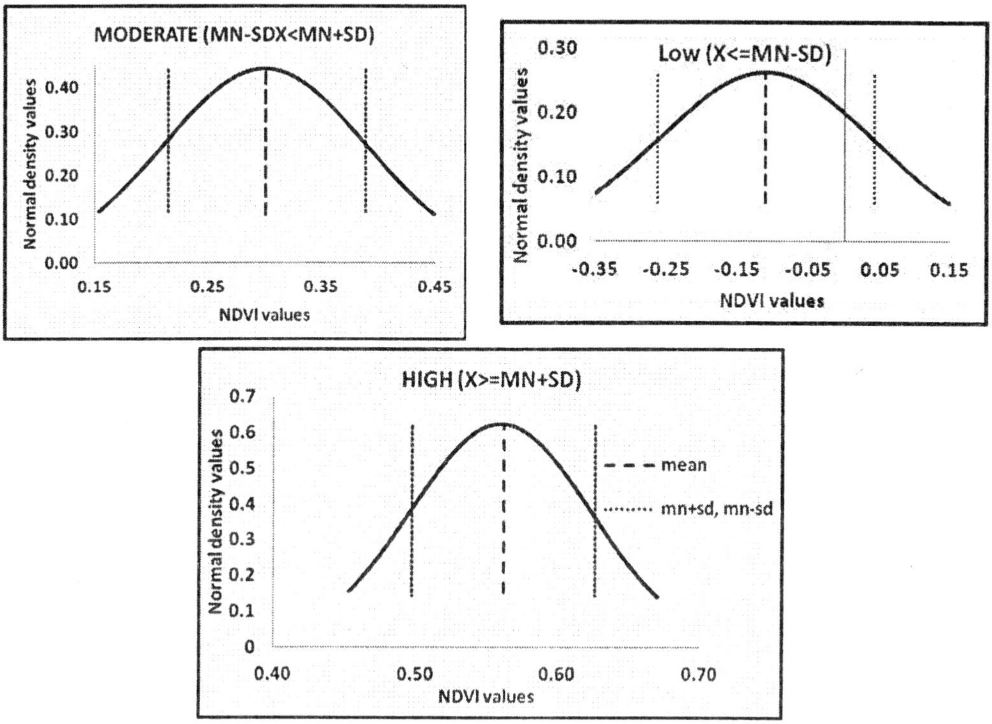

Fig. 17.6: Normal density function of individual classes of NDVI; i.e. 6.a Low, 6.b Moderate and 6.c High, showing membership of values in each respective class.

In class Low (Fig 17.6.a), the negative NDVI values represent the areas of exposed rock of Aravalli ranges, once covered with dense forest. The NDVI values from $\mu+\sigma$ of class Low (Fig 17.6.a) and $\mu-\sigma$ of class Moderate (Fig 17.6.b), i.e. areas with NDVI values 0.04 to 0.2, are representing a transitional zone. Immediate preventive action is required to increase the NDVI values of these area.

Similarly, the NDVI values from $\mu+\sigma$ of class Moderate and $\mu-\sigma$ of class High (Fig 17.6.c), i.e. areas with NDVI values 0.4 to 0.5 are also referring a susceptible zone of high risk. The values have a tendency decrease may be due to anthropogenic intervention. Thus these areas are in high risk of desertification due to deforestation and need combating measures to increase the NDVI values.

6. Results and Discussion

The above observations, as inferred from the normal density functions of individual classes of individual parameters, were potrayed spatially in GIS environment to actually know the areas under risk of desertification with their severity.

The interpretation reveals in southern part of Kota, in Ladpura and Sangod tahsil, the topography is prevailed by the residual mountain ranges of Aravalli with a height ranging from 350 to 450 metres with a slope range of 0.97 to 2.5 degree. Apart from few hill tops, the total range of Aravalli is under moderate risk of degradation (Fig 17.7). This area is under reserved forest, and once covered by dense forest as shown in toposheet

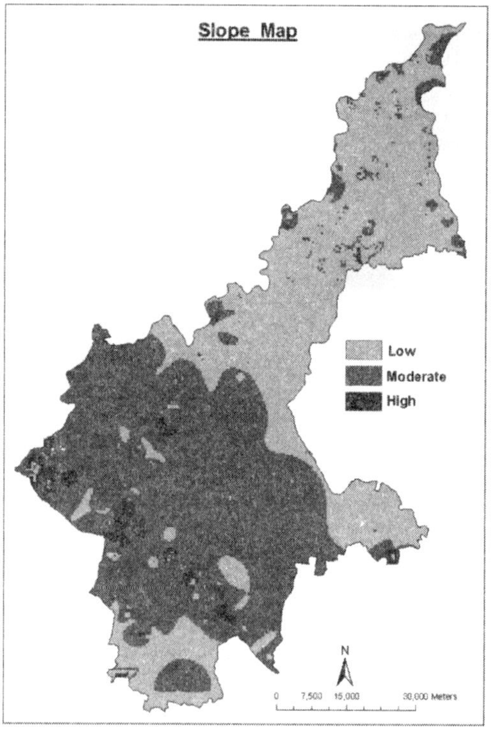

Fig. 17.7: Map showing three classes of slope, district Kota, Rajasthan

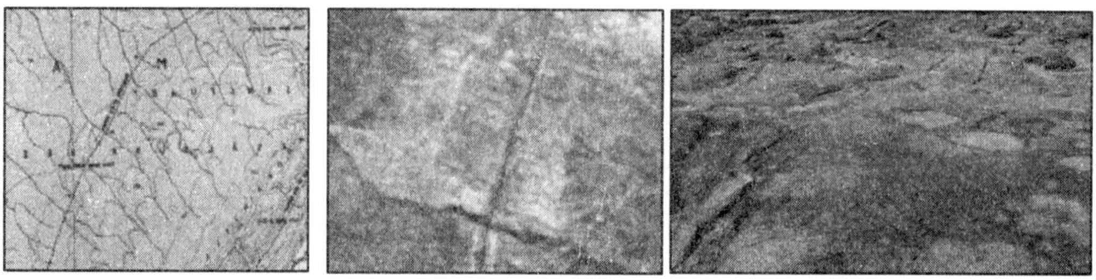

Fig. 17.8: The present condition of the eroded land, as depicted from satellite imagery (8.b) and field photo (8.c), once covered by dense forest as shown in toposheet (8.a).

(Fig 17.8.a). At present, as all the vegetation cover was removed by human intervention, the land surface is totally exposed to weather. Thus moderate to high slope coupled with surface exposure due to deforestation, accelerates the soil erosion specifically by rain splash and sheet erosion (Fig 17.8.b, c). Mostly all the soil layer has been eroded by vigorous continuous erosion process through surface runoff resulting in exposing the underlying parent rock. Consequently, the area is exhibiting negative NDVI (Fig 17.9.a, b, c) and low soil depth (Fig 17.10).

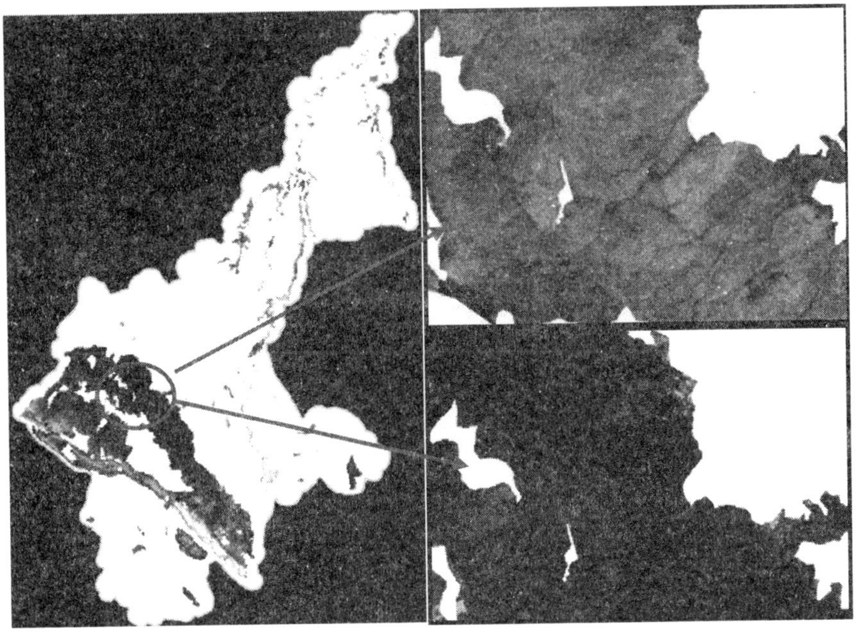

Fig. 17.9: The NDVI image of District Kota (9.a), highlighting certain part of exposed Aravallidue to deforestation, in satellite image (9.b) and NDVI image (9.c).

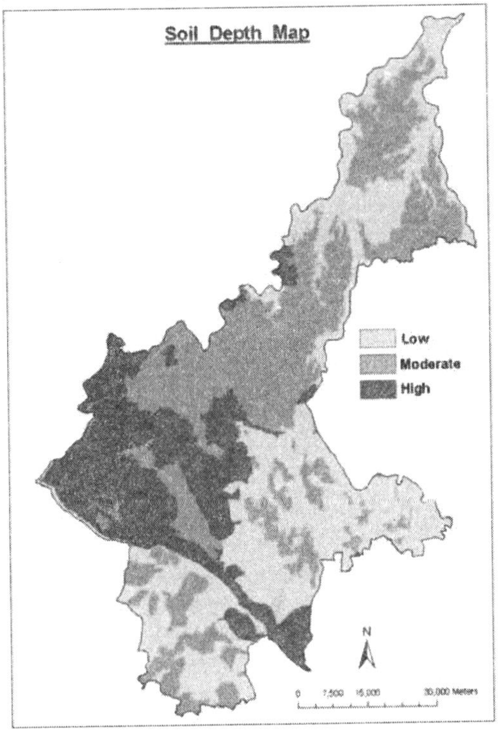

Fig. 17.10: Map showing three classes of soil depth, district Kota, Rajasthan

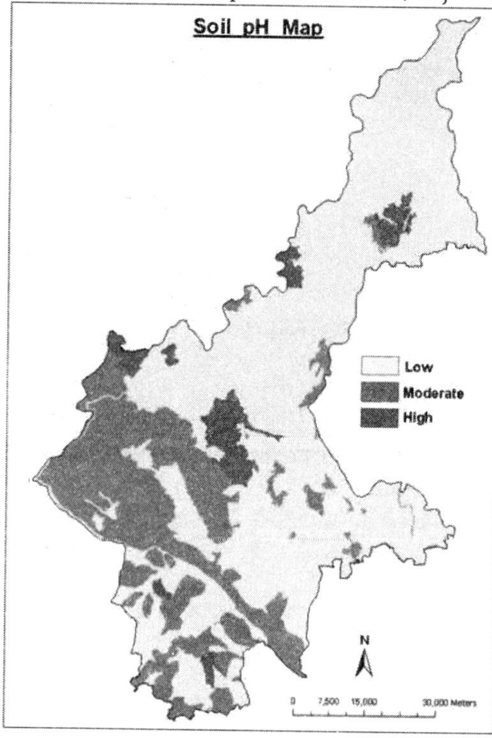

Fig. 17.11: Map showing three classes of soil pH, district Kota, Rajasthan

Additionally, in the middle part soil is getting eroded in and around the Kota city, near the lake region namely Kishor sagar and Sur sagar, and in the west of Kota Barage.

There are five areas (Fig 17.11) under the high risk of degradation in tems of pH. First one is in the northern part of the district in villages of Jharol, Ganeshganj, Mandil. Binayaka etc. This is mostly because of the soil type and the over-utilization of canal water. The second one is in the riverside of Chambal in Pipalda sand-Mandawara area, may be because of soil type and leaching process. Third one is in left side of the Kota barrage, Here the soil is extremely alkaline with the pH value of 7.9 to 8.3, because of waterlogging with the toxic swerage material coming from nearby factories. The fourth area is in the middle part of the district including the suburban areas of Rajpura, Kadihera, Galana, Balapura, Bhagwanpura, Anandkhera etc. here also the pH values ranges between 7.9 to 8.2, may be due to soil type, and dense populated area with high urban activities. The last significant area is in southern part of the district, near Ramganjmandi-Chachat area. Here soil is more acidic because of waste disposal of Kota-stone mining.

7. Conclusions and Future Scope

Space technology and Geoinformatics technique along with ancillary data, permits us to have a detailed analysis of desertification vulnerability by integrating all effective factors to take proper decisions.

Analysis of various observations made from different natural parameters results to identify the areas under different desertification risks.

The normal probability density function is suitable for membership analysis. The membership functions allow us to make out the areas presenting the transitional zone in between two successive classes. These are the vulnerable areas in two ways. These areas are susceptible to be altered into higher risk if unchecked. On the other hand these areas can be brought in lower risk with proper measures. In addition both the areas solely under positive or negative change were also identified.

The severity of risk has been recognized analyzing the individual response as well as their combined effect. The southern part of Kota, specifically Ladpura and Sangodtaluk are in high risk of desertification in terms of natural parameters like slope, soil depth and NDVI. The soil is exposed to weather due to indiscriminate deforestation and is continuously being eroded by sheet and rill erosion. Some explicit regions were identified as vulnerable in terms of soil pH. The area in and around Ramganjmandi-Chechat are noticeably in risk due to Kota-stone mining and waste deposit increasing the soil pH.

The future scope of the study is to predict the risk severity in future and to validate the model efficiency with future status. The assessment of socioeconomic parameters can also be incorporated into the study for more accurate appraisal of desertification.

[**Acknowledgement:** Authors are thankful to Dr. Irfan Khan, Director, TALEEM Research Foundation, Ahmedabad for his continuous support and encouragement.]

References

GLASOD, (1990). International Soil Reference and Information Centre, Wageningen, Netherlands, and UNEP, Nairobi, Kenya, 2004, http://lime.isric.nl/index.cfm?contentid_158

LADA, (2011). Final draft, Mapping Land Use Systems at global and regional scales for Land degradation Assessment Analysis, Version 1.1

Lillesand Thomas M., Kiefer Ralph W., (2000). "*Remote Sensing and image Interpretation*", Fourth edition, Chapter 7, 479-482 p., John Willey & Sons, Inc., USA.

Millennium Ecosystem Assessment, (2005). Ecosystems and Human Well-being: Desertification Synthesis, World Resources Institute, Washington, DC.

Reynolds, J.F., Smith, D.M.S., Lambin, E.F., *et al.*, (2007). Global Desertification: Building a science for dryland development, *Science*, Volume 316, 847-851p.

Sasikala K.R., (2001). Fuzzy Classification with A GIS As An Aid To Decision Making, *Journal of Fuzzy Sets and System*, ELSEVIER, Volume 118, 121-137 p.

Townsend W.N., (1973). "*An Introduction to the scientific Study of the Soil*", Fifth eition, Chapter 14, pp. 161-172, Edward Arnold, London.

UNCCD, (1994). Elaboration of an International Convention to Combat Desertification in Countries Experiencing Serious Drought and/or Desertification, Particularly in Africa (U.N.Doc.A/AC/241/27, 33 I.L.M. 1328, United Nations).

UNCCD, (2008). Desertification- Coping with Today's Global Challenges in the Context of the

Strategy of the UNCCD.

UNEP-UNO, (1992). Don't desert dry lands! Facts about deserts and desertification; www.unccd.int

UNEP, (1997). *The World Atlas of Desertification*, Arnold, Great Britain.

USDA, (1999). *Soil Taxonomy, A Basic System of Soil Classification for Making and Interpreting Soil Surveys*, second edition, http://soils.usda.gov/classification.

Velayutham, M., Mandal, D.K., Mandal, C., Sehgal, J., (1999). Agro-ecological Subregions of India for Planning and Development, NBSS&LUP, 88 p, NBSS 35, Nagpur.

Yang X. and Hodler T., (2000). Visual and statistical comparison of surface modeling techniques for point based environment data, *Journal of Cartography and Geographic Information Science*, Volume 27(2), 165-175 p.

18

Geospatial Modeling for Watershed Prioritization in Dahi Block, Dhar District (M.P.)

P.S. Dhinwa and *Madhavi Pore*

Introduction

The established procedure for watershed development planning in India is a top down approach by way of schemes to address specific problems and opportunities. These schemes are mandated by state and central government, and implemented by the sectorial institutions in a district. Existing decision methods are traditional, manual and biased in identifying the watershed sites for different schemes. Generally, base unit for these schemes is watershed, being a complete hydrological unit. Each scheme has a set of policies that are defined by legislation. Under these watershed management schemes, planners at district/sub-district level need to identify the priority watershed/ subwatersheds for preferential treatment/land use plans. Runoff is that portion of rainfall which moves down to the stream, channel, river or ocean as surface or subsurface flow. If the farmer can intelligently harvest the runoff from his field, store in a pond and recycle it for life saving or supplementary irrigation to crops, it will be possible to boost up his crop production and thus obtain good returns (Dhruva Narayana, 1993).

There are many reasons to develop a non-point source pollution prioritization system to determine which watershed requires the most attention. Once priority problems watersheds have been determined, an approach for directing specific actions is needed, i.e., targeting the specific areas of greatest potential erosion which can contribute to water quality degradation. Once targeted areas are identified, an optimizing management system can be used to examine trade-offs of the potential impacts from the implementation of various best management practice options. The best management practice options could be selected to minimize off–farm degradation of water quality and the farmers economic criteria.

Objective

The primary aim of the this study is to carry out watershed prioritization based on surface runoff and sediment yield models in Dahi Block, Dhar district (M.P.).

Study Area

The study area, Dahi block of Kukshi taluka is located in south-west part of Dhar district,

Madhya Pradesh. It lies between 22° 00′ and 22°15′ N Latitude and 74°30′ and 74°45′ E Longitude. The south and western boundaries of study area are marked by rivers Narmada and Hatni respectively. Total geographic area of this watershed is 470.46 sq. km. The major part of the study area is covered by Deccan traps of Lower Eocene to Upper Cretaceous Era. The study area is spread over two major geomorphic divisions, namely Vindhyachal range, and Narmada valley. Physiographically, Dahi block is characterized with high hills, inter hill uplands, interfluves, valleys and river terraces. The mean annual precipitation of the area is 395.86 mm with high degree of annual variability. The drainage density is moderate with predominantly dendritic to sub-dendritic drainage pattern.

Data Used

IRS 1D multi-date satellite data, ancillary information and collateral data have been used for preparing thematic maps to be used for prioritizing watersheds in Dahi Block Dhar district (M.P.) as given in Table 18.1.

Table 18.1: Data Used

	Satellite-ID	*Sensor*	*Date of Acquisition*
Satellite Data	IRS-P6	LISS-III	24.10.2007
	IRS-P6	LISS-III	28.01.2008
	IRS-P6	LISS-III	09.05.2007
Ancillary Information	46 J/12 (1:50,000 scale)		
Collateral Data	Daily Rainfall		
	Soil Texture and Depth Map		
	Cadastral Map		

Methodology

Various thematic maps, e.g. Land use/cover, groundwater prospect and slope were prepared through the use of multi date satellite data, ancillary information and collateral data as referred in Table 18.1. Later on, spatial database is created for these themes in GIS environment and subsequently these coverages are unionized and integrated in GIS environment for creating special themes for calculation of runoff and estimation of soil loss. For runoff estimation, Land use/land cover and soil maps were integrated and final runoff was predicted by using SCS model as shown in Fig. 18.1. For soil loss calculation was done based on sediment yield equation. Micro watershed prioritization has been done based on thematic inputs of Land Use/Land Cover, Soil, Slope and Hydro-geomorphology. These inputs were integrated in GIS environment to generate a map showing micro watershed wise priority areas for soil conservation as shown in flow chart in Fig. 18.2. Seasonal and month-wise runoff of the study area was calculated for the year 2007.

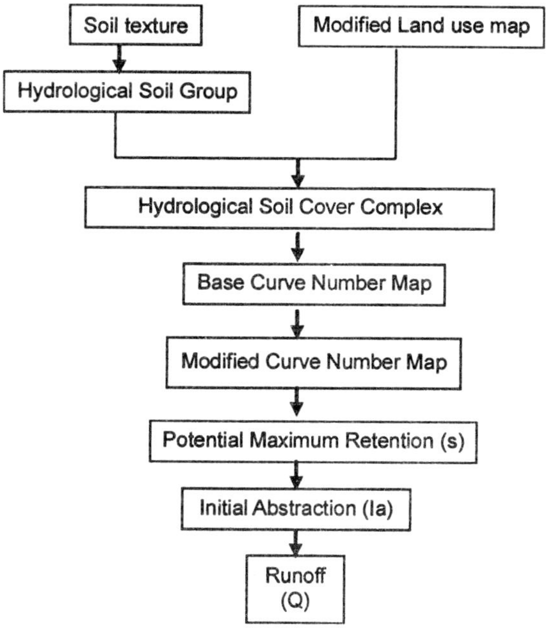

Fig. 18.1: Flow Chart of Runoff Estimation

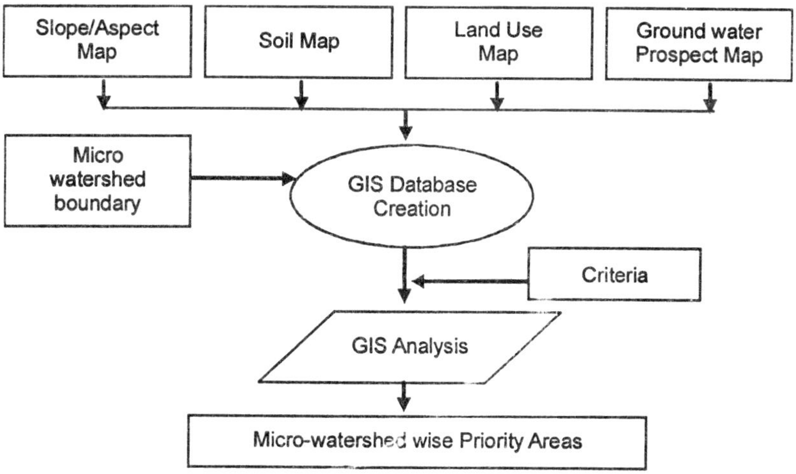

Fig. 18.2: Flow Chart for Micro-watershed Prioritization

Results and Discussion

Estimation of Runoff

Curve Number Method

Curve Number Method was developed by Ogrosky and Mockus (1957) for determining peak rate of runoff for small watersheds by synthesizing information about characteristics, physiographic factors and soil cover data. Runoff is the water in a river or stream that is derived from precipitation (Barry *et al.*,1988). In the present study, the model established by Soil Conservation Service (SCS), United States of Agriculture, is used to study the hydrologic response of the watershed to precipitation in order to prioritize watershed for adopting suitable water conservation measures. This method is used to estimate direct runoff volume from the rainfall depths. This method takes into account the parameters characterizing a watershed such as land use, hydrological soil cover and antecedent moisture condition for predicting yield from the watershed. The Land Use/Land Cover map that was generated based on IRS-P6 LISS III images was used as one of the inputs. There are 15 categories in Land Use/Land Cover, i.e., Built up - Villages (Rural), Agriculture - Kharif, Rabi, Double crop and Fallow, Wasteland-Land with scrub, Land with out-scrub, Barren Rocky/Stony waste, Mining (Industrial waste), Forest – Dense forest, open forest, scrub forest, forest blank, crop land in forest and water bodies. All these land use/land cover classes have been modified into three classes of poor, fair and good based on good crop, good canopy of forest/Land with scrub given in a modified land use/land cover map as given in Table 2. A NDVI map was prepared for the study area using satellite images. Then by using mean and standard deviation values under agriculture and forest poor, fair and good conditions were calculated as follows:

$$NDVI >= \mu + 6 \qquad \text{Good Condition}$$
$$\mu + 6 > NDVI > \mu - 6 \quad \text{Fair Condition}$$
$$NDVI <= \mu - 6 \qquad \text{Poor Condition}$$

Where, μ = Mean
6 = Standard deviation

Table 18.2: Modified land use/land cover classification system

Sl. No.	*Level II*	*Level III*	*Hydrological crop condition*
1.	Built-up	Village (Rural)	
2.	Agriculture	a. Kharif	Poor Agriculture
		b. Double crop	Fair Agriculture
		c. Cropland in forest	Good Agriculture
3.	Fallow	a. Rabi	Fallow
		b. Fallow land	
4.	Forest	a. Dense Forest	Good Forest
		b. Open Forest	Poor Forest
		c. Scrub Forest	Fair Forest
		d. Forest Blank	
5.	Wasteland	a. Land with scrub	Good Wasteland
		b. Land without scrub	Fair Wasteland
		c. Barren rocky/stony waste	Poor Wasteland
		d. Mining outcrop	
6.	Waterbodies	Streams/Reservoirs	

Antecedent soil moisture, as stated earlier is known to have significant effect on both volume and rate of runoff. This factor is duly considered by classifying moisture content in the soil into 3 antecedent moisture conditions. Antecedent moisture condition is five days moisture or wetness condition of the soil before occurrence of storm and hence, this is determined by the total in the 5 day period preceding a storm. To calculate AMC of a particular day, the cumulative rainfall values of the prior five days is calculated and decided with the help of Table 18.3.

Table 18.3: Classification of antecedent moisture classes (AMC) for the SCS method of rainfall abstractions

AMC Group	*Total 5 – day Antecedent Moisture Condition (Rainfall in mm)*	
	Dormant Season	*Growing Season*
I	Less than 12.5	Less than 35
II	12.5 – 27.5	35 -52.5
III	Over 27.5	Over 52.5

Example: If the cumulative rainfall for the prior five days calculated is 50 mm in the growing season then it is grouped into the AMC group-II.

Soil properties influence the generation of runoff from rainfall and they must be considered in the method of runoff estimation. The properties of the soil which influence the runoff are effective depth, clay in the surface layer, average clay in the profile, infiltration, permeability, soil texture etc. Soil map was prepared using soil texture information. Soil classification system developed by Soil Conservation Service (SCS) has been followed while classifying soil into different hydrological soil group, i.e., moderately low runoff, moderately high runoff and high runoff. Hydrological soil group map was integrated with modified Land Use/Land Cover to get Hydrological Soil Cover Complex Map (HSCC). Runoff numbers were assigned to the Hydrological Soil Cover Complex Map considering average antecedent moisture condition. For assigning of curve numbers for different combinations of HSCC table was used as given in Table 18.4.

Estimation of Direct Runoff Rainfall

The most generally available data in India are the amount measured by non-recording rain gauges and for such data rainfall runoff relationship was developed.

Rainfall-runoff relationship, in general form is given as:

$$Q = \frac{(P - I_a)^2}{(P - I_a) + S}$$

Where,

Q = Actual runoff in mm;

P = rainfall in mm;

S = potential maximum retention in mm

I_a = initial abstraction in mm

Black soil region AMC II and AMC III, Ia = 0.1 S

Black soil region AMC I, Ia = 0.3 S

All other regions Ia = 0.3 A

To show the rainfall-runoff relationship graphically, S values are transferred into curve number (CN) by the following equation:

$$CN = \frac{25400}{254+S}$$

Using the above, equations developed are:

$$Q = \frac{(P-0.3S)^2}{(P+0.7S)} \quad \text{... (1)}$$

$$Q = \frac{(P-0.1S)^2}{(P+0.7S)} \quad \text{... (2)}$$

The equation (1) is applicable to all soil regions of India except the black soil areas referred to Hydrologic Soil Groups. The equation (2) applies to black soil region.

After calculating runoff on daily basis it was calculated event-wise. Fig. 18.4 shows Total runoff for the year 2007.

Table 18.4: Runoff Curve Number for Hydrological Soil Cover Complex

Land use cover	*Treatment of Practice*	*Hydrological condition*	*Hydrological soil groups*			
			A	B	C	D
Fallow	Straight Row	Poor	77	86	91	94
Row crop	Straight Row	Poor	72	81	88	91
	Straight Row	Fair	67	78	85	89
	Straight Row	Good	65	76	84	88
Pasture or Range or wasteland		Poor	68	79	86	89
		Fair	49	69	79	84
		Good	39	61	74	80
Woodlands (farm wood lots or forest)		Poor	45	66	77	83
		Fair	36	60	73	79
		Good	25	55	70	77
Farmsteads			59	74	82	86
Roads (dirt)			72	82	87	89
(settlement)			74	84	90	92

Soil Loss Estimation

The sediment yield equation was used for estimating soil loss from each micro-watershed. To derive HLP map from the above four inputs, criterion analysis was carried out. The criteria used in this case were areas less than 1.5 sq. km., after union were deleted/dissolved and some categorical generalization was done considering the location and extent of units/polygons and the scale of mapping. As a first step, slope and soil coverages were intersected. Secondly, land use was intersected with derived layer of slope soil map. In the final step, the groundwater prospect map was intersected with slop soil land use map to derive HLP map. Later on weightages were assigned to HLP combinations as specified in Table 18.5. Soil erosion takes place at all the places but the

Table 18.5: Different Combinations of HLP and their Weightages

Sl. No.	HLP Combinations	SYI Calcode	Weightagess
1	Gentle slope categories and Deep soil categories and whole agriculture from Land use.	10	1
2	Gentle slope categories and Shallow soils and whole agriculture from Land use.	20	3
3	Gentle slope categories and Shallow soils and open forest, scrub forest and land with scrub	30	5
4	Steep slope categories and deep soil and whole agriculture	40	5
5	Steep slope categories and deep soil and open forest, scrub forest and land with scrub	50	4
6	Steep slope categories and deep soil and forest blank	60	4
7	Steep slope categories and shallow soil and whole agriculture	70	8
8	Steep slope categories and shallow soil and open forest, scrub forest and land with scrub	80	8
9	Steep slope categories and shallow soil and forest blank	90	8
10	Gentle slope categories and deep soil and open forest, scrub forest and land with scrub	100	2
11	Gentle slope categories and deep soil and forest blank and land without scrub	110	2
12	Gentle slope categories and shallow soil and land without scrub	120	5
13	Steep slope categories and whole soil categories and land without scrub	130	5
14	Gentle slope categories and shallow soil and forest blank	140	5
15	Water body in land use	99	0
16	Dense forest	77	1
17	Barren rocky/stony waste	88	2
18	Gentle slope categories and mining outcrop	661	4
19	Steep slope categories and mining outcrop	662	9
20	Village (rural)	55	0

intensity of erosion is different at different places. It is not possible to take measures for minimizing soil erosion at all the places at the same time due to constraints of money and technical manpower. Hence, areas which are eroding more in time and space are to be conserved first. In this study, an effort is made to identify priority areas for soil conservation by using empirical formula developed by All India Soil and Land Use Survey organization, Government of India, New Delhi (1977). The following sediment yield equation has been used for estimation of soil loss-

$$SYI = \frac{Aei \times Wei \times D \times 100}{Aw}$$

Where

SYI = Sediment Yield Index
Aei = Area of homogeneous land parcel unit
Aw = Area of micro-watershed
Wei = Weightage of erosion intensity, and
D = Delivery ratio

On the basis of SYI, the watersheds are categorized into very high, high, medium, low priorities for taking up soil conservation measures. In Dahi block 2715 ha of area in under low category, 855 ha. under moderate category, 2769 ha. under high category and 1379 ha. under very high priority category as given in Table 18.6 and shown in Fig. 18.5

Table 18.6: Priority areas soil conservation

Priority Class	*Area (ha)*
Low	27151.21
Moderate	855.8062
High	2769.017
Very High	1379.374

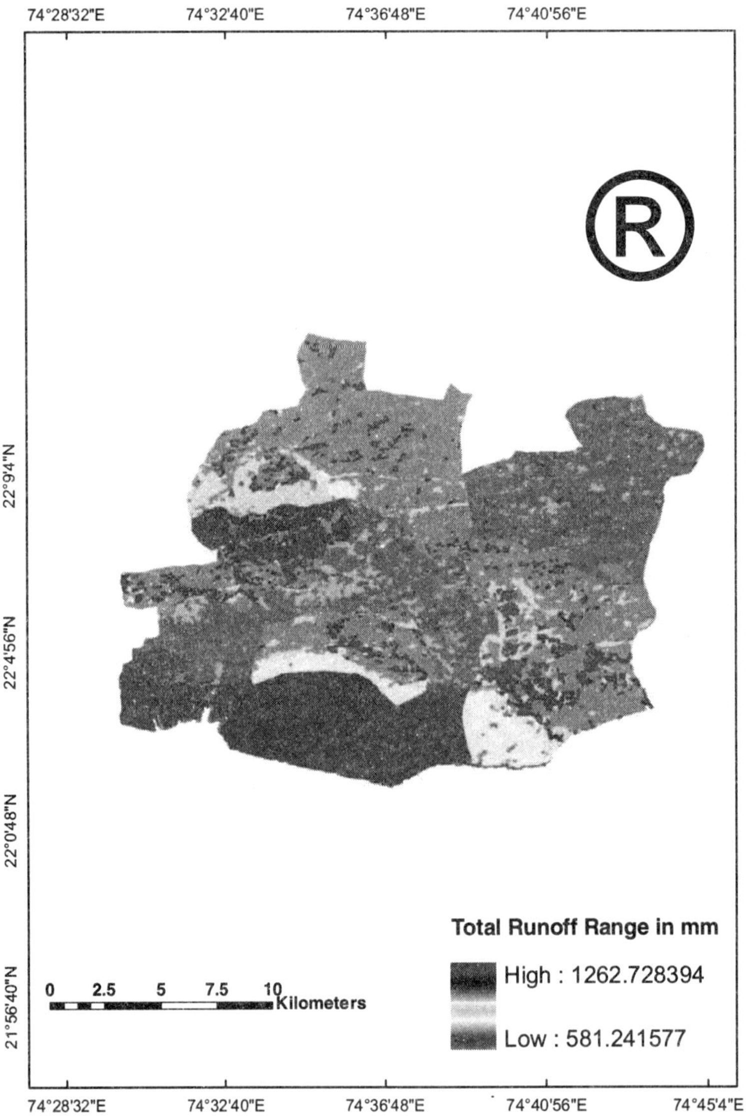

Fig. 18.4: Total Runoff Map for the Year 2007

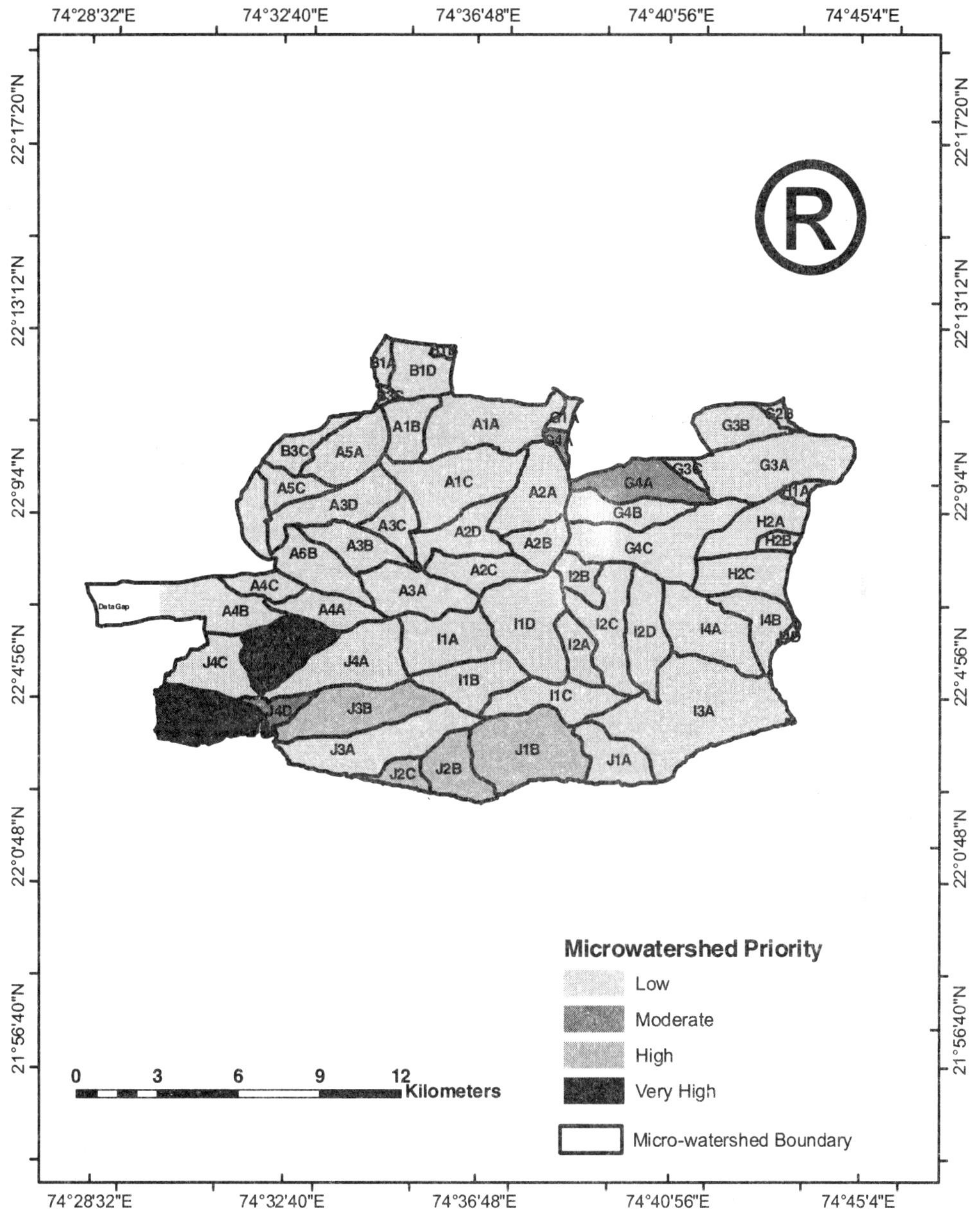

Fig. 18.5: Micro-watershed-wise Priority Areas for Soil Conservation

Conclusion

Watershed prioritization of selected study area of Dahi Block was carried out through estimating runoff and sediment yield using Soil Conservation Service model and

Sediment Yield Index Equation model. Geographic Information System (GIS) is proved to be an useful tool in spatial data base creation, for making data compatible from one format to another and retrieval of data as per requirements. Since input data is geographic in nature, GIS is found to be a potential tool while carrying out criteria based spatial analysis towards optimal management of natural resources. It is found that natural phenomenon such as hydrologic response of selected area of watershed to precipitation and its proneness to water erosion can be best simulated through GIS. SCS CN model and SYI model used for the estimation of runoff and soil loss respectively, have several advantages. They require few and easily available parameters, sensitive to changing land use condition/practices and easy to use by employing GIS techniques. Using GIS, it is possible to handle spatial and non-spatial data and it is important in runoff and sediment data handling. SCS CN model is very useful for prediction of runoff. Sediment Yield Index gives priority areas for soil conservation.

The analysis reveals that in Dahi block 2715 ha of area is under low category, 855 ha. under moderate category, 2769 ha. under high category and 1379 ha. under very high priority category. These inputs can be used for preparing proposals for taking up priority micro watershed for soil and water conservation by District Rural Development Agency, District Dhar, Dhar.

[**Acknowledgements:** The authors are grateful to Dr Irfan Khan, Director, TALEEM Research Foundation, Ahmedabad for his guidance and moral support.]

References

Anon. (1977). *All India Soil and Land Use Survey*, Governmnet of India, New Delhi.

Barry P.R., M.R. Church, Warren A.G. and David J. C. (1988), Relationship between Annual Runoff and Watershed area for the Eastern United States. *WRB* 24 (1): 35-41.

Dhruv Narayana (1993), *Advances in Soil and Water Conservation Research in India*. ICAR, New Delhi.

Orgosky, H O, Mockus, V (1957), The hydrology Guide. *National Engineering Handbook*, Section 4, SCS. USDEP, Agric.

19

Water Quality Index and Hydrogeological Interpretation to Assess Groundwater Quality using Geographical Information System in B. Ramaram and B. Pochampalli Watersheds

G. Praveen Kumar and *R. Pavanaguru*

Introduction

Groundwater quality is a none of the important factor in groundwater resource evaluation. Groundwater contains chemical ions which depend upon its interaction with aquifer material, rate of groundwater movement, ion exchange capacity and the source of groundwater. A critical analysis of existing water quality enables planning and development of water resources. Change in groundwater quality is due to the change in aquifer being tapped. The deepening of water levels to expose the deeper the aquifers more the change the water quality regime of the area. The area includes two watersheds, underlain by Granitoids, the aquifers represents weathered fractured granite. Though Hydrogeologically the two watersheds are similar, the groundwater regime is different in both the watersheds. The Musi river traverses through B. Pochampalli watershed, carrying treated effluent water from Hyderabad city perennially, and the groundwater levels in this area are quite shallow (< 5m. bgl). Where as in B. Ramaram watershed the surface water resources are very negligible thereby creating more stress on groundwater where the groundwater levels gone deeper most of the area (>20 m. bgl). The declining groundwater levels in the B. Ramaram watershed exhibits geogenic water problems. However, the anthropogenic activity is the main source of groundwater quality problem in B. Pochampalli watershed. The area located in the vicinity of peri-urban area of Greater Hyderabad Metropolitan City (GHMC), and the growth of industrial population may project the groundwater quality and quantity.

Geographical Information System (GIS) with various database can be integrated for evolving solutions for water resource problems, viz. assessing water quality, understanding the natural environment and developing management strategy for water resources. To understand the groundwater quality an integrated approach of traditional water quality analysis with GIS was applied to generate Water Quality Index map of the area.

The Water Quality Index (WQI) was first developed by Horton in the early 1970s. It is basically a mathematical means of calculating a single value from multiple test results.

WQI is an important parameter for demarcating groundwater quality and its suitability for drinking purpose (Tiwari & Mishra, 1985; Subba Rao, 1997; Vasantavigar *et al.*, 2010). Objectively it is to turn complex water quality data into information that is easily understandable and useful for the assessment and management of groundwater.

Location

The area extend over 657 Sq. km and lies between North latitudes 17° 14′ 51″ and 17° 41′ 34″ and between East longitudes 78° 40′ 05″ and 78° 55′ 49″ and included in Survey of India toposheet Nos. 56K/10, K/11, K/14 and K/15 (Fig. 19.1).

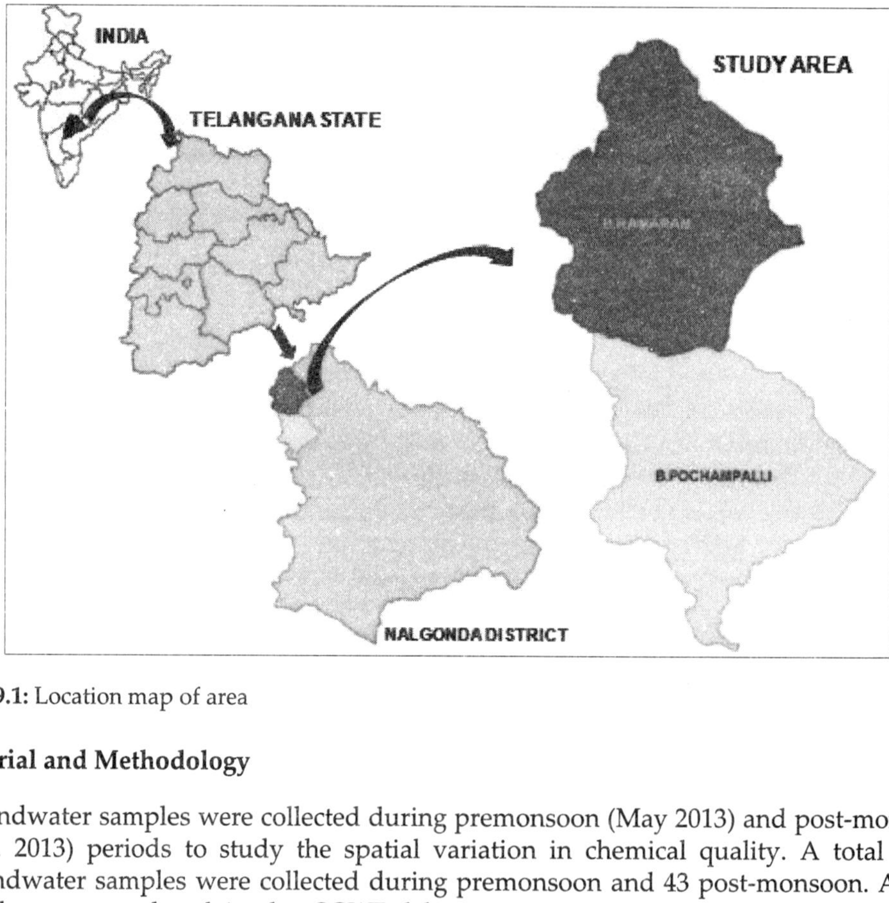

Fig. 19.1: Location map of area

Material and Methodology

Groundwater samples were collected during premonsoon (May 2013) and post-monsoon (Nov. 2013) periods to study the spatial variation in chemical quality. A total of 71 groundwater samples were collected during premonsoon and 43 post-monsoon. All the samples were analysed in the CGWB laboratory as per the standard procedures of NABL. All the parameters were expressed in mg/l, except EC and pH. Each parameter was compared to the desirable standard limit of that parameter stipulated for drinking water as prescribed by WHO and BIS.

The study was carried out with the help of topographic maps and interpreted by Mapinfo Discover. The maps were digitized from 1:50,000 scale maps and sampling locations were plotted using GPS and finally the parameter-wise analytical data was added to the locations. Various thematic layers on pH, Total Dissolved Solids (TDS). EC and ionic concentrations were generated using a spatial interpolation technique through natural neighbour method. This method of interpolation produces a comprehensive result by finding weighted averages at each interpolation point of the functional values. This is compared with the data of natural neighbours of each interpolation point. Based on this technique, Thematic maps for pH, TH, EC, TDS, Cl, SO_4, HCO_3, NO_3, Ca, Mg, Na and K were generated.

Hydrogeological Parameters

The area is composed of with the granitic rocks belonging to Archaean-Paleoprotorozoic Peninsular Gneissic Complex. The granites are intruded by basic intrusive rock (Dolerite). The depth, degree and the lateral extent of weathering of these rocks vary widely from place to place. Hydrogeological information, viz., well inventory was conducted over 76 wells. 75 key wells were considered as observation wells. The depth to water level (DTW) varies from 1.05 m to 44.50 m bgl with an average of 16.10 m bgl during pre-monsoon (Fig. 19.2a) and 0.5 m to 34.82 m bgl with an average of 7.15m bgl in Post-monsoon (Fig. 19.2b). In B. Pochampalli watershed the depth to water levels are mostly shallow (<5 m.bgl) whereas deeper water levels (> 20 m.bgl) are predominant in B. Ramaram watershed.

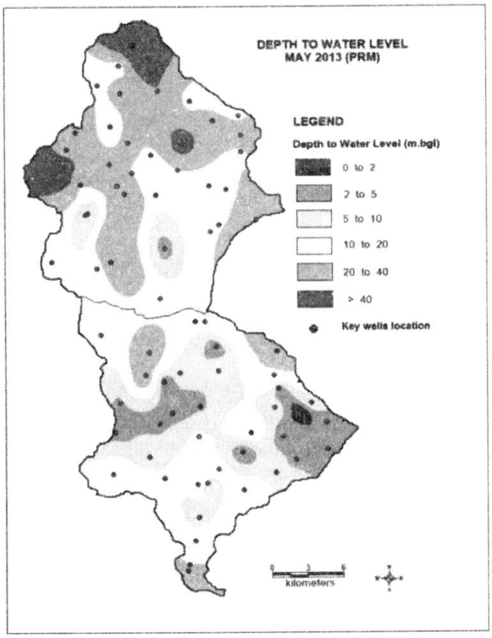

Fig. 19.2a: Depth to water level premonsoon

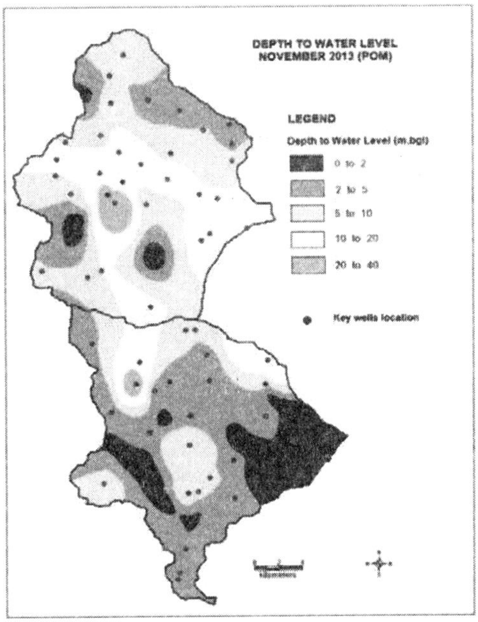

Fig. 19.2b: Depth to water level post monsoon

The depth of weathering in the area ranges from 3.0 to 24.4 m. bgl (Fig. 19.3a), the fractures in the hardrock encountered from 6.0 m bgl to191.5 m. bgl. In about 83 per cent of wells the weathering depth range is 5 to 20 m. bgl and the fractures encountered at 50 m.bgl are dominant (67%) compared to deeper fractures (Fig. 19.3b). It is also observed from the field data that the fractures upto 50 m. bgl are high yielding compared to deeper fractures.

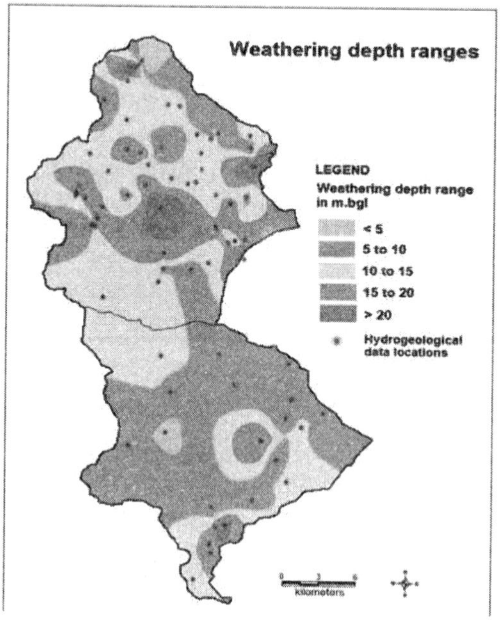

Fig. 19.3a: Weathering depth range

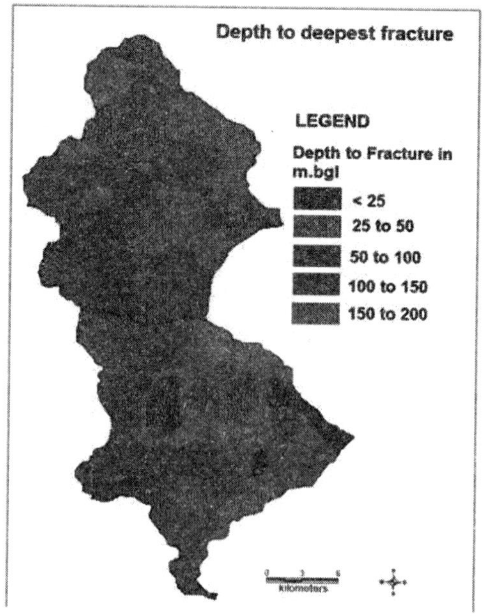

Fig. 19.3b: Fracture depth range

Results and Discussion

The analytical results exhibit abundance of anions and cations for both pre-monsoon and post-monsoon seasons: HCO_3 >Cl> SO_4> NO_3 and Na> Ca> Mg>K. The statistical comparision of groundwater analysis data is given in Table 19.1.

Table 19.1: Statistics of Groundwater samples

Parameters	*Premonsoon (Pre-monsoon)*				*Postmonsoon (Post-monsoon)*			
	Min	*Max*	*Avg*	*St.Dev*	*Min*	*Max*	*Avg*	*St.Dev*
pH	7.3	8.49	7.89	0.29	7	9.16	7.78	0.45
EC	390	3690	1587.94	771.13	307	4730	1733.16	887.59
TH	110	1140	482.61	218.88	85	960	462.33	221.57
Ca	32.064	296.592	117.19	56.19	16.032	328.656	115.11	74.62
Mg	1.216	145.92	46.26	34.92	4.864	126.464	42.59	29.62
Na	19.32	575	149.97	103.22	18.998	799.94	179.93	137.08
K	0.391	218.96	9.47	31.15	0	100	22.88	63.12
CO_3	0	6	0.34	1.39	0	60	2.23	9.80
HCO_3	152.55	610.2	340.42	115.97	79.326	610.2	330.22	123.13
Cl	10.635	907.52	270.57	207.59	10.635	779.9	283.11	205.52

SO_4	10.08	312	100.84	58.82	8.64	432	112.15	89.92
NO_3	2.48	399.9	63.81	74.83	1.24	479.88	110.90	103.44
F	0.22	3.9	1.30	0.71	0.22	2.6	1.06	0.60
TDC	241	2192	963.86	464.76	193.2211	3074	1067.16	563.80

The average concentrations of pre- and post-monsoon indicates that (Fig. 19.4) the concentration of EC, Na, K, CO_3, Cl, SO_4, NO_3 and TDS shows increase in average values whereas the pH, TH, Ca, Mg, HCO_3 and F shows the decrease in their concentration.

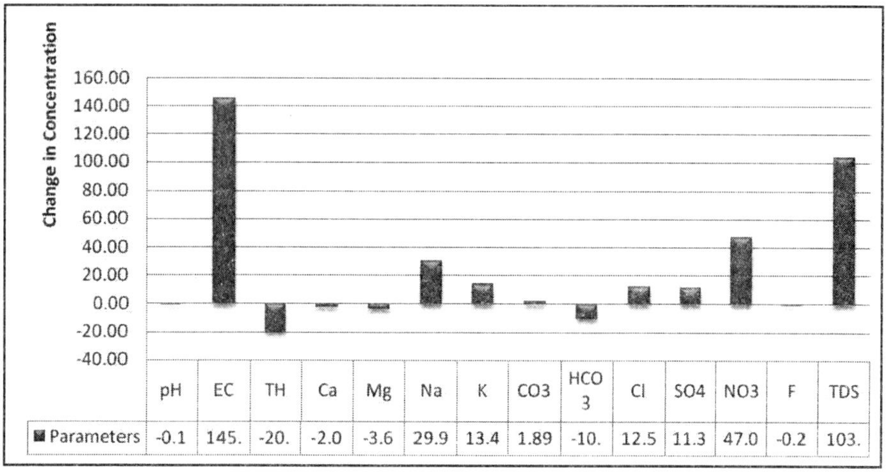

Fig. 19.4: Comparison of average concentrations of post-monsoon with pre-monsoon

pH becomes slightly alkaline during post-monsoon compared to pre-monsoon as 2 samples indicated > 8.5. The average EC increases during post-monsoon, 21 per cent of samples during pre-monsoon and 30 per cent samples during post-monsoon shows higher EC> 2250 µS.

The average value of TDS in the study area is 964 mg/L during pre-monsoon and 1067 mg/L during post-monsoon. According to Davis & De Wiest (1967), the average amount of TDS in the study area is near to the brackish range, 37 per cent during pre-monsoon and 29 per cent of the samples during post-monsoon are of brackish nature. TDS contour maps for pre-monsoon and post-monsoon are given in Fig. 19.5a and Fig. 19.5b.

Fluoride is electronegative and chemically reactive element of all halides. The desirable limit for fluoride in the drinking water is less than 1.5 mg/l. During pre-monsoon concentration of F range from 0.22 mg/l to 3.9 mg/l with an average of 1.30 mg/l and during post-monsoon it varies from 0.22 mg/l to 2.60 mg/l with an average of 1.06 mg/l. During pre-monsoon high concentration of F (>1.5 mg/l) is seen in northern parts of the area whereas during post-monsoon the area covered decreases, which can be attributed to the dilution of recharging water. Spatial distribution maps of Fluoride are given in Fig. 19.6a and Fig. 19.6b.

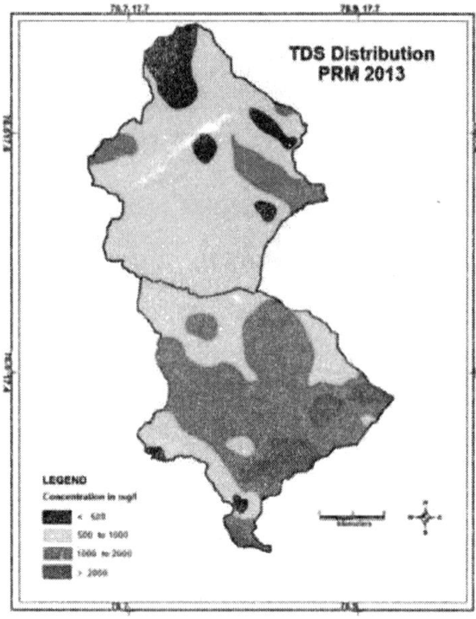

Fig. 19.5a: TDS distribution Pre-monsoon

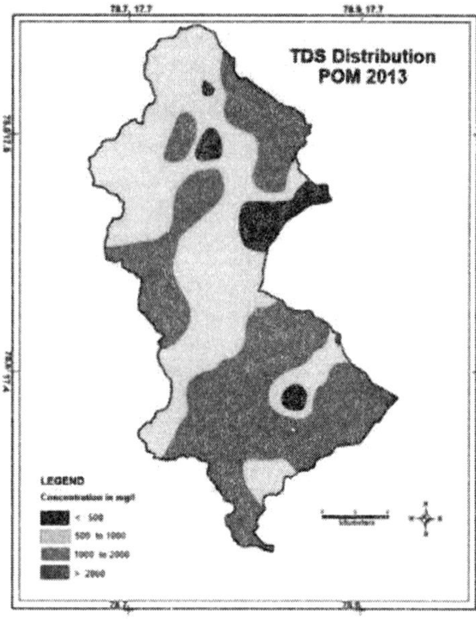

Fig. 19.5b: TDS distribution Post-monsoon

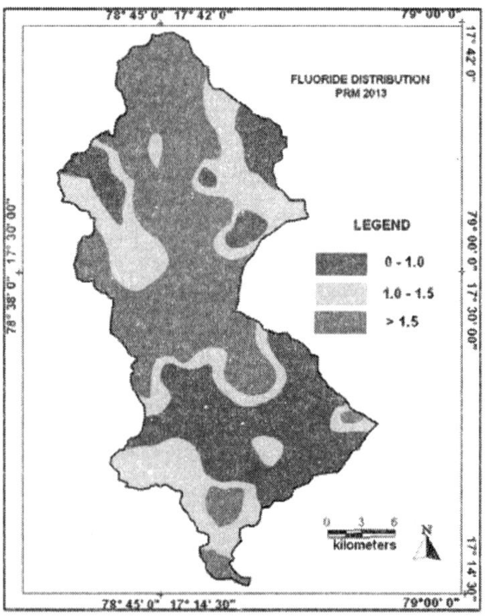

Fig. 19.6a: Fluoride distribution Pre-monsoon

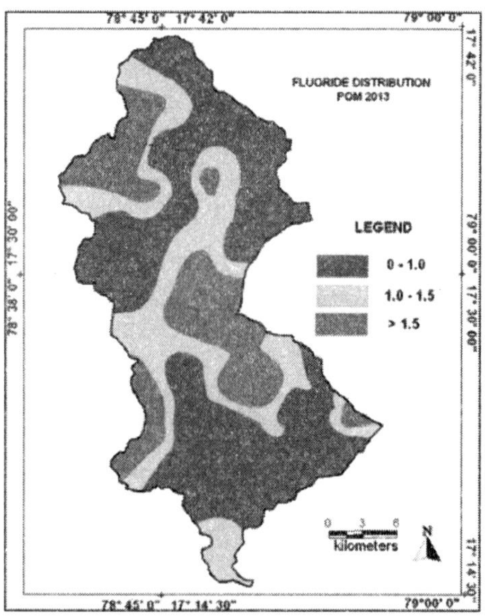

Fig. 19.6b: Fluoride distribution Post-monsoon

Bicarbonate is the predominant anion in the study area with 77 per cent of the samples exceeds the BIS limit (< 250 mg/L) for drinking water standards in both the seasons. (Fig. 19.7a and 19.7b).

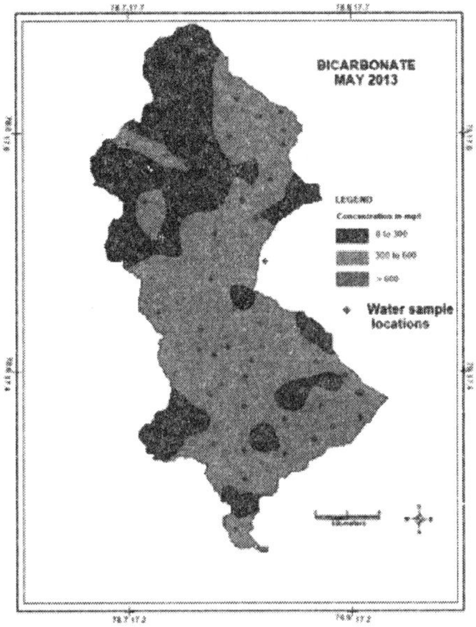

Fig. 19.7a: Bicarbonate distribution Pre-monsoon

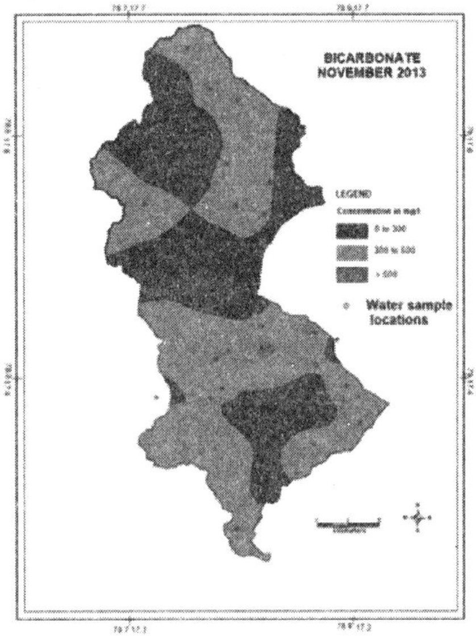

Fig. 19.7b: Bicarbonate distribution Post-monsoon

Chloride is the next dominant anion. As per the standards 41 samples during pre-monsoon and 51 samples during post-monsoon exceed the permissible limit (< 250 mg/l). Cl concentration between 250 to 1000 mg/L is dominant in southern parts during pre-monsoon, while during post-monsoon the concentration range of 250 to 1000 mg/l decreases in southern part and increase in northern parts (Fig. 19.8a and Fig. 19.8b).

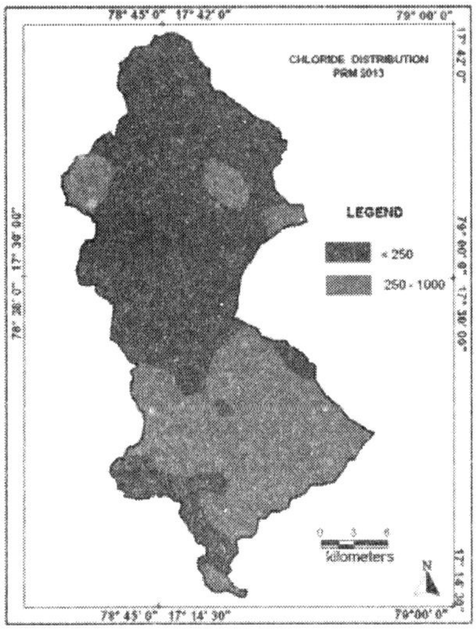

Fig. 19.8a: Chloride distribution Pre-monsoon

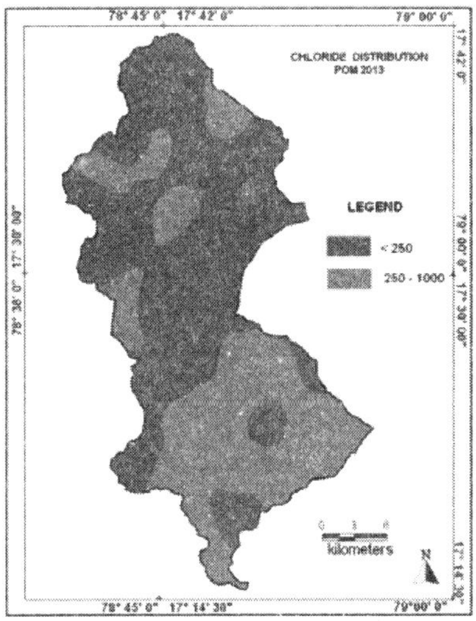

Fig. 19.8b: Chloride distribution Post-monsoon

Sulphate concentration exceeds the prescribed limit (< 400 mg/L) only during Post-monsoon in 5 per cent of the samples (Fig. 19.9a and 19.9b).

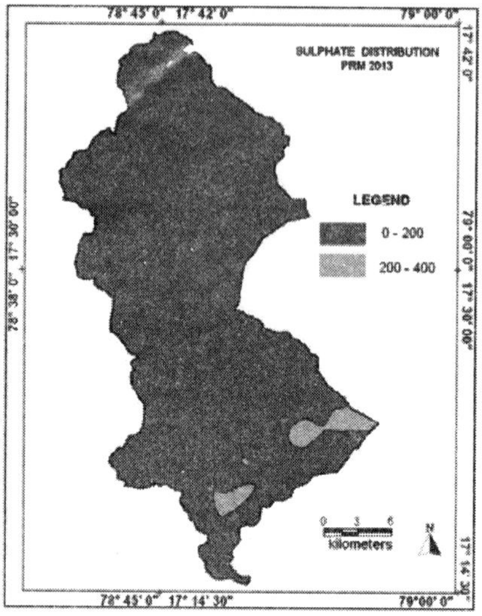

Fig. 19.9a: Sulphate distribution Pre-monsoon

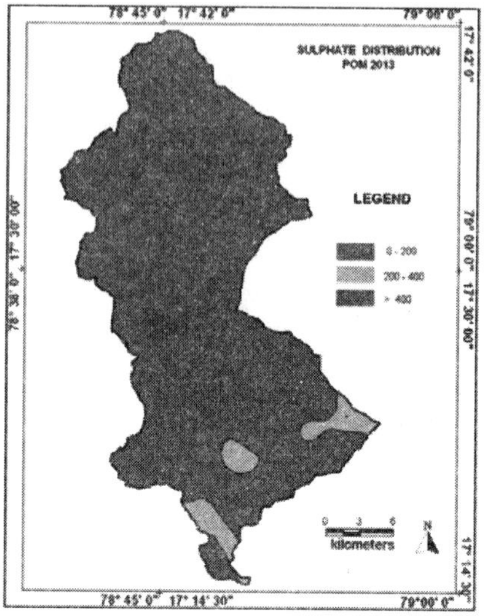

Fig. 19.9b: Sulphate distribution Post-monsoon

Leaching of nitrate with percolating water increases the nitrate concentration, 42 per cent of the samples during pre-monsoon and 65 per cent of the samples during post-monsoon shows the nitrate concentration above permissible limit (<45 mg/l) for drinking water. The concentration of Nitrate increases in most of the samples during post-monsoon, especially in B. Ramaram watershed where 75 per cent of the samples represent concentrations > 100 mg/l (Fig. 19.10a and 19.10b).

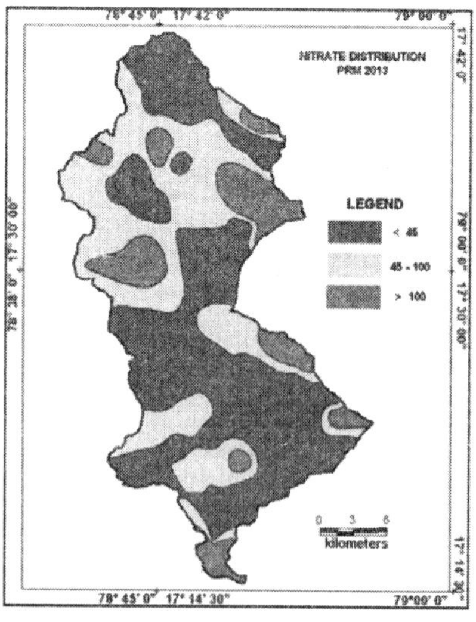

Fig. 19.10a: Nitrate distribution Pre-monsoon

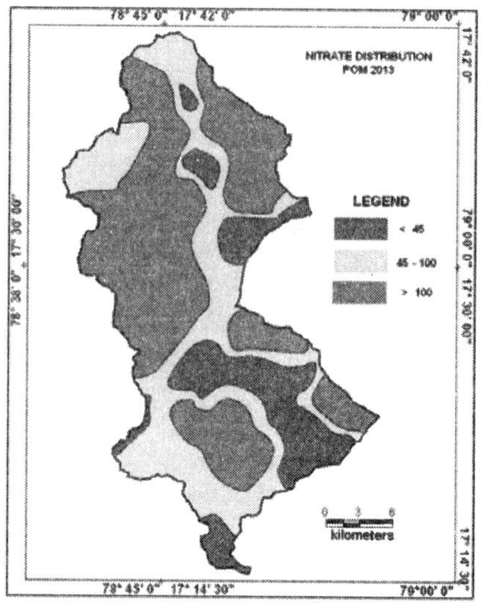

Fig. 19.10b: Nitrate distribution Post-monsoon

Among the cations, Sodium concentration increase from 15 per cent in Pre-monsoon to 21 per cent in Post-monsoon season (Fig. 19.11a and 19.11b).

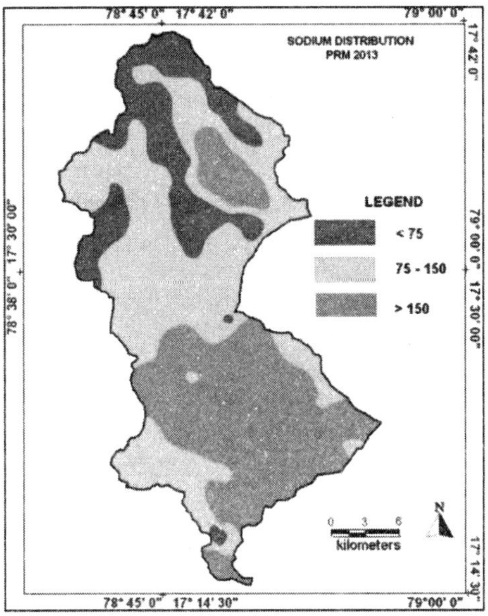

Fig. 19.11a: Sodium distribution Pre-monsoon

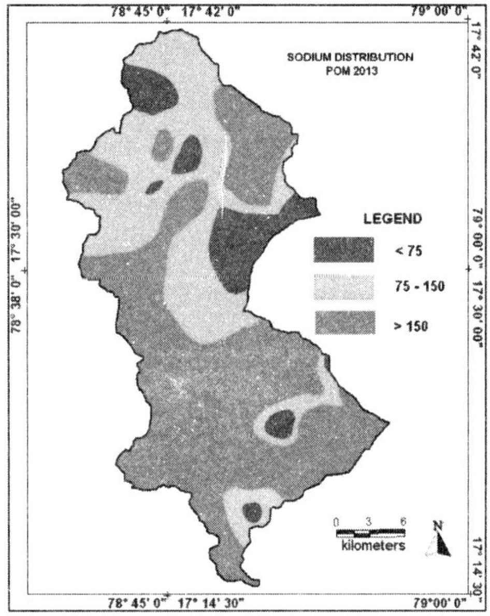

Fig. 19.11b: Sodium distribution Post-monsoon

Calcium concentration >75 mg/l is more dominant during pre-monsoon and concentration < 75 mg/l is dominant in post-monsoon. Small areas in both the watersheds represent calcium concentration > 200 mg/l (Fig. 19.12a and 19.12b).

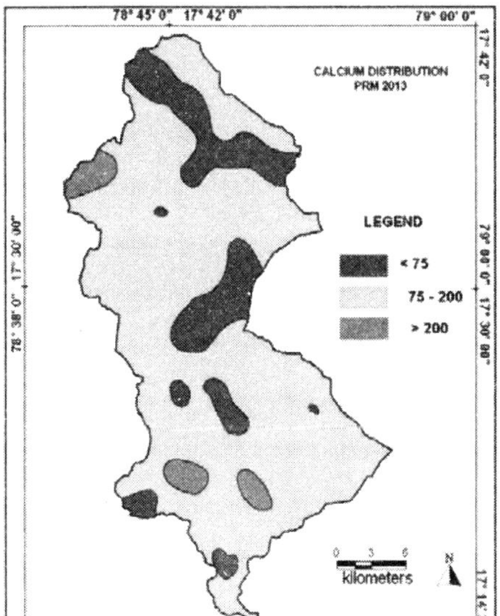

Fig. 19.12a: Calcium distribution Pre-monsoon

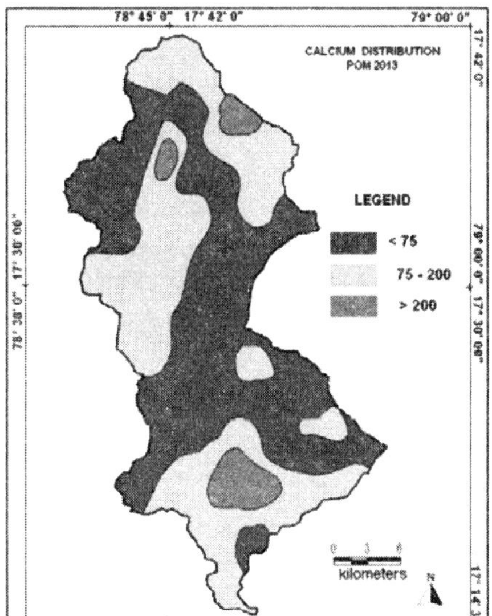

Fig. 19.12b: Calcium distribution Post-monsoon

Magnesium concentration is generally low compared to those of sodium and calcium, about 61 per cent in pre-monsoon and 63 per cent in post-monsoon exceed the permissible limit (<30 mg/l) as per BIS. (Fig. 19.13a and Fig. 19.13b). Concentration of magnesium >30 mg/l is more dominate in B. Pochampalli watershed whereas concentration < 30 mg/l is more dominant in B. Ramaram watershed during pre-monsoon. During post-monsoon, area is occupied by magnesium concentration > 30 mg/l decreased in B. Pochampalli where its spread increased in B. Ramaram watershed.

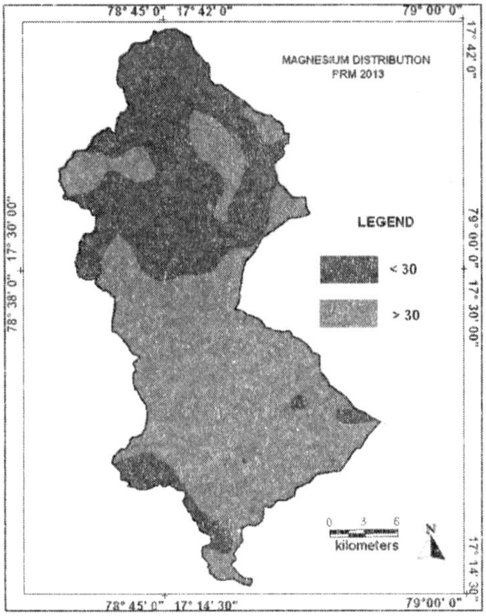

Fig. 19.13a: Magnesium distribution Pre-monsoon

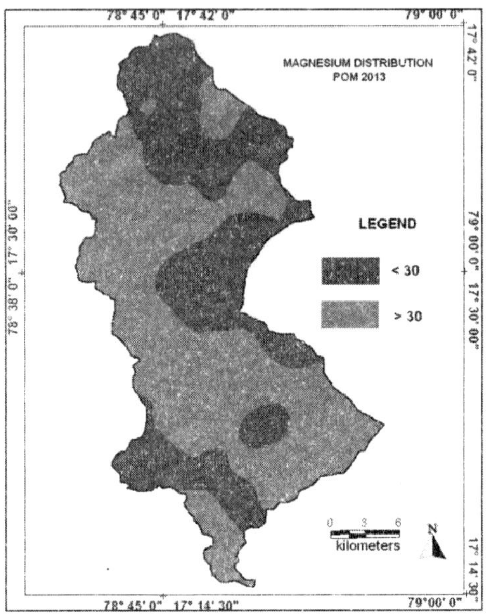

Fig. 19.13b: Mg distribution post-monsoon

The average concentration of potassium during pre-monsoon and post-monsoon are 9.47 and 12.66 mg/l respectively. Iso-concentration plots (Fig. 19.14a and 19.14b) shows that nearly 90 per cent of the area during pre-monsoon is occupied by potassium concentration < 12 mg/l, whereas concentration > 12 mg/l increase during post-monsoon. The increase in concentration can be observed in more areas in B. Ramaram watershed compared to B. Pochampalli watershed.

Water Quality Index WQI

Basically WQI attempts to provide a mechanism for presenting a cumulatively derived, numerical expression defining a certain level of water quality (Miller *et al.*, 1986). In order to calculate WQI, twelve important parameters i.e., pH, TDS (mg/l), TH (mg/l) Ca (mg/l), Mg (mg/l), Na (mg/l), K (mg/l), Cl (mg/l), HCO_3(mg/l), SO_4(mg/l), NO_3 (mg/l) and F (mg/l) have been considered which contribute to the maximum quality of the groundwater. In the first step, each of 12 parameters has been assigned a weight (w_i) according to relative importance in the overall quality of water for drinking purpose. The maximum weight of 5 has been assigned to the parameter TDS, NO_3 and Fluoride due to their significant

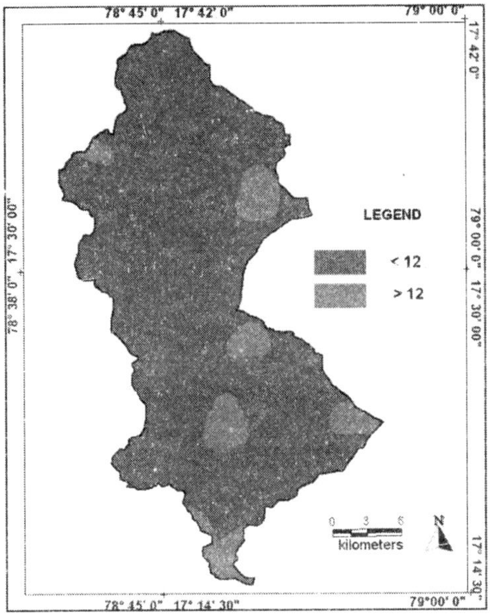

Fig. 19.14a: Potasium distribution Pre-monsoon

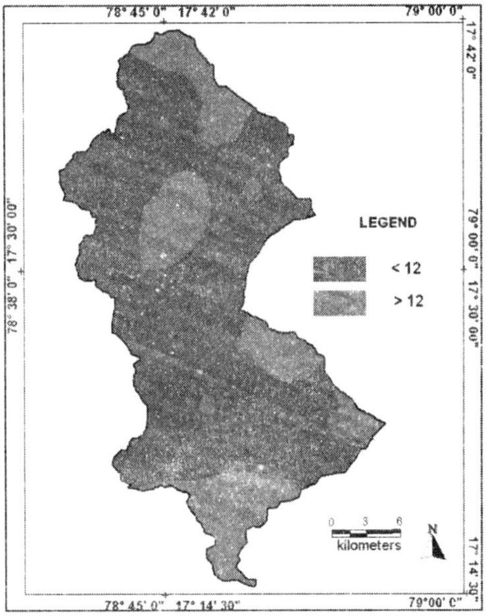

Fig. 19.14b: Potasium distribution Post-monsoon

importance in water quality assessment. Other parameters like Calcium, Magnesium, Sodium and Potassium were assigned a weight between 1 and 5 depending on their importance in the overall quality of water for drinking purpose (Table 19.2). In the second step, the relative weight (W_i) is computed for each parameter using Eq. 1

$$Wi = \frac{wi}{\sum_{1=0}^{n} wi} \quad \text{... (1)}$$

where, W_i = relative weight, w_i= weight of each parameter and n = Number of parameters.

Table 19.2: Parameter-wise standards and Weightage

Parameter	*Indian Drinking water standards*	*Weight (wi)*	*Relative weight (Wi)*
pH	6.5-8.5	4	0.103
TH	300-600	4	0.103
HCO_3	250-600	2	0.051
TDS (mg/L)	500-2000	5	0.128
Ca (mg/L)	75-200	2	0.051
Mg (mg/L)	30-100	2	0.051
Na (mg/L)	200	3	0.077
Cl (mg/L)	250-1000	3	0.077
SO_4 (mg/L)	200-400	3	0.077
NO_3 (mg/L)	45-100	5	0.128
F (mg/L)	1-1.5	5	0.128
K (mg/L)	12	1	0.026

The quality rating scale (q_i) for each parameter is assigned by dividing its concentration in each water sample by its respective standard according to the guidelines laid down in the BIS (1991) and the results are multiplied by 100. Permissible Sodium and Potassium values are not given by BIS the standard values are taken as per WHO standards.

$$q_i = \left(\frac{C_i}{S_i}\right)100$$

where, q_i is the quality rating
C_i is the concentration of each chemical parameter in each water sample (mg/L)
S_i is the Indian drinking water standard for each chemical parameter in mg/L.

For computing the WQI, the SI is first determined for each chemical parameter.

$$SI_i = W_i.q_i$$
$$WQI = \Sigma SI_i$$

where, SI_i is the sub-index for i^{th} parameter, q_i is the rating based on concentration of ith parameter. The computed WQI values are use to categorize the water into five categories from excellent to very poor (Table 19.3).

Table 19.3: Water quality categorization

WQI range	*Type of water*	*Pre-monsoon*	*Post-monsoon*
		No. of Sample	*No. of Sample*
<50	Excellent	2	2
50-100	Good	28	12
100-200	Poor	38	25
200-300	Very poor	2	3
>300	unsuitable	1	1

From this study it is understood that most of the water samples collected were good to poor for drinking water quality. The spatial distribution plots of WQI was prepared for both pre-monsoon and post-monsoon (Fig. 19.15a and Fig. 19.15b), the study of the plots reveal that during the pre-monsoon in most of the area the groundwater quality is good to poor, except small isolated patches in the eastern and southern parts where the quality is very poor and one sample at Malkapur village is unsuitable for drinking. During the post-monsoon the area covered by good quality decreases compared to the pre-monsoon, Isolated patches show very poor water quality for instance Suryapalli sample indicates unsuitability of the groundwater for drinking purpose.

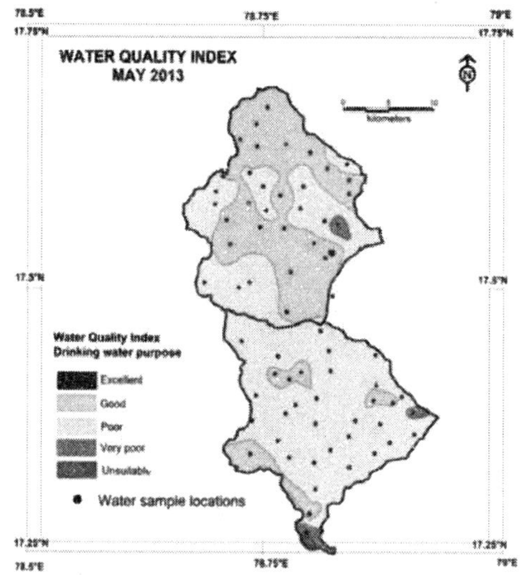

Fig. 19.15a: WQI map Pre-monsoon

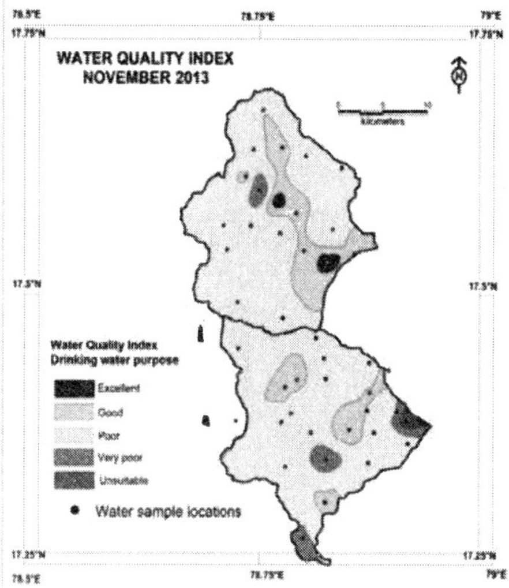

Fig. 19.15b: WQI map Post-monsoon

Comparison of WQI With Major Water Quality Parameters

Comparison of the Water Quality Index map with above permissible ranges of 3 major parameters indicate that poor to very poor WQI areas are associated with High TDS and Nitrate concentrations (Fig. 19.16a and Fig. 19.16b). One of the sample from Malkapuram in pre-monsoon is unsuitable for drinking, it has high concentrations of F (1.69 mg/l), Nitrate (400 mg/l), TDS (1898 mg/l) and one sample from Suryapalli in post-monsoon is unsuitable for drinking, it has high concentrations of F (1.80 mg/l), Nitrate (379 mg/l) and TDS (3074 mg/l).

The deeper water level during pre-monsoon in the northern parts contributed to high Fluoride concentration there by imparting poor water quality whereas high concentration of Nitrate and TDS imparted poor WQI in southern parts. The recharged water contributed to high concentration of Nitrate and TDS during post-monsoon lead to increase in poor WQI. The area covered details of chemical parameters are given in Table 19.4.

Conclusions

1. Hydrogeological information was collected from the irrigation wells to study the depth, degree and the lateral extent of weathering and deeper aquifers in the area. The data was plotted on GIS was effectively adopted to generate different thematic layers which understand the spatial distribution of various water quality parameters.
2. Hydrogeological data reveals the depth of weathering ranges from 3.0 to 24.4 m. bgl. The fractures in the hardrock encounter from 6.0 m bgl to 191.5 m. bgl. The average depth to water level (DTW) is 16.10 m bgl in pre-monsoon (May'13) and an average of 7.15 m bgl during post-monsoon (Nov'13). It is observed that the water level fluctuation is very less in the Musi command area ranging between 0 to 4 m., where as it is very high in the recharge areas in B. Ramaram watershed. (ranging between 10 to 36 m).

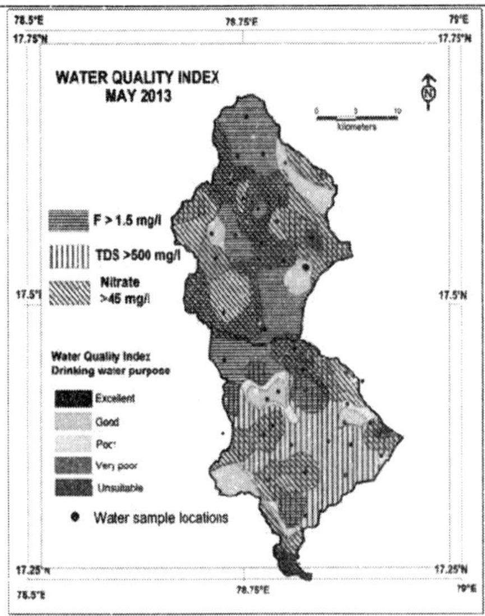

Fig. 19.16a: Pre-monsoon WQI comparision

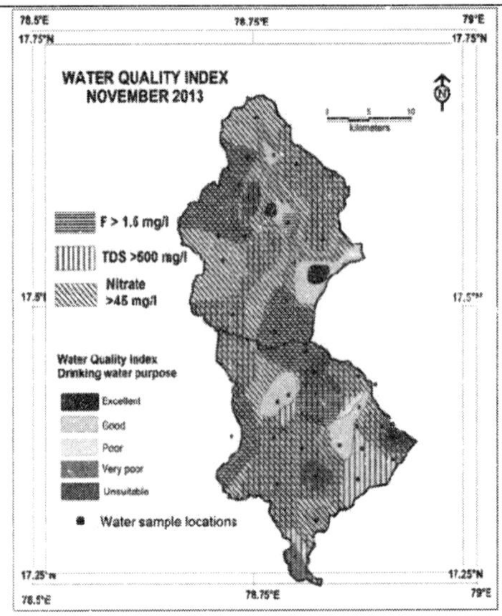

Fig. 19.16b: Post-monsoon WQI comparision

Table 19.4: Season-wise area occupied by major constituents

Parameters > permissible limit	*Area occupied in Sq. km*	
	Pre-monsoon	***Post-monsoon***
F	276	143
Nitrate	326	535
TDS	236	308

3. The average value of pH in the study area is 7.80 during pre-monsoon and 7.59 in post-monsoon. The EC varies between 390 μS to 3690 μS during pre-monsoon and 307 μS to 4730 μS during post-monsoon. The average value of TDS in the area is 964 mg/l during pre-monsoon and 1067 mg/l during post-monsoon.
4. The average concentration of F in the groundwater samples in the area is 1.30 mg/l in pre-monsoon and 1.06 mg/l in post-monsoon season. High concentration of Fluoride is observed in deeper water level areas, Dilution of F is identified in the post-monsoon samples. Concentration of Bicarbonate ranges between 152 to 610 mg/l during pre-monsoon and 79 to 610 mg/l during post-monsoon. Chloride concentrations range from 10.63 to 907 mg/l during pre-monsoon and 10.60 to 780 mg/l during post-monsoon. The concentration of Nitrate increase during post-monsoon in most of the samples especially in B. Ramaram watershed where 75 per cent of the samples show high concentrations (>45 mg/l).
5. Increase in Sodium concentration was observed during post-monsoon and extends into the northern part of the area whereas Calcium concentration decreases in most of the area from pre-monsoon to post-monsoon.
6. In order to calculate WQI, eleven important parameters viz., pH, TDS (mg/l), TH (mg/l), Ca (mg/l), Mg (mg/l), Na (mg/l), K (mg/l), Cl (mg/l), HCO_3(mg/l),

SO_4(mg/l), NO_3 (mg/l) and F (mg/l) were selected. The maximum weight of 5 has been assigned to the parameter TDS, NO_3 and Fluoride due to their importance in water quality assessment.

7. WQI considered for water samples collected fall under 'good to poor' during the pre-monsoon and decreases further during post-monsoon period, occupied by good quality index.
8. The study helped in transforming complex water quality data into simple information that is easily understandable and adoptable.
9. Spatial distribution maps with quality attributes generated for WQI exhibits its ease to understand and facilitate the policy planners and decision makers for initiating groundwater quality utilization and development not only in the area but in other areas where this methodology is adopted.

References

1. BIS. (1991). Indian standard specifications for drinking water, B.S. 10500.
2. C.R. Ramakrishnaiah *et al.* (2009). Assessment of water quality index for the groundwater in Tumkur district, Karnataka State, India. *E-Journal of Chemistry* 6(2): 523-530.
3. Davis, S.N. and De Wiest, R.J.M. (1966). *Hydrogeology,* John Wiley Inc., New York. 463 p.
4. Hem, J. D. (1985). Study and interpretation of the chemical characteristics of natural water, 2nd edn. *US Geological Survey Water Supply.* Paper-2254.
5. Mouna Ketata-Rokbani *et al.,* (2011). Use of Geographical Information System and Water Quality Index to assess groundwater quality in El Khairat Deep Aquifer (Enfidha, Tunisian Sahel). *Iranica Journal of Energy and Environment* (2): 133-144
6. Ramesh K. and BhuvanaJagadeeswari, P. (2012). Hydrochemical characteristics of groundwater for domestic and irrigation purposes in Periyakulam Taluk of Theni District, Tamilnadu. *I Research Journal of Environment Sciences,* Vol. 1(1), 19-27.
7. Swarna Latha. P., Nageswara Rao, K. (2010). Assessment and Spatial Distribution of Quality of Groundwater in Zone-II and III, Greater Visakhapatnam, India Using Water Quality Index (WQI) and GIS. *International Journal of Environment Sciences.* Vol. 1. No. 2
8. WHO (1992). *International Health Regulations,* 3rd annotated ed. Geneva, World Health Organization.
9. Tiwari, T.N. and Mishra, M. (1985). A preliminary assignment of water quality index to major rivers. *Ind J. Protect.* Vol. 5, 276.
10. Vasantavigar, M. Srinivasamoorthy, K. Vijayaraghaven, K. Rajivaganthi, R. Chidambaram, S. Anandhan, P. Manivannan, R. and Vasudevan, S. (2010). Hydrogeochemical assessment of groundwater in Thirumanimuttuar sub-basin, implication of water quality degradation by anthropogenic activities. Recent trends in water research, Hydrogeochemical and Hydrological prospectives. I.K. International Publishing House Pvt. Ltd., New Delhi. 234 p.
11. Elango L, Kannan R. (2017). *Rock-Water Interaction and Its Control on Chemical Composition of Groundwater.* Chapter 11. Vol. 5. Developments in Environmental Science: Elsevier.
12. Miller, W.W. Joung, H.M. Mahannah, C.N. and Garrett, J.R. (1986). Identification of water quality differences in Nevada through index application. *J. Environ. Quality.* Vol. 15, pp. 265-272.

13. Raghunath, H.M. (1987). *Geochemical surveys and water quality. Groundwater,* Wiley Eastern Limited, New Delhi, India. pp. 343-347.
14. Subba Rao, N. (1997). Studies on water quality index in hard rock terrain of Guntur district, Andhra Pradesh, India. National Seminar on Hydrology of Precambrian Terrains and Hard Rock Areas, pp. 129-134.

Index